BEYOND UKRAINE
DEBATING THE FUTURE OF WAR

TIM SWEIJS and
JEFFREY H. MICHAELS (*Editors*)

Beyond Ukraine
Debating the
Future of War

OXFORD
UNIVERSITY PRESS

Oxford University Press is a department of the
University of Oxford. It furthers the University's objective
of excellence in research, scholarship, and education
by publishing worldwide.

Oxford New York

Auckland   Cape Town   Dar es Salaam   Hong Kong   Karachi
Kuala Lumpur   Madrid   Melbourne   Mexico City   Nairobi
New Delhi   Shanghai   Taipei   Toronto

With offices in

Argentina   Austria   Brazil   Chile   Czech Republic   France   Greece
Guatemala   Hungary   Italy   Japan   Poland   Portugal   Singapore
South Korea   Switzerland   Thailand   Turkey   Ukraine   Vietnam

Oxford is a registered trade mark of Oxford University Press
in the UK and certain other countries.

Published in the United States of America by
Oxford University Press
198 Madison Avenue, New York, NY 10016

Copyright © Tim Sweijs, Jeffrey H. Michaels and the Contributors, 2024

All rights reserved. No part of this publication may be reproduced,
stored in a retrieval system, or transmitted, in any form or by any means,
without the prior permission in writing of Oxford University Press,
or as expressly permitted by law, by license, or under terms agreed with
the appropriate reproduction rights organization. Inquiries concerning
reproduction outside the scope of the above should be sent to the
Rights Department, Oxford University Press, at the address above.

You must not circulate this work in any other form
and you must impose this same condition on any acquirer.
Library of Congress Cataloging-in-Publication Data is available

ISBN: 9780197790243

Printed and bound in the United Kingdom by Bell and Bain Ltd, Glasgow

To our daughters Liztophe and Carla
May they be spared the horrors of war

CONTENTS

Acknowledgements   xi

List of Figures and Tables   xiii

Introduction   1
    Tim Sweijs and Jeffrey H. Michaels

PART I
BOUNDING THE IMPACT OF THE WAR IN UKRAINE

1. Revisiting Putin's 2022 Invasion of Ukraine: 23
   Implications for Strategic and Security Studies
       Antulio Echevarria

2. The Futures of War: A Recent Western History   41
       Frans Osinga

3. After Russia's Invasion of Ukraine: Is the Decline of   71
   War a Delusion?
       Azar Gat

4. The Next War Would Be a Cyberwar, Right?   85
   Lessons from the Russo-Ukrainian War
       Paul Ducheine, Peter Pijpers and Kraesten Arnold

vii

PART II
LANDSCAPES OF FUTURE WAR

5. Four Faces of War 107
   Frank Hoffman

6. People's War vs Professional War: Which Has the Future in Europe? 125
   Jan Willem Honig

7. Urbicide and the Future of Civil War 147
   David Betz

8. Living with Denial? Great Power Competition Over Economic and Military Access 163
   Paul van Hooft

PART III
MILITARY INNOVATION AND THE FUTURE OF WAR

9. Military-Technological Innovation in the Digital Age 183
   Audrey Cronin

10. The Rising Dominance of the Tactical Defense? 201
    T.X. Hammes

11. Artificial Intelligence and the Nature of War 223
    Kenneth Payne

12. Assembling the Future of Warfare: Innovating Swarm Technology within the Dutch Military–Industrial–Commercial Complex 241
    Lauren Gould, Linde Arentze and Marijn Hoijtink

PART IV
ANTICIPATING THE FUTURE OF WAR

13. The Past as Guide to the Future 265
    Beatrice Heuser

14. Forecasting the Future of War 287
    Collin J. Meisel

15. The War in Ukraine and the Apocalyptic Imaginary   305
    Jeni Mitchell

16. War as Becoming   329
    Antoine Bousquet

Afterword   349
    Christopher Coker

Notes   355

Index   433

ACKNOWLEDGEMENTS

This volume emerged from papers presented at The Future of War Conference that took place in Amsterdam from 5–7 October 2022. The conference brought together many scholars from different parts of the world whose research focuses on the study of war. Set against the backdrop of the war in Ukraine, it offered an opportunity for these scholars to reflect on the ways in which their ideas about the future of war were affected by the most devastating conflict Europe has witnessed since the Second World War. Occurring after a prolonged period when no such in-person gatherings could take place due to the Covid pandemic, the event provided a useful venue for these scholars to present and debate a diverse range of perspectives. So valuable did we find these interactions for our own understanding of the topic that we decided a collection of the papers presented would make an important contribution to the future of war literature, particularly as the war in Ukraine had already led many scholars to revise the views they held prior to the war.

The conference was hosted by the War Studies Research Centre of the Netherlands Defence Academy in collaboration with the Changing Character of War (CCW) Centre of the University of Oxford. We would like to express our gratitude to the team that made this conference possible, including to Frans Osinga, who helped initiate it and allocated the necessary funds; to Martijn Kitzen, who served as co-chair of the conference alongside Tim Sweijs; to Robert Johnson, for the CCW partnership; to Dorith Kool, who

ACKNOWLEDGEMENTS

was the point guard and producer of the event; to Vibeke Gootzen, Claire van Voorst tot Voorst, Maggie van Heesch, Martijn Rouvroije and Daan van den Wollenberg, who all made pivotal contributions in organizing the conference, and to all other members of the Netherlands Defence Academy staff who played a role.

Special mention goes out to Ivor Wiltenburg, who passed away in September 2023. Ivor, despite his illness, stepped in and stepped up in the final weeks leading up to the conference to make sure everything clicked and everyone landed on their feet. He is and will be missed.

We would like to thank Harm Roelant van de Plassche for his equally diligent and meticulous research assistance and valuable contribution to the preparation of this manuscript. We also would like to thank Tom Connolly and Tom Draaijer for proofreading the manuscript. Finally, we would like to express our gratitude to the great team at Hurst, especially to Lara Weisweiller-Wu and Mei Jayne Yew, for their keen focus, valuable input, kind tolerance for our dithering, and ultimately firm hand, as well as to Michael Dwyer for seeing potential in the project when we first pitched it and for pushing it forward in record time.

We dedicate this book to our daughters Liztophe and Carla, sources of endless joy and unceasing amazement. We hope they'll be spared the horrors of war.

LIST OF FIGURES AND TABLES

Figure 3.1: GDP per capita (2023), current prices US dollars

Figure 3.2: The Long Peace phenomenon between the great powers since 1815

Figure 3.3: GDP per capita (1820–2018)

Figure 5.1: The Four Faces framework

Figure 11.1: ChatGPT exchange

Table 8.1: Impact of military and economic denial on interactions between great powers, middle powers, and small powers/non-state actors

Table 10.1: Projected balance between offense and defense in each of the domains of warfare

Table 14.1: Summary of patterns of (un)knowability, their relation to war, and partial remedies for analysis

Table 15.1: Apocalyptic threat typology

INTRODUCTION

Tim Sweijs and Jeffrey H. Michaels

Typically, the question of what the future of war will look like looms large only in the minds of military officers and a relatively small circle of defence planners, industrialists, strategists, futurists and academics. At some moments in history, however, it captures the attention of a much larger swathe of society. Russia's large-scale invasion of Ukraine in 2022 sparked such a moment. Politicians, pundits and concerned citizens are now grappling with what the future of war holds in store, still adapting to the unexpected reality of the return of large-scale interstate war in Europe. In this sense, interest in the future is really about present-day decisions: how do we prepare now for what is coming sometime in the future? Often our assumptions are wrong. Or to be somewhat more charitable, the answers we have today always seem to be plausible, but developments occur within a handful of months or years that seem to invalidate these visions and generate demands for new ones. The war in Ukraine has once again proved this point.

The war in Ukraine was not expected. Even when the possibility of war became a probability in late 2021 and early 2022, it was still difficult to admit that the war we've since witnessed—the most intense in Europe since 1945—could really occur. In large part, this inability to foresee such a war was due to incorrect assumptions not only about the future of war, but also about what was still possible in the present. What transpired in Ukraine was a type of war believed to be long dead. Surely, we'd given up on *that* type of war, learned our lessons from the past, become more 'civilized', more technologically sophisticated, and moved on. Surely such a war couldn't happen *now*. Yet, despite all the advanced technology that has played such a large part in the war, many of the war's most visible features—trenches, the dominance of artillery, the mobilization of a nation for total war, the requirement to sustain heavy rates of fire over months and possibly years, the enormous losses on a battlefield stretching hundreds of kilometres—are more reminiscent of the First World War than they are of the twenty-first-century visions of war that had previously dominated the thinking in many Western defence establishments. The ongoing war also contains features that are seemingly timeless: political leaders blundering into long wars they assumed to be short ones; their willingness to sacrifice the lives of others; their disregard for international law when it suits them. It also demonstrates the readiness of ordinary people to defend their homeland rather than bend the knee for an aggressor or become a refugee.

Amidst the carnage of Ukraine, lest it be forgotten that Russia embarked on the war following years of indecisive conflict in Afghanistan and Iraq, in which the US-led forces, possessing overwhelming resources and the most advanced technology, were unable to defeat poorly armed insurgents. After so many years of bloodshed in Central Asia and the Middle East, the resort to large-scale war hardly seemed an attractive prospect for any sensible state leadership. Even if Putin's war had initially succeeded in destroying Ukraine's armed forces, there should have been every expectation that irregular conflict would continue, drawing the Russian military into a quagmire lasting many years. So much for the cautionary lessons of recent history.

Many Western armed forces, particularly those of the United States, were quite content to leave the 'small wars' behind and focus instead on preparing for 'big wars' again, just as they had during the Cold War. The idea of the future of war as a 'war amongst the people'—a notion popular in the decade after 9/11—quickly gave way to a very different vision. The wars in Afghanistan, Iraq, Libya and Syria became synonymous with the sort of violent conflict in which Western military power was ill-suited. A return to preparing

INTRODUCTION

for future wars against other states provided a preferable alternative, somewhat reminiscent of the US military's 'return' to the European theatre in the 1970s following years of conflict in Vietnam. China's more assertive behaviour in the South China Sea and Russia's 2014 annexation of Crimea served to facilitate this transition.

Running parallel to the conduct of recent and ongoing wars, germinating in the background far from the front lines, were visions of future war largely based on expectations of what the next generation of advanced weapons might be able to achieve. This is a vision of future war that is dominated by technology, but technology that is so sophisticated and expensive that very few countries can afford the entry ticket. For those that cannot afford it (which is most of them), visions of future war are necessarily very different. Notably, these technological visions of future war downplay or ignore the political, economic, societal and organizational context of war; that is, both what wars will be fought for and what limits will be placed on how they are fought. Instead, they focus on the advantages that one technological system will have over another. Despite the limitations of this approach, technology remains integral to war. Consequently, visions of future war demand a technological component, though preferably one that is integrated within a broader context.

At the time Russia launched its large-scale invasion of Ukraine on 24 February 2022, mainstream Western conceptions about the future of war had been in flux for about a decade, with the dominant conception being one of wars *between* states rather than *within* them. With a large-scale war finally underway, these pre-Ukraine conceptions of what a large-scale war (and specifically a large-scale war against Russia) would consist of had to be reevaluated and adapted based on the practice of war since. As the war drags on, the practice of war will continue to evolve, those changes will be closely observed, and visions of future war will be adapted accordingly. Even so, we should not discount the possibility of other major outbreaks of violent conflict erupting in the meantime, such as occurred in the Middle East following the 7 October 2023 attack by Hamas against Israel. These too will affect how future war is conceived.

The study of war—with the explicit objective to anticipate and prepare for future wars—has occasionally been frowned upon, especially in academic circles. With some justification, futures work has been disparaged as theoretically light, historically shallow and insufficiently robust. How, after all, can the future possibly be reliably studied if it hasn't happened yet? Scrutiny of the performance of expert judgement from the 2000s and 2010s casts serious doubts on the ability of scholars and strategists to accurately anticipate the outbreak, geographical location or character of future war. Looking at a breadth of examples of incorrect predictions over the past century and a half, Sir Lawrence Freedman, in his 2017 book *The Future of War: A History*, convincingly argued that our track record for anticipating the future character of war is weak. He therefore advised to take work on the future of war 'seriously' but also cautioned to 'treat it skeptically'.[1] These admonishments are extremely relevant, but they do not address the question as to *how* the study of war can fruitfully help to inform anticipation of the future of war in order to prevent it, to prepare for it, or both. At the same time, a sizeable research community of practitioners and analysts are involved in efforts to understand the causes and the character of future war. Their understanding of what war is or isn't, what actors will be involved, for which political objectives they will fight, what means they will use, and in which domains they will clash, feeds into real decisions that shape the way future wars will be waged.

Against this background, it bears noting that there has been great progress in future anticipation across different walks of life—from weather prediction to consumer behaviour forecasting—based on a combination of theoretical progress, the explosion of information, and the adoption of new analytical techniques. War, however—and especially interstate war—is a more pernicious phenomenon. We have a rather poor track record when it comes to predicting the onset of interstate war, its geographical location, its scope and magnitude, as well as its very incarnation. War is a true chameleon that is hard to pin down. This is also explained by the fact that war in its very essence is an interactive process in which strategic actors act and react in complex ways. Whilst undeniably coming a long way from the time when the Greeks sought counsel from the Priestess of Pythia at the Oracle of Delphi, or when Roman auspices studied the patterns of bird flights to divine the will of the gods,

INTRODUCTION

there are inherent difficulties that neither the advances of time nor sophistication of technique have been able to completely overcome, which applies equally to war as it does to warfare.

To illustrate the difficulties, one need only consider some of the challenges of answering the question of what sort of *history* of war is useful to inform thinking about future war. The first problem is determining what period in history one should look at to gauge patterns and variations in armed conflict. Should we look at the three decades that have passed since the end of the Cold War? The two centuries following the Congress of Vienna in 1815? The emergence of the modern state system, often dated to the Treaty of Westphalia in 1648? Or do we go further back and examine war since *homo sapiens* started settling down approximately 12,000 years ago? Obviously, there are lessons to be gleaned about the dynamics of war from each of these periods, but clearly these dynamics will manifest themselves differently in today's world, let alone in the world of tomorrow. Even if we are confident that we know this history of war because it already happened, the ability to capture its complexities in an accessible, written form presents enormous difficulties. Naturally, the further back one goes, or the more types and geographies of war one examines, the more difficult it becomes to capture the nuances of war as a social phenomenon. In this sense, there are natural limitations to the study of war, past, present and future.

Another problem with any analysis of past wars is their diversity. Within a specific period, wars may be similar in some respects but vary in other respects; they may fit within certain categories, such as interstate or intrastate, yet follow very different patterns across geographic regions. Attempting to generalize is a fraught endeavour. As any database of wars and conflicts will indicate, there are often multiple wars occurring simultaneously. Some of these will be large-scale wars, others low-intensity conflicts; some wars will last years, others only days or weeks; some will involve alliances or coalitions, others will be limited to two adversaries facing each other on their own. Among these wars, there are those that receive attention, those that are ignored, and those that seemed important at the time but were then forgotten when a more important war came along. The degree of attention is often a function of one's direct participation:

5

we care most about those wars in which we are directly involved, and care much less about wars in which we have little or no stake. We also seem to care more about wars that are deemed significant, such as those wars that traumatized our societies or, alternatively, those wars in which great victories were achieved and of which we are proud. Conversely, because some wars were so traumatic or didn't go so well, societies may try to forget them. Regrettably, there are too many wars and only so many we can realistically digest.

The reasons why individuals and groups continue to fight, and whether or not these reasons will remain valid over the coming decades, is yet another factor that must be accounted for when contemplating future wars. War may continue to be about fear, interest and honour, but these elements have traditionally manifested themselves differently, and often in unpredictable ways, across eras and across regions. Wars have been waged to conquer territory, to exploit natural riches and to subjugate populations, but they have also been launched to gain control of strategic geography, to promote an ideology, to demonstrate the power of a state, and for personal ambition. Regardless of motive, mobilizing societies to sacrifice for war requires narratives designed to cast the adversary as evil and dangerous, and to promote one's own cause as noble and just. For the 'enlightened' outsider, it can prove very difficult to divine the degree to which such narratives, which can often border on the absurd, still resonate amongst a sufficiently large audience to facilitate large-scale armed conflict. Many analysts were reluctant to believe that the Islamic State in Syria's apocalyptic visions of a caliphate would hold any wider appeal until foreign fighters from across the world travelled to the Middle East to bulldoze the Sykes–Picot borders. Similarly, they refused to believe the possibility that dreams of the restoration of the Russian Empire, even though openly professed by Russia's despotic leadership, would actually factor into any decision to invade Ukraine. Currently, many observers in Europe seem out of touch with the way in which visions of regional or global hegemony in both the US and China may play a key role in decisions about war and peace in the Indo-Pacific region.

Thus, the 'objective' analyst of future wars will be cognizant that war will come in many shapes and sizes. In contrast, practitioners will

INTRODUCTION

be more targeted. In crude terms, a US military officer's view of the future of war is likely to be very different from an Ethiopian view, a Chinese view different from that of a Colombian view, a Serbian view different from a Saudi view, and so forth. There are those that prefer to focus on the high-tech because they can afford to, and others that only care about the low-tech because that is all they can afford. To the extent one can speak in terms of 'Western', 'Eastern' or 'Global Southern', clearly there will be a great diversity in outlook. Even within the North Atlantic Treaty Organization (NATO), an alliance that stresses 'standardization' among its members, especially when it comes to equipment and doctrine, there have always been different perspectives on future war; or rather, which types of wars some countries will prioritize and those where they let other members take the lead. In today's NATO, there are those members that want to focus on the Russia threat, those that want to focus on instability and terrorism along NATO's southern periphery, and those that wish to strike a balance among both while also addressing the rise of China. Similarly, during the Cold War, NATO members envisaged many types of armed clashes with the Soviet Union and the Warsaw Pact, ranging from a seizure of West Berlin, incursions along the Northern or Southern Flanks, a full-scale invasion of Western Europe, and conventional and nuclear wars of different shapes and sizes. Among these members, some had to prepare for conflicts outside Europe, whereas others were exclusively focused on Northern, Central or Southern European war scenarios.

Apart from the content of past wars—who fought whom, with what means and for what reasons—one must also consider more general trends in the evolution of war as a social phenomenon. Some scholars have pointed to a decline in war over the centuries, particularly the numbers of wars, the involvement of states in wars, and their lethality relative to the size of populations and combatants. In some cases, this is narrowed to a decline in wars directly fought among the great powers, whereas wars fought among lesser powers, lesser powers fighting directly with great powers, or the great powers meddling indirectly are recognized as an ever-present reality. Putting aside the frequency and intensity of interstate wars, civil wars and insurgencies must also be accounted for.

7

BEYOND UKRAINE

Prior to the 2022 invasion of Ukraine, many decision-makers had come to view war as an illegitimate instrument to attain political objectives. Undeniably, war's de-glorification and its stricter codification in international law had become more deeply and more widely entrenched. Meanwhile, the attractiveness of territorial conquest for economic riches and national glory had diminished. Yet, if that applied to some states, it was certainly not true across the board, and certainly not for polities in wholly different political, socio-economic and cultural environments. These developments should give pause and require us to revisit other assumptions pertaining to the future of war. Is progress inevitable? Do civilizations progress towards some peaceful end-state, or do they go astray, auguring in conflagrations of staggering proportions?[2] Will the tradition of nuclear non-use continue into the indefinite future? Will the great powers return to direct conflict with one another? And if they do, what will such wars entail? Will they be short, intense and decisive, or sustained, subdued and indecisive? Will they involve the mobilization of tens of millions of citizens taking up arms? Are such wars even contemplatable? One can speak of laws and norms of war, but one can also point to the regular violation of laws and shifting of norms depending on societal mood, political will, new technological developments and the nature of the issues being contested.

From the vantage point of 2023, scholars looking ahead are quite limited in the period they feel somewhat confident predicting, usually measured in no more than a few years (and even that is a stretch). Practitioner demand, usually construed in terms of military procurement lead times or military and societal long-term planning, tends to look ahead no more than a couple of decades. Typically, there is little demand for predictions about the future of war in 2123, much less 2223. Put slightly differently, today's practitioners demand future visions of war that concentrate on the next generation of Starlink rather than the sort of visions associated with Star Wars or Star Trek.

In this context, it is somewhat paradoxical that many military organizations value science fiction in their intellectual preparations for future war. Science fiction now features as compulsory reading on the syllabi of numerous Western staff command courses. Defence

INTRODUCTION

planners have enlisted the services of scenario specialists who collaborate with design specialists to imagine different future conflict environments. The US Army's Training and Doctrine Command (TRADOC), for instance, hosts a 'Mad Scientist Initiative'. In France, the Defence Innovation Agency has created a team of science-fiction analysts to help identify threats that fall outside the purview of regular horizon-scanning exercises. In Germany, the Federal Ministry of Defence has funded the academic research project Cassandra, in which works of literature are utilized as an early warning tool. These are clear attempts to decipher the writing on the wall for a wide range of factors that shape the future of war, which can then be used to inform preparations for it.

Yet, with all these plausible and not so plausible perspectives to account for, what confidence can we have that our ideas about the future of war will bear a ballpark resemblance to what these wars will be like five, ten, twenty, or more years from now? The short answer, unsatisfying as it may be, is that we will probably get some aspects right and some aspects wrong. But in one key respect, it may not matter very much anyway. Such is the unceasing demand for answers that even ones we are highly sceptical of are still deemed useful for the here and now. In any case, we need to start somewhere, and given our ostensible familiarity with the present and recent past, we can always try to project forward as best we can while accounting for how various geopolitical, economic, technological, social and related trends affect future war. We can always attempt to identify potential wild cards—those radical developments that will blow our projections wildly off course. Yet as history has demonstrated, the key game-changers may have less to do with technology and more to do with how organizations make clever use of the technology; less to do with clever generalship and more to do with how political and societal standards are transformed. At the same time, it would be foolish not to recognize that amidst changes, even radical ones, considerable continuity remains. Change may be inevitable, but changes occur at different paces.

Likewise, when looking back on the history of war and then attempting to look forward, we can identify a spectrum of different types of wars that have remained fairly consistent. Changes will occur

9

within these categories, and sometimes the categories will overlap, but it is rare that new categories outside this spectrum are created. At the end of the day, even the diversity of war has its limits. We can also observe that most countries focus on a handful of categories, yet only a small minority of countries are even hypothetically capable of engaging across the full spectrum (though in practice, even with this small subset of capable states, one category will receive the bulk of attention at any one time while the others will be placed on the back burner). It is precisely due to the political and military requirement of identifying which category to prioritize and which to place on the back burner, along with all the military procurement and organizational changes that accompany this prioritization, that demand for ideas about the future of war remains consistently at a premium.

It is on the basis of these ideas, whether genuinely believed or simply used as a convenient justification, that billions of dollars, euros or yuan may be spent on one military project rather than another, spent on building up one military service or branch to the detriment of another, or simply spent on guns instead of butter. Such are the stakes involved that future of war thinking is highly prone to self-interest rather than disinterest. Careers as well as contracts are often on the line. And even the most disinterested scholars cannot completely escape their natural biases. Given these difficulties, perhaps the most effective method of providing a well-rounded perspective on the future of war is to bring together multiple insights on multiple aspects of the subject to inform our thinking. Although our volume tries to do precisely this, we are cognizant that many voices could not be heard, many topics not covered, and that it is mainly the 'Western' (read: North American and Western European) lens through which the future of war is being viewed. Presented here are American views of the role of technology in future wars rather than Brazilian, Iranian or Indonesian views; Western European views of military innovation rather than Ghanian or Bangladeshi; Western perspectives on the war in Ukraine rather than Chinese or Ecuadorian, and so forth. Admittedly, it is simply impossible to capture such a diverse range of voices and perspectives in a single volume. Nor would a set of volumes be likely to fare much

INTRODUCTION

better. Despite this limitation, such is the diversity and novelty in the approaches taken by the chapter authors gathered in this volume to study the future of war that they make a highly valuable contribution to our understanding of it.

Since the full-scale invasion of Ukraine, the future of war question has gained renewed urgency, particularly as many ideas about the conduct of war previously held—especially those pertaining to Russia's ability to wage different types of hybrid and conventional wars, to say nothing of Ukraine's ability to offer effective resistance—have been shown to be seriously compromised, if not outright wrong. However, it is not our intention to suggest that the war in Ukraine should necessarily be a major turning point. For instance, even if the new awareness of Russia's limitations in waging war needs massive adjustment from the previously high expectations, this single case can only shed limited light on the conduct of war more generally. In other words, we can probably assume that other wars will be fought in the years and decades ahead, even if few are likely to resemble the war in Ukraine in terms of its size, scale, international participation, duration, weaponry employed, and so forth. Moreover, although many militaries are observing the ongoing war and will adapt their posture, doctrine, procurement, etc., to some extent, this Ukraine-inspired change should not be over-hyped.

Prior to 24 February 2022, many visions of future war existed. Some of these will have been adapted to account for the war in Ukraine, whereas for others it will have little relevance. Some analysts believe that the future of war will be radically different from wars in the past. They conjure up images of robots doing battle on isolated fields, cyber warriors crafting weapons from zeros and ones, and mosaic units doing the fighting. Others believe that the more things change, the more they remain the same. They argue that mass and scale will continue to be decisive factors in wars of attrition throughout the twenty-first century. Such wars will require the full industrial weight of wartime economies, as well as coalitions willing and able to sustain the war effort. Then there are also those who argue that wars are more likely to be intrastate rather than interstate, possibly to be fought within the major powers rather than between them. While planners in the Pentagon envisage a future of

11

algorithmic war fought at hypersonic speeds, and while many armed forces around the world continue to purchase tanks, drones, cruise missiles, frigates and fighter aircraft, more stringent budgets and limited industrial capacity elsewhere place important restrictions on the means to fight. But where there's a will, there's is a way, and creative approaches to fighting wars can always be found to make the most out of the cheapest and simplest lethal weaponry. It is because of these competing visions and realities that it is important to ensure the war in Ukraine is accounted for where appropriate but to remain principally focused on the bigger picture.

We have therefore selected four broad themes, divided over four sections, each consisting of four chapters that will allow for useful engagement with the key contemporary debates. In this context, the sections assess the impact of the war in Ukraine on strategic studies and visions of future war, survey the landscapes of future war, examine how technological innovation shapes future war, and scrutinize our ability to anticipate the future of war.

Bounding the Impact of the Russian War against Ukraine

The first theme is an attempt to bound the impact of the Russian war against Ukraine. It considers its impact on ideas about future war and identifies lessons for adjacent disciplines more generally. The section starts with Chapter 1 by Antulio Echevarria, who takes stock of the war's implications for strategic and security studies. He argues that the war casts in a different light six prevalent explanations for the waning of interstate war, which include the proliferation of weapons of mass destruction, the spread of democracies and democratic values, the growth of multilateral institutions, increasing economic integration, the influence of international law and the laws of armed conflict, and the spread of anti-war norms. He argues that in the case of the war against Ukraine, instead of dampening the odds of war, they either fuelled its onset or severity, or were less relevant than previously thought. In addition, he shows how the war harnesses important insights for the study of strategic coercion, including the relevance of deterrence by denial when faced with a determined aggressor, and the opportunities and limitations afforded by non-

INTRODUCTION

military compellence. Echevarria concludes that the Russo-Ukrainian war is an interstate war that is both a war against the people and a war amongst the people—a fact that Western military strategy and doctrine ignore at their own peril.

His assessment is followed by Chapter 2, in which Frans Osinga surveys the evolution in Western visions on the future of war over the past thirty years. He traces their roots to recent military operational experiences, traumatic events, national security priorities of the moment, the specific needs of individual services, the perceived revolutionary potential of emerging technologies, the rise of new adversaries, and strategic culture. Osinga then distinguishes five visions of future war (to wit: humanitarian wars, sophisticated barbarism, immaculate war, cool war, and major war). Proposing that '[w]ars are educational moments that serve to gauge the extent to which prior views on future war hold water', he highlights how each of these five visions were manifest in the Russo-Ukrainian war, albeit in ways different than initially anticipated. Osinga suggests that the five visions he presents are akin to the constituent elements of 'string theory in physics' that depict reality as composed of multiple dimensions, and argues that strategists will need to take these into account as they prepare for future wars.

In Chapter 3, Azar Gat revisits his own thesis that peace is profitable, and asks whether the decline of war is a delusion. Although the decline of war thesis continues to be met with disbelief, he argues that the decline is real, with one major qualification: it only exists in the 'zone of peace', populated by socio-economically advanced countries, as opposed to the 'zone of war', which includes those less fortuitous countries. Here, war is alive and kicking, both intrastate as well as interstate. He argues that although nuclear weapons have played a role, free trade, economic growth and the spread of liberal democratic values largely account for the process of modernization that has led to the dwindling of war in the 'zone of peace'. Nevertheless, even though the proliferation of peace is real, Gat prophesizes it is 'far from being foolproof and free from challenges'. Elsewhere, it rests on fragile pillars and whether above the surface, as in Ukraine, or beneath it, the possibility of war continues to cast an ominous shadow over the future.

13

The section is capped by Paul Ducheine, Peter Pijpers and Kraesten Arnold, who examine the extent to which pre-war visions of cyberwar were borne out by the Russo-Ukrainian war and reflect on what that implies for our understanding of future war. They establish that cyber operations have been an integral element in the modus operandi of both sides in the lead up to and during the war, while showing that these operations had a different impact than was anticipated prior to the war. They argue that the war clearly shows that contemporary wars are fought across multiple domains, including the physical, virtual and cognitive dimensions of the information environment. This will remain true for the wars to come, and policymakers and professionals should prepare accordingly.

Landscapes of Future War and Warfare

The second theme asks: what does the future of war and warfare look like? In Chapter 5, Frank Hoffman examines how a diverse assortment of social, political, economic and technological developments shape future warfare. Recognizing that many aspects of future wars are extremely hard to predict, he offers a useful framework to think about future types of warfare based on two axes that run from state to non-state, and from direct and kinetic to indirect and non-kinetic. He thus distinguishes between four 'faces of warfare': societal, cognitive, proxy and conventional. Societal warfare is aimed at the critical infrastructure of the opponent; here, the goal is to dislocate and paralyse, and to achieve objectives through coercion. In cognitive warfare, the goal is to manipulate the minds of specific audiences, especially in the general population. To achieve political objectives, 'the message is the munition'. In an age of heightened interstate competition, proxy warfare will also be prevalent. In this type of warfare, major powers use proxy actors in the indirect pursuit of their objectives. By utilizing covert means that offer plausible deniability, the risks of escalation can be reduced. Finally, conventional warfare absorbs the lion's share of the budgets of armed forces, and features the use of traditional and modern military instruments to occupy land and to 'attrit or outmaneuver' the adversary. Hoffman explains the

INTRODUCTION

current and likely future incarnations of these faces, and assesses the risk associated with each of them based on their likelihood and impact.

As all war is political, we ignore the *why* of war at our peril. Polities and people will undoubtedly continue to fight for fear and honour, and to compel their enemies to do their will. But the ostensible purposes and the concrete objectives for which they fight are bound to change, as is the role of ordinary people in war and warfare. In Chapter 6, Jan Willem Honig considers the involvement of people rather than professionals in future wars. Honig provides an historical analysis of the role of populations in interstate war over the past two centuries—not just as victims, but also as supporters and warriors in their own right. He shows how the involvement of people was the subject of continuous debate until the end of the Cold War, when European states and their societies suddenly shifted to full professionalization. In his historical survey, Honig shows that the people not only have an 'immediate stake' but also a 'deciding vote' in determining the outcomes of war. He observes that the war in Ukraine, specifically 'the re-appearance in a developed country of the people on the field of battle', requires us to reconsider the question whether 'reliance on regulars' will suffice for wars in the twenty-first century.

In Chapter 7, David Betz argues that while images of high-tech weaponry tend to dominate Western visions of future war, the reality is likely to be radically different. He thinks that in the coming years, 'the West's armed forces will be exhausted and demoralized, their arsenals drained, and their economies too weakened to rebuild them'. He sketches an apocalyptic landscape of 'militias stabbing, shooting and bombing each other in the burning rubble of Western civilization'. Based on an analysis of structural trends that render technologically advanced societies vulnerable to disruption, and an examination of the views and plans of discontented individuals and groups on anonymous chat boards, he describes a dark future of what civil war will look like in Western societies. His argument is provocative and disconcerting, arguing that 'we are in for a wild ride'.

The section is capped by Paul van Hooft in Chapter 8, in which he considers the emerging nature of competition and conflict among the great powers. He argues that despite globalization, the ability to deny military access to conflict theatres and to restrict access to advanced technologies is distributed unevenly among great powers and the rest of the world. Whilst minor powers will increasingly have the ability to deny access at the tactical and operational level, great powers will have the ability to deny access at the strategic level and restrict access to advanced technologies. This will mean that invasion and occupation of territory is likely to be more costly for all powers. However, great powers can make use of stand-off capabilities in coercing weaker powers, and restrict their access to advanced technologies. Additionally, only the great powers possess the resources and production capacity to support a sustained war effort, which cannot be easily cut off by external states.

Military Innovation and the Future of War

The third theme turns to military innovation and the future of war, and investigates how the future of war is created. As both an elaboration and a counterpoint, it offers a nuanced discussion of the role of new technologies on the character of future conflict. War is and has always been a reflection of the technological, economic, social and political context in which it is waged. Unfortunately, visions of future war are often tech-centric. Nonetheless, if there is one thing that is clear from history, it is not the technologies but rather the ways polities and their societies are able to innovate and exploit technologies that allow them to create wealth and prevail in war. Throughout history, different political, social and economic configurations have been conducive to achieving this evolving dual purpose. As we find ourselves at the beginning of what is sometimes referred to as the Fourth Industrial Revolution, with an assortment of emerging and disruptive technologies changing the nature of both wealth creation and military innovation, their implications for war and warfare deserve examination.

In Chapter 9, Audrey Cronin discerns the arrival of a new age of tinkering, in which individuals rather than large corporations shape and steer technological innovation, similar to how Guglielmo Marconi (radio), the Wright brothers (flight) and Alfred Nobel

INTRODUCTION

(dynamite) did so in the late nineteenth and early twentieth centuries. She argues that the convergence of mature technologies in computing, robotics and biology, which are readily accessible to the many instead of the few, reshapes not just how war is waged on the battlefield but also how the security of our societies is threatened. It necessitates the rethinking of state-driven, top-down models of military innovation that remain a legacy of the Cold War.

Partly building on Cronin's argument and drawing on battlefield developments in Ukraine, T.X. Hammes argues in Chapter 10 that the proliferation of advanced technologies such as satellites, 3D printing, and unmanned and semi-autonomous systems is providing pervasive surveillance, long endurance and cheap, fast and increasingly distant strike capabilities to most conflict actors. He analyses how this changing of the tactical offense–defence balance is reflected in the land, sea, air, space and cyber domains.

In Chapter 11, Kenneth Payne offers a compelling exploration of the implications of artificial intelligence for the very nature of war. There are reasons why groups of humans engage in war, reasons that are encapsulated in the narratives about identity and purpose that bind them together and steer them to a common cause. These narratives also inform and shape how they prepare for war and how they fight. Engaging with the long-running debates about war's enduring nature versus its ever-changing character, Payne forces us to consider what the emergence of large language models such as ChatGPT imply for war's future nature. Yes, war has always been that 'human thing', but as these forms of technology increasingly penetrate all aspects of human existence, including warfare, they are bound, he argues, to change not only war's character, but also its nature.

Concluding this section in Chapter 12, Lauren Gould, Linde Arentze and Marijn Hoijtink demonstrate that it is humans that create the future of warfare, both on the battlefields of Ukraine as well as within what they refer to as the military–industrial–commercial complexes elsewhere. In making this argument, they analyse how different constituencies assemble that very future in an in-depth case of how swarm technologies are being developed in the Netherlands within an intricate constellation of policymakers, practitioners,

17

researchers and corporations, each with their own different yet aligning interests. Their contribution serves as a powerful reminder that the future of war is created now, and that those present at the creation can and should engage with that act critically.

Future War Anticipation

The fourth and final theme looks at anticipating future war. It asks the question: what insights and approaches can give us a more sophisticated appreciation of war's uncertain future? Given the enormous impact of war on our societies, it is important to continue work on attaining a better and more differentiated understanding of war and warfare in its past, present and future incarnations. This is not just about actors, geography, size, scale and frequency, but also concerns the tools, technologies, strategies and organizational structures relevant to war and warfare. Note that when it comes to understanding future wars, analysis of the changing character of war should be rooted in deep historical knowledge and familiarity with war's fundamental principles. But understanding also derives from other forms, such as system dynamic modelling, structured multimethod exercises, tech watch, net assessment, what-if and what-if-not approaches, the analysis of war's depiction in computer games, the use of literature in early warning, and finally also science fiction. In key respects, the future of war is partly a function of prediction, partly a function of imagination, and partly a function of engineering. Given that all thinking about the future of war is speculative, this section considers the promises and perils of future war anticipation.

In Chapter 13, Beatrice Heuser examines 'what the past does and does not tell us about the future'. She examines how factors such as financial hardship brought about by economic crises, climate change and the history of war shape societies' receptiveness to future war, recognizing that concerted action by governments can either exacerbate or mitigate their effects. In doing so, she rejects a determinist view, observing that the future is what we make of it. She warns of the dangers of presentism, pointing out how forecasts of war are shaped by the experiences of the last war as well as

INTRODUCTION

shallow historical frameworks of reference applied by analysts and decision-makers alike. History does not provide a crystal ball, but a good understanding of the dynamics of war in previous eras, in combination with in-depth expertise of potential adversaries, strengthens anticipatory abilities.

In Chapter 14, Collin J. Meisel examines the techniques of war forecasting. He assesses our ability to predict the onset of various types of armed conflict based on a review of existing track records, showing progress with respect to intrastate armed conflict as well as significant limitations when it is applied to interstate war. He then reflects on five factors of (un)knowability related to complexity, measurability, equilibration, stochasticity and tractability, and suggests measures to at least partially address the challenges associated with them.

In Chapter 15, Jeni Mitchell considers the utility of apocalyptic imagery in thinking about future war. She offers a concise history of the apocalypse in human thought, and makes a strong argument that we need to start taking apocalyptic thinking seriously, through sober engagement not just with science fiction, but also with the existential risk literature. To do so, Mitchell offers an analytical typology to differentiate between different apocalyptic threats and their effect on policy and strategy. She then examines the effect of the war in Ukraine on apocalyptic thinking, arguing that although it may not have generated a new 'genre of apocalyptic fears', it certainly strengthened 'an already existing brew of existential anxieties' that are becoming more prevalent in today's *zeitgeist*.

In Chapter 16, Antoine Bousquet bypasses the epistemological questions and focuses more on an ontological exploration of the phenomenon of war. He questions the widely accepted notion that war's nature is enduring. He argues that rather than possessing timeless qualities, war is in a constant process of becoming, and its nature is therefore changing too. He proposes that our understanding of war will be served if we stop trying to divine war's very essence. Anticipating the objections of 'the card-carrying Clausewitzians', he argues that Clausewitz himself agreed that war changes its nature and is both composite and variable, based on humanity's malleability and adaptivity. That, however, should only compel us to try and

19

understand war in its various incarnations to shape a future in which humans rein in the risks of war, rather than succumb to them.

Prior to his unfortunate recent death, Christopher Coker prepared a review of this book for Hurst that essentially constituted a review essay in its own right. In honour of his memory, we are republishing it here as an Afterword. In it, Coker offers a lucid appraisal of why the study of war, and particularly its future, is as relevant as ever, despite recent attempts by some well-known scholars to cast it as a moribund topic.

* * *

As we write these words, the war in Ukraine rages on, with no end in sight. This war, like so many others, has resulted in staggering levels of physical destruction and human suffering. Despite aspirations for humanity to move beyond war, it remains an integral part of the human condition, and this shows no signs of changing. So long as this is the case, there will be a need to study the war phenomenon— whether to prepare for its future incarnations, prevent its future occurrence, or simply to enlighten the interested onlooker. In the following chapters, multiple viewpoints are presented. Some of these are generally consistent with and complement one another; others offer contradictory perspectives. Taken together, they offer the reader, regardless of their motivation for wanting to study war, a diverse range of scholarly insights into how we might think about a deadly phenomenon, the end of which we are unlikely to see.

PART I

BOUNDING THE IMPACT OF THE WAR IN UKRAINE

1

REVISITING PUTIN'S 2022 INVASION OF UKRAINE
IMPLICATIONS FOR STRATEGIC AND SECURITY STUDIES

Antulio Echevarria

The defense establishments of the United States, its strategic partners and their rivals have already extracted numerous operational and strategic lessons from the Russian invasion of Ukraine on 24 February 2022.[1] How well these lessons will hold up from 2023 onwards remains to be seen; for just what a 'lesson' is and how it is 'learned' remain contested issues. Of necessity, the interrelated fields of strategic and security studies, beholden as they are to rolling academic interests and rigorous research methods, move more slowly. That characteristic has both advantages and disadvantages, as it affords academics opportunities to arrive at defensible conclusions. Military thinkers, on the other hand, would like academia to move more quickly so that they can put scholars' good research to practical use.

This chapter is an effort to encourage some of that good research by drawing attention to several themes important to strategic and security studies. This conflict—the largest interstate clash of the twenty-first century to date—has seriously challenged some of our assumptions about military strategy and major war. To begin with, it calls into question half of the six principal explanations for the apparent waning of major war since 1945. It also adds important new insights to the growing body of literature on strategic coercion, particularly with respect to the criticality of information flows and the effectiveness of extensive financial and cultural sanctions. Similarly, the war reveals much about the limitations of the popular paradigm 'war amongst the people,' as advanced by British general Rupert Smith some two decades ago. Essentially, this conflict shows at least some modern wars are not only 'amongst the people,' but 'against' and 'by' them as well. Furthermore, this conflict tells us a great deal about those forces—enmity, chance, political purpose—commonly associated with the Clausewitzian model of war's nature, especially the power of enmity as a strategic multiplier. This special commentary offers some initial thoughts about each topic in turn. But it is important to make clear that this list is hardly exhaustive.

The Waning of Major War

Despite what some pundits and scholars have argued, large-scale, interstate wars have not become *passé*.[2] Since at least the beginning of the twentieth century, peace has generally been more profitable than war, largely due to the avoidance of market disruptions.[3] In any case, such wars rarely occurred more than once or twice per century in the modern era. Nevertheless, half of the six explanations many scholars have given for the decline in the frequency of large-scale conflicts (detailed below) have acted instead as accelerants in the case of the war in Ukraine.[4] The insecurities and fears autocrats such as Vladimir Putin have felt, in other words, may indeed be real (at least to them), which in turn suggests other autocrats may also act aggressively in the future. As such, we may see more major wars before we see fewer. That does not mean major wars will become an epidemic, but it does suggest that allowing ourselves to feel sanguine about the possibility of a major war is unwise.

1. The proliferation of weapons of mass destruction

According to this explanation, major wars have declined in number due to the risk such conflicts pose with respect to nuclear escalation, which could well lead to unparalleled devastation. Instead, states have opted to pursue limited conflicts that do not present existential threats to other regimes or to compete within the so-called 'gray zone'—the realm of aggression short of war. As some sources have noted, Putin chose to launch large-scale operations against Ukraine precisely because his previous invasions led only to 'frozen conflicts' in the Donbas and Luhansk oblasts, and these gray-zone activities did not yield the results he desired.[5]

2. The spread of democracies and democratic values

This explanation suggests the decline of major wars has occurred because the number of democracies worldwide is increasing, and democracies purportedly do not go to war with one another.[6] Yet, as multiple accounts have indicated, Putin perceived Ukraine's movement toward a fully democratic and representative government as a threat to his style of autocratic rule, and so he opted to arrest that progress with military force. In this case, the spread of democracy and democratic values increased, rather than decreased, the likelihood of a major war. Given the fact that autocratic regimes frequently see democracies as threats, the spread of democracy itself appears likely to cause more wars before it can be said to cause fewer of them.

3. The growth of multilateral institutions

Multilateral institutions such as the North Atlantic Treaty Organization (NATO), the United Nations (UN), and the European Union (EU) are believed to have reduced both the number and scale of armed conflicts by increasing security more collectively and by creating 'new normative standards, communication channels, and institutional practices.'[7] These new alternatives and customs have provided states with opportunities to enhance their security and to channel their competitiveness in less belligerent ways. Unfortunately, Putin saw at least one of those multilateral organizations—namely NATO—as a threat to his own nation's security. In 1946, George

Kennan described the Russian mind as perennially suspicious and insecure, a characterization we may hope will one day be overcome by events.[8] But that day is not yet here. In terms familiar to students of Thomas Schelling, even an alliance built merely to deter must, by definition, be intimidating.[9]

4. Increasing economic integration

According to this explanation, governments refrain from choosing armed conflict to settle their grievances because war in general—and interstate war in particular—causes a high degree of economic disruption. Armed conflict clearly benefits some sectors of the global defense industry; however, it disrupts commerce and financial markets, driving up prices and increasing other costs even for parties not directly involved in the conflict. Even though the Russian economy is relatively small compared with many Western economies, the sanctions imposed on it by the West have started a ripple effect that some experts warn might halt globalization and separate the world's economy into three spheres: a China-led one, a US-led one, and a European one divided between the other two.[10] Whether or not the effects extend that far, fears over the negative impact a major war might have on an integrated global economy are at least partially founded, as second- and third-order economic effects are notoriously difficult to predict. For his part, Putin gambled in two ways: first, that Russian financial institutions would find sufficient workarounds to remain effective; and secondly, that the campaign in Ukraine would conclude before sanctions could take full effect. On the first gamble he was correct, as projections by the International Monetary Fund for Russian economic growth in 2023 are at 0.3 per cent, rather than a negative number.[11] However, it remains to be seen how much longer the Russian economy, the eleventh-largest in the world, with a GDP of US$1.70 trillion in 2019, can endure such pressures as the conflict becomes more protracted.[12]

5. The influence of international law and the law of armed conflict

This rationale suggests the influence of international law and the law of armed conflict have restricted the reasons states may legally go to war and how they may wage it. To be sure, to have legal restraints on the conduct of war is useful. But for this explanation to be persuasive, prosecutions of war criminals must occur in a timely fashion.[13] Historically, this has not been the case. For example, 'It took two decades for the Nazi Adolf Eichmann to be called to account. It was two and-a-half decades for former Chilean President Augusto Pinochet, and four decades for Kang Kek Iew, Nuon Chea, and Khieu Samphan.'[14] Clearly, the existence of the International Criminal Court and the promise of post-conflict investigations into possible war crimes did not dissuade Putin from invading Ukraine or from allowing his troops to attack non-military targets. In fact, attacking noncombatants appears to be one of the Russian army's primary tactics.

6. The spread of anti-war norms

Quite distinct from the legalist explanation is one concerning changing cultural norms, which no longer buy into the 'glamorization' and 'heroization' of war. This explanation says the expansion of anti-war norms has made it much more difficult to 'sell' a contemporary populace on the need to participate in an armed conflict. To be sure, anti-war norms have ebbed and flowed throughout modern history. Nonetheless, they represent an important measure of national will (or international will in some cases). They also have a critical downside in that aggressors can leverage such attitudes to bully states into policies of appeasement. Putin has successfully employed that tactic throughout much of his presidency. Fortunately, the situation reversed itself after his invasion of Ukraine. Most of the free world, with the assistance of a brilliant Ukrainian information campaign, bonded emotionally with President Volodymyr Zelensky and the Ukrainian people, and came to see the Russian state as having brutally victimized its peace-loving neighbor.

To be sure, all six explanations offer plausible reasons for the purported decline of major conflicts since the Second World War. However, as mentioned above, none proved sufficient to dissuade Putin's invasion of Ukraine, with the first three functioning more as accelerants, or encouragement, for that aggression.

The fourth—economic integration—is essentially neutral. The globally integrated economy affects aggressors, defenders and

BEYOND UKRAINE

nonaligned parties alike, though certainly not equally. It is worth noting that sanctions and economic embargoes have become 'weapons of first resort,' at least for pro-Western democracies with robust economies. In this case, however, sanctions proved disappointing, as they did not deter Russia, which claimed to have become inured to them. There was some truth to that claim: having been subjected to Western sanctions several times before, the Russians had developed some workarounds. Russia also retaliated economically by attempting to wage an energy war against Ukraine and the West by restricting the flow of gas and oil, or by shifting it all together. Both sides have had successes and failures. In any case, sanctions and embargoes clearly require time to work, and Putin wagered he could achieve his objectives in Ukraine before such strictures took effect. As recent reports indicate, he almost did: the initial assault on Kyiv, designed to neutralize the Ukrainian government, was a near-run thing.[15] In addition, Putin initially had the acquiescence of China, India and several other countries, who kept much of their markets open. As such, the volume of Russia's imports in Western markets has 'plunged,' but the volume of its exports to Asian markets has doubled or tripled.[16]

Sanctions and embargoes clearly require the cooperation of other parties to be effective. However, such cooperation cannot be assumed in a global economy, as those states imposing them will also suffer from the potential loss of valuable goods and services. Russia retaliated against the sanctions and embargoes imposed on them, yet other states not directly involved in the conflict or with weak economies have also suffered hardships as a result. By most accounts, Russia's economy has been severely damaged by the combination of these restrictions alongside their involvement in a major war. But observers also indicate the Russians have adapted economically by streamlining their production lines, and thus are girding themselves for a long war of exhaustion with the West. As such, if Russia manages to break the West's willingness to continue supplying Kyiv with war materiel, the war might develop into an insurgency or some form of hybrid conflict.

Not surprisingly, the influence of international law and the law of armed conflict neither dissuaded Putin nor his top advisors. It

28

is unrealistic to expect Putin to be deterred from taking an illegal action simply because it was illegal. The Russian attitude, in any case, appears to be that laws are meaningless unless they can be enforced, and it is not clear whether the West will be able to apprehend those Russians who have committed war crimes (and survived the war). Nonetheless, the existence of legal conventions offers some hope of exacting at least partial justice for similar crimes in the future, under different circumstances. That might in turn deter would-be war criminals.

The last explanation—the spread of anti-war norms—is interesting because it offers aggressors advantages during peacetime, but then quickly works against them in wartime. Anti-war sentiments transformed almost overnight into antipathy for the Russians and sympathy for the Ukrainians. Before the invasion, Putin's bullying tactics gave him a distinct advantage in dealing with heads of state who wanted to avoid war. But he lost that edge once the conflict started and antipathy grew, which led to a host of cultural sanctions, such as barring Russian athletes from competing in international events.

This list is instructive for what it includes as well as what it omits. Oddly, a seventh potential explanation for the low incidence of interstate wars since 1945 is the relative balance of military power, especially regionally. Heads of state might have feared nuclear escalation and therefore decided to avoid going to war; however, they might also realize that they do not have enough military advantage over their rivals to take the risk of going to war. This contemporary 'balance of power' is not the 'balance of nuclear terror' that existed between NATO and the Warsaw Pact countries during the Cold War, but it might be as effective. After all, until recently, Russia chose to compete within the 'gray zone' (that is, below the threshold of war) not because of Ukraine's military power vis-à-vis its own, but because of NATO's. Obviously, miscalculation is always possible (as Putin's difficulties so far have shown) because Russia is facing much of the military (and economic) power of NATO and other parties, just not their troops. The term 'balance of power' is a misnomer in any case as military power is usually in a state of imbalance. The question is what type of imbalance is needed. By the time this conflict in Ukraine comes to an end, NATO will need to have determined how much imbalance is necessary between Russia and Ukraine to prevent another miscalculation.

Strategic Coercion—Re-examining the Concept

The conflict in Ukraine both confirms and challenges our understanding of strategic coercion. Although there are many definitions of strategic coercion in the academic literature, the best is still the one developed by Sir Lawrence Freedman; he describes it as the 'deliberate and purposive use of overt threats to influence another's strategic choices.'[17] Strategic coercion is generally thought to consist of two elements: compellence—making adversaries do something we want them to do; and deterrence—making adversaries refrain from doing something we do not want them to do. For the latter, the academic community has settled on two basic types of deterrence: by punishment and by denial. Analysts generally agree that the first type—deterrence by punishment—which consisted largely of economic sanctions, was not serious enough to deter Putin. The Russians have experienced Western economic sanctions before and believed they could weather the storm, so to speak. To put it in Thomas Schelling's terms, the Russians did not consider the 'pain yet to come' from further sanctions to be sufficient to deter them from attempting to accomplish their goals.

It is not clear, however, whether deterrence by denial—that is, increasing the odds that aggressors will not be able to accomplish their objectives—might have worked. The West was clearly one step ahead of the Kremlin, releasing critical intelligence about Moscow's planned 'false-flag' operation before it happened, for instance, all of which ought to have indicated the Russian military's chances of achieving its operational goals were dubious at best. Yet those efforts failed. It may well be that deterrence by denial requires something more, such as a viable defense concept, demonstrated publicly through training exercises and so on, to demonstrate more persuasively that the attackers will fail to achieve their objectives. The Ukrainians were moving in the direction of such a concept with President Zelensky's announcement in early January 2022 of a policy of 'total defense,' which essentially amounted to a modern-day form of *levée en masse*.

The policy was not funded, however, and was enacted too late to be truly viable. In the end, it resulted in thousands of Ukrainian citizens being issued a Kalashnikov assault rifle and two magazines, and being told to defend their cities and towns.[18] Despite the confusion, the frictions and the inevitable fratricides this mobilization caused, the question is whether the participation of Ukrainian citizens in their own defense, both as combatants and as intelligence sources, accomplished just enough to offset the Russian timetable. Or were armed citizens more of a nuisance, especially since the Russians had also armed sympathizers operating within Ukraine? After all, how does one distinguish friend from foe under such circumstances? Details about Putin's plan of attack remain sketchy, but enough information has surfaced to suggest it was not based on conventional Russian military doctrine, but rather on an attempted decapitation strategy (or *coup de main*) engineered by the Kremlin's intelligence services. A combined thrust of Russian armor, air and special forces was to seize Kyiv, neutralize the Ukrainian government, and transition to stabilization operations by D+10.[19] NATO's leaders, meanwhile, needed four or five of those ten days to assess the full extent of the Russian attack and to decide how they would respond. The bottom line is that the amount of time NATO needs to assess the situation and to determine how to respond represents a window of opportunity for an aggressor and a window of vulnerability for a defender. Unless that window can be closed, perhaps by a well-organized and well-led territorial defense force, aggressors may be tempted to exploit it by attacking. A better-prepared citizen force— one that is trained, equipped and funded—might help close that gap in a future war and thereby bolster a concept of deterrence by denial. Nonetheless, Putin simply might not have been deterrable in this case. All the more reason, then, for a well-integrated defense concept should deterrence efforts fail.

Since deterrence of any type depends on the flow of credible information to the would-be aggressor, the war in Ukraine has exposed another dilemma for strategic coercion: the irrational actor versus the ignorant actor. The former term has become controversial since it first appeared in the nuclear era. Actors who would willingly work against their own interests, especially those who might risk nuclear suicide, were considered irrational. But the emergence of violent non-state actors and terrorist groups willing to commit suicide for their cause introduced a new sense of the term 'rational actor.' Consequently, the term 'irrational' has lost its utility. However, the conflict in Ukraine has raised the question of the extent to which actors are genuinely aware of their situation, including friendly and enemy capabilities and losses. If target leaders are receiving accurate information but choose to reject it, perhaps because it is unpleasant, that is one type of problem. If target leaders are not receiving accurate information from their advisors, again perhaps because it is unpleasant, that is another problem. The latter assumes target leaders might consider compromising or complying if they receive reliable information. Rational but ignorant actors, unlike irrational actors, have thresholds they do not wish to cross; these parties can be convinced or compelled. Research regarding strategic coercion has not yet separated the irrational actor problem from the ignorant actor problem. To illustrate the point, Putin's advisors may not have given him accurate information initially, perhaps due to fear, or simply as a byproduct of the Russian military culture of 'overly optimistic reporting.'[20] Perhaps aware of this culture himself, Putin took steps to obtain better information and evidently succeeded. In the interim, he was surrounded by false or unreliable information, and therefore in a state of ignorance about the military situation. Accordingly, our attempts to coerce him (assuming he could be coerced) with our information may not have seemed credible to him, and hence had little chance of succeeding. Put differently, the theory and practice of strategic coercion would be enhanced by efforts to analyze Putin's decision-making not simply from the traditional dichotomy of a rational–irrational actor, but from the standpoint of an ignorant–informed actor.

Strategic coercion would also benefit from research into how the concept's two essential complements—compellence and deterrence—function together synthetically. While separating the two has some value, particularly when simplifying them for educational purposes, it tends to obscure their complementary nature. Compellence is implied in deterrence, and vice versa. Both aim to make our adversaries do what we want and to be unable to do something else. Achieving that aim requires pushing our adversaries down some avenues while blocking them from moving down others. A perfect military strategy, in fact, would be a combination of push and block, compellence and deterrence. Carl von Clausewitz's use of a wrestling metaphor when he defined war as 'an act of force to compel our adversaries to do what we want' is illustrative because a proper wrestling hold makes it impossible for our opponent to do anything but what we want.[21] Nonetheless, Schelling's bargaining model of war still applies because no matter how excellent the wrestling hold, our opponent still has to concede. The bargaining model applies to more situations than Schelling himself appears to have realized. On the other hand, his model of 'brute force,' which entails forcibly taking something such as territory from a foe, applies less frequently than he realized.[22] The reason is that even forcibly taking territory from a population with brute force requires them to concede by opting not to fight for that territory by indirect or unconventional means, such as an insurrection. As a nuclear strategist and theorist of limited war, Schelling can be forgiven for not giving the element of public consent its due in armed conflict. As the United States' foremost game theorist at the time, he was preoccupied with dichotomies. The bargaining-versus-brute force model was one such dichotomy. However, brute force is not an equal element in that dichotomy: it can only occur in situations where consent is immaterial, where aggressors decide to eliminate a defender or forcibly depopulate an area and are militarily strong enough to do so. We are seeing the Russians carry out a policy of depopulation in some locations within Ukraine; fortunately, they are not strong enough to do so everywhere. Nevertheless, it is a characteristic of the Russian way of war, as evidenced by their tactics in Chechnya, Georgia and Syria. The West needs to find better ways of countering it.

For their part, Kyiv wants to compel Putin to give up his aggressive intentions, deter other states from supporting him, and avoid bleeding its own military white in the process. That is a difficult task. But strategies of control, which emerged in the United States during the 1950s and 1960s through the works of J.C. Wylie, Henry Eccles and Herbert Rosinski, can be useful in this situation.[23] Kyiv has gained control of several important dimensions in this war: it has taken control of the information dimension of the war with a masterful public information campaign; it has retaken key military and political terrain via the cities of Kharkiv and Kherson; it has denied the Russians control of the air dimension through effective (and improving) air defenses; it has secured materiel support from the West capable of outpacing Russian production capacity; and it has obtained sanctuaries among NATO states where its troops can be trained and its equipment refitted free from Russian long-range strikes. Less is known at this point about the cyber and space dimensions of the conflict. Control of these dimensions (and others not mentioned) is critical in a long war, especially one in which it is impractical for the defender to advance on the aggressor's capital. The center of gravity in this war is Western materiel support for Ukraine, which is partly contingent on public opinion in the West, but also on how strong the West's strategic interests are vis-à-vis Russia and other threatening actors. In short, the conflict in Ukraine is providing new data capable of illuminating how strategies of control might facilitate strategic coercion, and vice versa. These data, moreover, should justify fusing compellence and deterrence together more formally, rather than informally or accidentally.

To return to the topic of financial sanctions mentioned above, Putin's invasion of Ukraine has shed additional light on the coercive value of economic power vis-à-vis a modern state with a relatively small economy but a large military. Sanctions have become the weapon of 'first resort' for the West's policymakers for a variety of reasons.[24] But they are also controversial, since many experts do not believe they are effective. The sanctions the West has imposed since 24 February 2022 fall somewhere between outright economic warfare—which aims to ruin an adversary's economy—and economic coercion—which brings pressure to bear by applying (or withholding) foreign aid, monetary power, financial power and trade.[25] The Russians countered the West's economic pressure by finding workarounds, such as sailing their merchant vessels under false flags, and by attempting to wage an energy war against Ukraine and the West by restricting the flow of gas and oil, or shifting it entirely. Both sides have had successes and failures. The volume of

Russia's imports in Western markets may have 'plunged,' but the volume of its exports to Asian markets has doubled or tripled, giving the Kremlin a bona fide lifeline.[26] The Russian economy has endured longer than expected, suffering considerably less than analysts predicted.[27] But it might still collapse. In any event, this conflict is providing a wealth of data that may move the debate forward over the strategic efficacy of economic pressure.

On a related note, the war in Ukraine has revealed an important weakness in the West's ability to wage a major war, namely that the production capacity of its industrial base is insufficient to sustain large-scale combat operations for extended periods while also maintaining robust deterrence postures abroad. One of the things the war in Ukraine has thus far shown rather distinctly is the rate of material consumption, especially in ammunition, which has been much higher than prewar thinking anticipated. The Ukrainian Armed Forces have reportedly consumed an average of 100,000 rounds of 155mm artillery ammunition per month, yet the US defense industry can replace only 14,000 rounds per month (though that number should increase to 20,000 rounds per month over the next two years).[28] The story has been similar for Stingers, Javelins, high-mobility artillery rocket systems, and other critical weapons.[29] Moreover, several of the components and subcomponents needed for these weapons are manufactured outside the United States, which compounds the problem. The West, particularly the United States, has been rapidly partnering with private industry to close the gaps, but the fact that the gaps occurred in the first place reveals a lack of strategic imagination and a sense of complacency about the types of wars the West was going to have to fight.

War Amongst, Against and By the People

In the early twenty-first century, British general Rupert Smith attempted to introduce a new paradigm of armed conflict, which he referred to as 'war amongst the people.'[30] This paradigm was intended to shift defense thinking and procurement in the West away from its preoccupation with force-on-force conflicts, or what Smith refers to as 'interstate industrial wars,' to contemporary wars. According to Smith, these wars are characterized by six major trends. First, the ends for which wars were fought have changed from the 'hard absolute objectives of interstate industrial war to more malleable objectives to do with the individual and societies that are not states.' Second, wars were now fought 'amongst the people,' as exemplified by the 'central role of the media,' which bring armed conflicts into 'every living room,' even as they are being fought in streets and fields far away. Third, modern conflicts 'tend to be timeless,' since they center on establishing conditions that must be maintained until treaties or peace agreements are reached, which can take years or decades. Fourth, fighting takes place in a manner designed 'not to lose the force,' rather than employing the force and expending it as necessary in pursuit of the overall aim of the conflict. Fifth, 'old weapons' designed for industrial war were out of necessity being adapted to 'new uses' so as to accommodate 'war amongst the people.' And finally, the sides in contemporary conflicts consist mostly of non-state actors (meaning multinational groupings such as alliances or coalitions) pitted against parties that were not states.

Smith can be faulted for attempting to use Thomas Kuhn's framework of conceptual paradigms to describe different types of wars.[31] Paradigms are better at describing the systems of thought or the ways individuals and groups think about things than the things themselves. Wars are notorious for the 'contemporaneity of the non-contemporaneous,' a phrase once popular among French and German sociologists to describe generational overlap (that is, when individuals of two or more generations occupy the same space and time). An example is Western society in the 1960s, when a younger generation embracing anti-establishment values clashed with an older, more conservative one hewing to traditionalism. In short, classifying phenomena according to periods can be problematic because things can be *in* an era without being *of* that era. So it is with wars. Industrial-age, interstate conflicts such as the First and Second World Wars occurred temporally with many of the United States' 'Banana Wars,' for instance, in which the US military often had to deal with violent non-state actors. Yet the two types clearly differed. (The two world wars, incidentally, were fought by alliances, which Smith and others classify as non-state actors.) The United States has participated in at least ten times more non-interstate, non-industrial-age wars than it has interstate, industrial-age wars.[32] Nothing about the twenty-first century thus far suggests this ratio will change in favor of interstate wars. While his attempt to classify wars is problematic, Smith should not be faulted for having tried to persuade defense establishments in the West to develop better tools for fighting non-industrial, non-interstate wars. That dream remains both noble and urgent. Western defense establishments continue to resist investing in the capabilities and skill sets needed to deal with such wars, while ironically not investing enough in the logistical requirements necessary for large-scale combat operations. Traditionally, larger profits have come from producing military hardware for interstate wars, and that hardware has become increasingly exquisite and expensive over the years, probably because it has not had to survive in a highly lethal environment. Nor has it been necessary to mass-produce it.

Most of the trends Smith identified are not true trends. His description of the 'absolute nature of political aims,' for instance, was true for the Second World War, but not really for the First World War. Nor does it hold up when we consider the Korean and Vietnam wars, both of which featured heavy fighting as part of a Schelling-like bargaining process. He is correct about the second trend, but not only have the media shaped our perspectives of a conflict, but we have also shaped how the media represent the conflict to us. Moreover, for decades now, the explosion of social media has enabled us to participate directly in the representation or misrepresentation of conflict; it has empowered us to 'spin' events almost as we please, irrespective of whether we believe what we post.

His third trend—that modern conflicts tend to be timeless—was also true in the early modern period, as evidenced by the Hundred Years' War and the Thirty Years' War, but the world wars were short by comparison. The conflict in Ukraine, meanwhile, may take on various incarnations, become many different types of wars over time, like the conflicts in the Middle East. The Russian invasion of 24 February 2022, for instance, was preceded by Moscow's seizure of Crimea and territories in the Donbas in 2014.

His fourth trend—that military operations are conducted in such a way as not to lose their force—has been true for many militaries in

Iraq and Afghanistan. In the latter especially, a state's political objectives might be met by simply 'showing up.'[33] Taking casualties, meanwhile, would have left governments vulnerable to criticism at home. This type of reaction reflects the attitude of anti-warism discussed earlier in this chapter, but it also reveals the sense that the voting publics in many Western countries simply did not believe the goals in Afghanistan were achievable or worth risking their soldiers' blood. Western governments are, after all, accountable to their constituencies.

Trend number five—adapting weapons designed for conventional warfare to irregular warfare (for which they were presumably ill-suited)—is true. But it is an age-old story and an outcome of the inelasticity of the modern defense industrial base, which sees little profit in meeting the low demand for materiel for small-scale operations.

The sixth trend—that modern wars are carried out less by states and more by multinational coalitions or alliances against non-state actors—is also true, but there is more to the story. States are key players in galvanizing coalitions and alliances into action, partly because there is comfort in numbers and partly because alliances like NATO need their raison d'être renewed periodically. Such realities have existed before the modern era.

The salient characteristic Smith rightly ascribes to new wars—counterinsurgencies and peacekeeping operations—is that they occur amongst the people. But as we have seen, that is also true of major wars today. As of 5 April 2022, the UN International Organization for Migration reported some 11 million people had been displaced within Ukraine and more than 4 million had fled the country.[34] Refugees would have impacted any conflict that might have broken out in Central Europe during the Cold War, though Smith's point is that military doctrine and training exercises at the time rarely took account of the refugee flow and how its presence might impede operational maneuver. But the plight of refugees and displaced persons in Ukraine means this war is not only *amongst* the people, it is also *against* the people in the most egregious way. The Russians have directly targeted noncombatant populations in Ukraine, and video evidence and personal testimonies have documented many of those incidents for adjudication later. While the targeting of noncombatant populations might provide limited tactical advantages, it comes with heavy strategic disadvantages as people across the globe watch the conflict play out on their television sets, iPads and computer screens, and call upon their governments to do more to support the Ukrainians.

In short, noncombatants are reluctant participants in this war just as much as Ukrainian and Russian military personnel. This war is therefore as much a war amongst the people as it is a war against the people in every sense, despite being interstate and multinational in character. Western military strategy and doctrine must account for this phenomenon as it may well manifest itself in similar ways in other theaters.

War's Changing Character and Dynamic Nature

The fact that the conflict in Ukraine is a war amongst and against the people raises an important point about the relationship between war's character and its nature. Western militaries tend to refer to a chameleon analogy when discussing war's character, which they borrow from Clausewitz's *On War*. In the analogy, war's character—the institutions that participate in armed conflict, the weapons and doctrines employed, and the whole process of warfare itself—changes like the skin of a chameleon, while war's nature—political motives, human emotions and the element of chance—does not. The chameleon remains a chameleon, and war remains war. However, the fact that war has common denominators such as political motives, human emotions and the element of chance does not mean that they do not fluctuate and interact with each other. After all, these forces are dynamic, perhaps even more so than the institutions that make up war's character. What this dynamism means is that war's character could remain relatively stable while important fluctuations are occurring within war's nature that will transform the chameleon into a dragon. This dynamism is also at work in the war in Ukraine, where enmity for the aggressor has motivated the defenders to resist fiercely. Clausewitz would have recognized this conflict as a war of national resistance or of national liberation wherein the citizenry takes up arms in its own defense.

Of course, it is impossible to predict how long this spirit will last, though it is clearly a strategic multiplier while it does. It could fuel an insurgency. It could fade without material support. Or it could turn against its leaders if they are perceived to have given up on the cause. These possibilities underscore how important it is to understand the forces that make up war's nature and the bearing they might have on a war's direction and intensity. Important as it is, it is not enough to study war's character. Such studies are literally only skin-deep. War's nature and its character should be understood together; for each could have serious implications for integrated deterrence and defense.

Conclusions

In sum, research into the conflict in Ukraine already offers a wealth of answers to some fundamental questions in the field of security and strategic studies. Paradoxically, it will also create more questions for academics to ponder. Moreover, each of the topics discussed above informs the general context of the war in Ukraine in important ways. Three of the explanations for the decline of major war, for instance, also contributed to shaping Putin's justifications for the 2022 invasion of Ukraine. In turn, theories of strategic coercion have influenced the quality of each side's thrusts and parries. 'War amongst the people' is still a valid way to frame modern conflict, though it includes many more dimensions than its author originally conceived. Finally, the motivational element of war's nature has proven quite powerful indeed in favor of one side and to the obvious detriment of the other. Any war in the not-too-distant future is bound to share some of the characteristics of Putin's invasion of Ukraine in 2022. But there will likely be enough differences to warrant caution with regard to military lessons. Strategic and security studies enjoy a luxury that is not necessarily shared with military thinking, which is the ability to self-correct over time. We should therefore expect, if not welcome, the implications discussed in this chapter (as well as those not discussed) to change in interesting ways in the years ahead.

2

THE FUTURES OF WAR
A RECENT WESTERN HISTORY

Frans Osinga

Introduction: The Recent History of the Future of War

The future has a long history. Visions of future war often reflect recent, sometimes traumatic experiences, cater to the specific security concerns of a nation, address the ambitions and needs of specific services, focus on the revolutionary potential of emerging technologies, provide warnings for the rise of new types of actors or specific peer competitors, or reflect the strategic culture of a specific nation. Certainly over the past three decades, there has been no shortage of visions on what the future of war might look like. Those visions often reflected the experiences of the recent wars Western militaries had been engaged in, and the complex operational dynamics they encountered during these interventions. While such visions often emphasized continuity in war and warfare, others have predicted discontinuities or even revolutionary changes.

Prior to the 2022 Russian invasion of Ukraine, three decades of societal dynamics, operational experiences, new technologies, tactics employed by non-state actors and perceived changes in the security

41

environment inspired five different yet plausible perspectives on the future of war. These were:

1. Humanitarian wars;
2. Sophisticated barbarism;
3. Immaculate war;
4. Cool war; and
5. Major war.

All of these arose within Western military, policy and academic circles. When published, some seemed relevant for the US, whereas others would be more in tune with European security cultures. Most of these inevitably suffered from presentism: they either emphasized continuities, or offered speculative expectations about the possible impact of new technologies and insisted on disruptive innovations. Other views had a normative slant. With the war unfolding in Ukraine at the time of writing, analysts have once again suggested that this particular war paints the landscape of future war. This chapter aims to capture the trajectory of Western thinking on the future of war prior to the Russian invasion of Ukraine on 24 February 2022, and concludes that the future of war after the invasion cannot simply be reduced to a single one of the five above-mentioned perspectives. Instead, we should expect the overlap of several different perspectives simultaneously.

The Future in the 1990s

A Revolution in Military Affairs

The end of the Cold War in the early 1990s heralded a 'New World Order', especially for the US, whose military power formed a key foundation for maintaining and expanding the liberal world order. The ideological struggle between East and West was over, and democracy, liberalism and capitalism seemed destined to gradually but steadily spread across the globe. In 1991, Operation Desert Storm seemed to demonstrate that US military power was vastly superior to any potential competitor. Indeed, the American political scientist Eliot Cohen suggested that an embryonic revolution in warfare was underway.[1] Technologies such as precision weapons, stealth aircraft,

THE FUTURES OF WAR

conventional cruise missiles, electronic jamming devices and new generations of sensor platforms, all connected through data links and coordinated in an operational headquarters in which the traditional and dysfunctional divisions between the armed forces had been removed, resulted in an unprecedentedly effective force that was capable of defeating Iraq (at that time the fourth-largest army in the world) without suffering major losses.

Some considered this new epoch to represent an 'information revolution' since the US-led coalition had information dominance while Iraq was constantly lagging behind, their command centres 'blind and deaf' after targeted attacks. Others labelled it a 'precision revolution' and predicted 'precision-age warfare'; small targets could now be attacked with unprecedented accuracy, a single bomb being enough to destroy a building or a tank, whereas previously several bombs had to be dropped with corresponding collateral damage. As Cohen suggested, everything that can be seen could now be hit. Finally, some analysts mainly saw an 'air-power revolution' because the technological developments had mainly benefited the effectiveness of air operations.[2] Air power was no longer to be regarded merely as a means of support for land operations; with the offense now dominant in air warfare, in the future it would also be possible to wage and decide a war without large troop formations in the area of operations.[3]

Despite intense inter-service debates, the US Army and Navy, like the US Air Force, embraced the new technologies. Whilst the Army proceeded to digitize their armoured vehicles, tanks and operational headquarters,[4] the Navy concentrated investments in information and communications technology (ICT) and data links. The Air Force, meanwhile, tried to shorten the time between observation of a small mobile target and the moment of attack—the so-called sensor-to-shooter time—through data links between sensors and offensive weapon systems. US Admiral Arthur Cebrowski stated that his country's future conflicts resided in 'network-centric warfare',[5] with units interconnected through data links and with support systems such as fighter jets that could 'swarm' and coordinate attacks over far greater distances than was previously possible. The introduction of networks made other flatter, decentralized forms of organization possible, which would enhance operational tempo.[6] Subsequent successful operations in the early 2000s (such as Operation Enduring Freedom and Operation Iraqi Freedom) seemed to confirm the validity of this 'New American Way of War', and suggested wars could be waged with only limited political risk, as well as in terms of risk for American ground troops, civilian casualties and collateral damage.[7]

The European Perspective: The Humanitarian Impulse

In Europe, strategic history and futurizing largely evolved differently from that in the US. Critics argued that military-technological revolutions only provide military superiority temporarily. Moreover, as this revolution was based on civilian ICT, this dominance would be relatively short-lived because other countries could quickly catch up.[8] Indeed, visions of future war, if at all present, were not shaped by the notion of a technology-driven revolution in military affairs (RMA), but rather by questions concerning where, and for what purposes, the military would ever be deployed now the enemy had disappeared. Towards the end of the Cold War, it had become subrationally unthinkable in most Western European societies that war could happen, and was no longer considered a useful and legitimate tool of statecraft, as John Mueller noted in 1989.[9] War had become 'obsolete'; useless as an instrument because of its destructive effect in the atomic age; superfluous because of the growing interdependence and the binding effect of international organizations; and, moreover, normatively no longer conceivable and no longer appropriate in the policy instruments of highly developed Western societies.[10] 'Soft power'—the positive effects of globalization, international treaties and organizations—was considered more important. Europe, including its armed forces, had become 'postmodern'.[11] As Lawrence Freedman observed, the future would be one filled with 'wars of choice' rather than 'wars of necessity'.[12]

The Balkan crisis reinforced the conviction that peace operations in civil wars represented the dominant role for European militaries. These needed to have the capability to provide what Mary Kaldor called a 'cosmopolitan law enforcement' in an era that would be

THE FUTURES OF WAR

marked by civil wars. The utility of force, as Rupert Smith argued, was no longer the achievement of a decisive military victory, but the management of force; the creation of a condition in which a political solution can be sought.[13] The use of military force was only considered legitimate when it concerned humanitarian interests. Huge investments in RMA capabilities relevant for high-intensity warfare seemed irrelevant because:

> War no longer exists ... war as cognitively known to most non-combatants, war as battle in a field between men and machinery, war as a massive deciding event in a dispute in international affairs; such war no longer exists.[14]

Frustrating peacekeeping experiences informed much of the military and academic debate. When civil war broke out in Yugoslavia in the early 1990s, the European Communities—later the European Union (EU)—and subsequently the North Atlantic Treaty Organization (NATO), decided on a peacekeeping operation to contain the violence, force the warring factions to agree to ceasefires and thereby alleviate the humanitarian suffering of the civilian population. The operations were initially set up according to the then current UN Blue Helmets model, which assumed several prerequisites: (1) an interstate conflict with functioning political and security authorities; (2) that said political and security authorities actually exercised effective control over the military units; and (3) warring parties consented to the presence of the peacekeeping force, who would remain neutral in the conflict and only be allowed to use force in self-defence.[15] In Bosnia in 1992, these assumptions turned out to be incorrect. In this intrastate war, the warring factions (consisting of Serbs, Croats, Bosnian Muslims, Bosnian Croats and Bosnian Serbs) violated temporary ceasefires when it suited them and showed little interest in initiatives by the United Nations (UN), the Contact Group of Diplomats, or the European Community. UN observers were threatened, UN aid convoys were blocked, and so-called UN Safe Areas turned out to be extremely vulnerable. Air support requested by Blue Helmets frequently resulted in UN observers being taken hostage as a reprisal measure, neutralizing the UN threat.[16] Studies as early as

45

1992 thus argued for a greater use of force, with suggestions for both 'wider' and more 'robust' peacekeeping.

The horrors of the Srebrenica massacre in July 1995 forced European states to switch from peacekeeping to peace enforcement; conduct effective coercive action by means of an air offensive, and adopt robust rules of engagement accordingly. Operation Deliberate Force—eighteen days of air assaults coupled with artillery and mortar fire by the 10,000-strong ground contingent—finally managed to bring the Serb and Bosnian Serb leaders to the negotiating table.[17] Similar complexity and vicious dynamics plagued NATO in 1998–9 when it became clear that Slobodan Milošević had no intention of stopping the purge of Kosovo. After seventy-eight days of bombing, NATO's Operation Allied Force succeeded in forcing Milošević to withdraw his troops from Kosovo.[18]

Strategic analysts subsequently explored which theories of interstate coercive diplomacy and nuclear deterrence could be applied within the constrained contexts of peace enforcement using only limited military means. In the mid-1990s, these forays identified preconditions such as escalation dominance; a credible and rapidly deployable military capability that can inflict the threatened damage; clearly formulated requirements; a strong, united coalition; the creation of a sense of urgency to meet those requirements; and a reputation that threats are fulfilled.[19] Debates on military strategy revolved around the question of whether the mere threat of bombing would suffice, would symbolic attacks be required, and should bombing be intensive or follow a gradual escalation.[20] By this point, precision weapons provided new options to strike at strategically relevant targets—even those in the middle of cities—and thus exert pressure, introducing 'decapitation' as a new coercive mechanism next to punishment and denial.

Humane Warfare

The superiority of the West in terms of military technology reinforced this shift in Western strategic culture. Moving forward, industrial-style warfare was *passé*, and RMA now provided unchallenged maritime and air dominance.[21] The resulting power-projection capability promised assured access to the 'global commons' and

THE FUTURES OF WAR

low risks in conducting interventions.[22] This became a valuable development according to Christopher Coker, who coined the term 'humane warfare' to describe the Western way of war in the 1990s and societal attitudes towards the use of force. It was 'humane' because warfare was accompanied by an unprecedented respect for the law of armed conflict, and because the West only waged war when it revolved around humanitarian interests (or at least when it could be framed that way).[23] For Western militaries, the bar was now set very high precisely because of an increasing aversion to collateral damage and sensitivity to casualties.[24] The high-tech way of warfare therefore became the political and ethical norm. Other scholars cynically labelled Western humanitarian interventions as tools for reducing risks for Western states as these aimed to contain distant conflicts, and prevent regional destabilization and refugee flows that might reach Western borders.[25] Martin Shaw referred to this as 'risk transfer warfare'. In this concept, risks for Western societies, military personnel and politicians must be excluded, and the risks of warfare, if any, are to be passed on to the targeted society and the innocent in the form of euphemisms such as 'collateral damage'.[26]

New Wars, Clausewitz out?

According to Robert Kagan, this strategic culture and vision of the future of war was a denial of history, warning that 'Americans are from Mars and Europeans are from Venus'.[27] Colin Gray was similarly sceptical, giving his book looking at conflict in the twenty-first century the title *Another Bloody Century*.[28] According to various authors, wars of the 1990s demonstrated the future of conflict would be epitomized not by revolutionary weapon systems, but new types of actors employing vicious tactics, rendering Western military superiority irrelevant. These 'new wars' revolved around identity, and featured new dynamics and a fundamentally different conceptualization of war. As such, Martin van Creveld and Mary Kaldor both argued that the instrumental Clausewitzian understanding of war no longer applied.[29]

War had become an end in itself, and everyone with a different identity was an enemy. New types of warlords were, in a parasitic-symbiotic relationship with rivals, intent on continuing local wars in

47

order to maintain their position of power. Identity wars were total in their societal impact and horrific tactics, which included ethnic cleansing, rape, political assassinations, killings of civilians and the razing of houses. The distinction between combatant and civilian was meaningless.[30] Studies on the 'social construction' of political violence showed how leaders of ethnic factions mobilize myths, cultural artifacts and religion to create enemy images and gradually legitimize extreme violence against 'the other'.[31] Participating in this struggle gives meaning to their existence, as exemplified in the title of Chris Hedges' book, *War is a Force That Gives Us Meaning*, which examined how the ethnic conflicts of the 1990s exposed the existential experience of war.[32]

By 1989, Martin van Creveld had already predicted a future featuring such vicious tactics, and Bill Lind had warned that such violent non-state actors would wage 'fourth-generation warfare' which circumvented Western conventional military superiority. This eventually took the form of jihadist terrorists, who would strike deep in the heartland of Western societies to achieve their strategic effects. In doing so, Western casualty sensitivity would be fully exploited, undermining public and political perceptions about the legitimacy of Western action.[33]

The Future after 9/11[34]

The Future is Counterterrorism

Francis Fukuyama's bestseller *The End of History and the Last Man* seemed to promise that the end of the Cold War heralded the end of geopolitics. According to Fukuyama, the demise of fascism and communism meant there was no longer an ideological rival to the liberal-democratic model.[35] Those who rejected this idea, such as Samuel Huntington in his book *The Clash of Civilizations*, in which he argued that wars could still develop along cultural and religious dividing lines across the world, were conveniently criticized, as were those who envisaged rapidly spreading zones of turmoil and warned for the coming anarchy.[36] This sense of complacent security in the West was destroyed on 11 September 2001 with the horrific attacks on the Twin Towers in New York and the Pentagon in Washington, DC, and in 2004 and 2005 by the terrorist attacks in Madrid and London respectively. It had become clear that conflict and violence—still present in so many parts of the world—would not pass by Western societies.

These terrorist attacks spawned a flood of studies on the apparent new form—the fourth wave—of catastrophic fundamentalist terrorism, on radicalization processes, the logic of suicide terrorism, the role of religion and the possible strategies to counter this problem. Studies on al-Qaeda, the Taliban, Hezbollah, and, more recently, the Islamic State in Syria (ISIS) tried to gain insight into the motives, objectives, organizational forms and tactics of these fundamentalist movements.[37] These studies suggested that while the Western instrumental perspective on war may partly explain the behaviour of such groups, there are also eschatological perspectives in play, as demonstrated by al-Qaeda's vision of an eternal 'cosmic', metaphysical and existential religious struggle.[38]

The initial US response to 9/11, Operation Enduring Freedom, demonstrated the value of high-level expeditionary capabilities and the merits of the 'new American way of war'. With relatively little effort, US special forces, in cooperation with abundant air power and the Northern Alliance, defeated the Taliban, giving rise to predictions that this 'Afghan Model'—air power combined with special forces and proxy forces—would be a new strategic feature in future wars. The RMA thesis also seemed validated when the Iraqi army was rapidly defeated in 2003, and Saddam Hussein overthrown in the initial stages of Operation Iraqi Freedom. The US proclaimed a Global War on Terror: in light of US military capabilities, 'rogue' and terrorist leaders now needed to worry for their survival. Future Western security now hinged on assisting failing states to combat criminal groups, armed bands and terrorists, and enhance their own national security to prevent those groups from conducting hostile activities against the West. The 'logical' answer and future role of Western (mostly US) militaries was therefore counterterrorist operations in the so-called 'Arc of Instability'—that region of failing and failed states where violent non-state actors found sanctuary.

The Future is State-building

Looking for more long-term structural solutions to the problem of failing states, the UN and affiliated think tanks suggested 'lines of operations'—development paths—based on the 'Washington Consensus', to enhance the power, legitimacy and effectiveness of local and national governments.[39] This involved the recovery of the economy, the implementation of the 'rule of law' (i.e. building a legal system according to a Western model, with Western values and norms), an effective and representative democratic political system, and restoring security in these countries. The EU told its member states in its 2003 strategic vision that Europe needed to become more interventionist so as to establish a ring of well-developed countries around the continent. Such liberal state-building operations required an interagency approach and not just a military one.[40] From 2005 onward, the 'comprehensive approach' (also referred to as the 'interagency approach', the 'whole of government approach', or the 3D approach) became the dogmatic perspective on future coalition operations within NATO, the UN and the EU.

While sound in theory, the practical problems Western troops encountered in Afghanistan and Iraq quickly led to intense discussion about the validity of liberal state-building.[41] It required a prolonged presence of a sizeable force to ensure security which never materialized, and the inevitable use of military force against militant and criminal elements subsequently eroded the legitimacy of the operation. Local power brokers proved corrupt and disinclined to demonstrate an ability to do without large financial support, while Western legal norms and democratic principles encountered local resistance.[42] In the power vacuum that developed in Iraq, a civil war between the Shiites and Sunnis developed from 2003 onwards. In Afghanistan, the Taliban quickly managed to fill the security gap that had arisen when it turned out that the international force was far too small to exert influence throughout the whole of the country. Successful state-building requires a sufficient degree of security, and gradually politicians were forced to acknowledge that Western forces were now fighting insurgencies.

THE FUTURES OF WAR

The New Face of Future Irregular Warfare

Some felt that the solution to these issues lay in a rediscovery of counterinsurgency (COIN) strategies, and an awareness took hold that irregular warfare would be the most likely feature of future war.[43] Classic works on COIN were scrutinized for their current relevance, and concepts such as 'hearts and minds' campaigns and 'clear, hold, build' were rediscovered. However, they were also criticized, as these insurgencies differed (unsurprisingly) from the insurgencies classical COIN theorists studied. While the West harped on about winning the hearts of locals, the use of force to convince the minds—historically an integral part—was no longer an option.[44] David Kilcullen argued that it is during this competition for control where groups such as the Taliban and ISIS intimidate the population and local leaders, and, if necessary, use brute force.[45] Other dissimilarities are that insurgents were not necessarily out to take over the state anymore, but consisted of multiple rival groups and opportunistic criminal actors, rogue militias, child soldiers and fundamentalist terrorists who operated increasingly in urban environments rather than the countryside.[46]

The intifadas that Israel faced between 1987–93 and 2000–05, along with the Second Lebanon War of 2006, indicated the limitations of the traditional Western categorization of war types. Hezbollah, often designated as a terrorist organization, suddenly employed an arsenal of medium- to long-range missiles, and was able to run a capable command-and-control system using commercial ICT. The group also applied tactics for different types of warfare: standard guerrilla tactics, but also positional defence of villages and the ability to disable Israeli tanks. It thus combined characteristics of an insurgency along with those of a regular army, including a sophisticated media organization. In 2007, Frank Hoffman introduced the label 'hybrid conflict' to draw attention to this category-breaking aspect.[47] The emergence of the Islamic State in Iraq and Syria attested to this development, with the group brandishing tanks and artillery while also deploying suicide terrorists in its assaults on Iraqi villages.

The wars against these non-state actors also highlighted that with Web 2.0—a shift towards user-generated content and participatory culture on the Internet—anyone with a laptop and a network connection could now instantly and cheaply send, edit and store evocative images, sound and text worldwide, reaching a global audience. Cyberspace, in addition, offered unprecedented access to funding sources and opportunities to recruit, plan and coordinate actions from afar. As such, a physical infrastructure and proximity to members of the organization is no longer required to plan and execute operations. Future war became one of ideas and narratives being played out virtually on YouTube, Twitter, Telegram and (much later) TikTok.[48] Borders had become meaningless, border control impossible. The Internet provided a refuge, a platform, an audience, an organization and a battlefield.

Indeed, Hezbollah in 2006 managed to claim a victory over Israel not because of military success, but through a targeted media campaign. In Iraq and Afghanistan, groups such as al-Qaeda, the Taliban and ISIS used social media to give the impression that they had a much greater position of power than was actually the case.[49] Suicide bomber attacks, gruesome killings, beheadings and ambushes on patrols of Western troops were planned and carried out not for their immediate tactical military effects, but rather to produce propaganda material.[50] The aim was to erode local support for Western troops, to intimidate both those troops and the local population, to suggest the futility of Western efforts, and thus influence the political processes in the capitals of Western countries.

In response, Western military forces developed 'influence operations' doctrines, which were conceptually merely a rediscovery of traditional psychological warfare. Through so-called 'strategic narratives' and coherent reporting (i.e. strategic communication), an attempt was made to explain both to home audiences and the population in the mission area what the intentions and legitimacy of the Western military presence were,[51] and to de-legitimize the actions of the insurgents or terrorist movement.[52] This, however, turned out to be an asymmetric battle. Insurgents can ground their actions within the historical and religious context of the local population,[53] and can also freely distort facts and spread lies, which for Western units would directly undermine their credibility.[54]

The Search for Low-risk Warfare

Not surprisingly, the highly ambitious yet frustrating operations in Iraq and Afghanistan inspired critique.[55] The West had rediscovered the myriad challenges COIN campaigns pose for democracies: the requirement of long-term commitment and large numbers of troops; the severe political constraints; the inevitably high number of Western military casualties; and the problems of demonstrating tangible success.[56] The future would instead reside in alternative low-footprint, low-risk operational concepts.[57] The success of Operation Enduring Freedom in 2001 highlighted the strategic value of the so-called 'Afghan Model', or proxy warfare.[58] Yet some doubted whether the success of this concept could be replicated outside Afghanistan.[59] The US application of this concept during Operation Iraqi Freedom in northern Iraq, however, where American special forces teams worked with Kurdish fighters to bind several Iraqi divisions in the north, suggested otherwise.[60] In 2011, proxy warfare was again successfully employed in Operation Unified Protector— the NATO operation against Muammar Gaddafi's regime in Libya— when a UN mandate precluded deploying NATO ground forces. The fight against ISIS similarly demonstrated the strategic utility of proxy warfare, which was deemed an important new policy instrument for Western leaders, as it promised success with minimal investments in lives of Western soldiers and thus limited political risk.[61] However, success depends on the extent to which the interests, objectives, endurance and risk assessments of the Western coalition are aligned with those of the proxy forces.[62]

The use of proxy warfare has energized fierce societal and academic debate. Some have argued that drone strikes and proxy warfare were just forms of 'surrogate warfare' and another Western method for avoiding the risks of war, albeit with questionable effectiveness, ethics and legitimacy. Drones, for example, with their ability to carry out long endurance observation and kinetic engagement with leaders and bomb experts of terrorist movements, initially offered an alternative future mode of waging low-risk asymmetric warfare.[63] However, their use has involved 'extra-judicial killing' marred by inadequate assurance of 'accountability' in the targeting process.

Moreover, moral disengagement and dehumanization were deemed inevitable when operators experienced these strikes as video games, as demonstrated in the rising numbers of civilian casualties.[64] Others warned of the 'dronification of foreign policy' and 'everywhere war' in which violation of sovereignty became too easy and risk-free for Western politicians. Subsequent empirical studies, however, discounted many of the initial concerns and criticisms as unfounded and demonstrated their significant impact on the targeted groups. The International Committee for the Red Cross even concluded that, in principle, carefully deployed drones enhanced accuracy and discrimination of attacks.[65] Other RMA technologies also worked against violent non-state actors.

2014 and the Return of the Future of War as We Knew It

The Future is Hybrid Conflict

Russia's annexation of Crimea in 2014 seemed to radically invalidate the pre-2014 predictions that proclaimed sustained low-intensity operations against violent non-state actors would be the norm. As a result, the West was now forced to recalibrate what the future of war would look like. Russian Prime Minister Dmitry Medvedev told the West in 2016 that a new Cold War was a fact.[66] Dispelling *The End of History* myth, the ideology of authoritarianism now presented itself as a competitor to liberalism. Russia, keen to restore its status as a superpower, believes in a unique and dominant civilization with almost mythical (and certainly orthodox religious) foundations in which there is no room for Western liberal values and legal principles. Behind the rhetoric lies a deep mistrust of 'the West' that inspires its aim to restore the spheres of influence and strategic buffer between Russia and Western Europe.[67] The West was caught by surprise by Russia's rapid success in Crimea, insufficiently aware of the many subversive actions that preceded the operations and the non-military face of it. Absent of tank battles, the annexation of Crimea was not war in the traditional sense, but rather seemed to suggest the West needed to prepare for a future filled with 'hybrid conflict'.[68]

The 'hybrid warfare' label denoted the orchestrated Russian deployment of conventional military means, irregular methods of combat, subversive activities, deployment of paramilitary units, incitement, psychological warfare, propaganda, media manipulation, deception activities, deployment of unmarked special forces units, cyberattacks and control of the media. Studies concluded that hybrid warfare amounts to a long-term, gradual strategy that suffices with the achievement of small incremental political successes—a *fait accompli*; individually, these are probably not *casus belli* for the West, but can, over time, cumulatively create a new situation. It is a strategy that aims to exert influence without directly using armed force; in this way, it is hoped it will ultimately achieve its political goals or create the conditions to quickly achieve a local victory by military means. Through continuous actions—intimidation of political leaders and opinion leaders, bribery, media manipulation, sanctions, military threats along the border in the form of exercises, cyberattacks, etc.—it seeks to test political resistance, sow social unrest, increase the degree of international commitment and undermine the credibility of Western deterrence.

Troll armies, as well as official state-controlled media, manipulate public opinion through targeted social media influence operations, and undermine the legitimacy of an incumbent government and/or the outcomes of democratic elections. The views of Western authorities and the reliability of Western media are questioned through frames that suggest that Europe is weak and decadent, violates traditional Christian values, and has no answer to problems such as migration, Islamic terrorism, drugs and organized crime.[69] Emerging out of a close study of the high-tech 'Western way of war', hybrid warfare purposefully seeks to stay below the violence threshold of the Western understanding of war, so that the geographically and time-dispersed actions of diverse assets are not recognized as acts of war, and effective decision-making on countermeasures becomes difficult (and may not always be forthcoming). In doing so, it becomes the grey zone between peace and war.[70]

Critics noted that the label of, and focus on, hybrid conflict was a manifestation of both confusion and ignorance.[71] The combination of military and various non-military means has long been the norm in warfare. The Second World War, for example, in addition to clashes between armies, also saw continuous guerrilla actions, economic blockades, industrial mobilization and propaganda campaigns. In Crimea, we merely saw a repeat of Russian Cold War practice with the deployment of Spetsnaz units deep into the opponent's territory, incitement of local political radical elements, application of *maskirovka* (use of deception, camouflage and covert action) and 'reflexive control' methods (spreading confusing information in such a way that the target audience no longer knows which source to trust, what is true and what is fiction). In 1999, two Chinese colonels reminded the West of such a conceptualization of war in their book titled *Unrestricted Warfare*, arguing that a multidimensional strategy, with many non-military methods of influence, was the way to circumvent and compete with the military superiority of the West.[72]

Rediscovering Deterrence and State-on-State War

The annexation of Crimea also seemed to demonstrate a surprising level of professionalism and discipline of Russian military units, and an impressive modernization of combat systems.[73] With these resources, Russia executed 'snap exercises' in which large numbers of equipment and personnel (sometimes more than 100,000 men) traversed quickly over strategic distances to produce an intimidating threat along a European border under the guise of an exercise.[74] Western military security was further challenged by Russia's 'Anti-Access/Area Denial' (A2/AD) capabilities, which precluded future assured Western access to Eastern Europe and the use of its airspace, roads and sea transport.[75] This undermined NATO's ability to defend the Baltic states and hence the credibility of the West's conventional deterrence posture.[76] NATO responded with enhanced readiness measures, positioning of 'tripwire' forces, air-policing operations, strengthening its cyber capabilities, and calls for increasing defence spending. Member states started to conduct high-intensity warfare exercises to relearn Article 5 operations. Suddenly, the future of war for the West looked to be quite different from what they had experienced during the Cold War.

THE FUTURES OF WAR

This future also included the nuclear dimension. While Russia was testing new cruise missiles, positioned nuclear capable missiles in Kaliningrad and was prepared for a (limited) nuclear war,[77] Europe after 1990 produced hardly any discourse on nuclear deterrence and had long harboured serious doubts about the relevance of these types of weapons. Despite NATO's Warsaw Summit declaration in 2016 that 'any employment of nuclear weapons against NATO would fundamentally alter the nature of a conflict', analysts concluded that the organization lacked an adequate response to a Russian escalation, and by default had to rely on a relatively low credibility 'deterrence by punishment' strategy.[78]

Emerging Technologies and the Future of War

The 'New Cold War' also brought attention to the increasing threat posed by cyberattacks. The first article on cyberwarfare had been published in the US as early as 1993, and increasing emphasis was placed throughout that decade on the importance of information operations, particularly gaining and maintaining information dominance, disrupting the enemy's ability to effectively command units by disabling its headquarters, sensors and data links (so-called command-and-control warfare), and the rapid sharing of information within one's own ranks (a core theme within network-centric warfare). Furthermore, the protection of critical infrastructure, such as telecommunications facilities, energy networks, financial networks and transportation infrastructure, gained prominence.[79]

The Russian denial-of-service attack on Estonia in 2007, however, demonstrated the actual susceptibility of open modern societies to cyberattacks.[80] Increasing Chinese cyber activity, the 2010 Stuxnet attack on Iran and the 2012 Flame attack in the Middle East also propelled cyberattacks into the security-political and military realm.[81] In June 2011, then US Secretary of Defense Leon Panetta warned that the next Pearl Harbor could be a cyberattack, echoing Richard Clarke's 2011 use of this metaphor as a warning in a much-discussed book.[82] British general Sir David Richards warned that the West had to rethink its perception of security, war and warfare because while tanks, fast jets and fleet alliances had been the dominant weapons of war ten years earlier, future attacks would be carried out semi-anonymously via cyberspace.[83]

Here, too, theorizing and debates quickly followed incidents and technological developments. Given that cyber weapons are not the prerogative of nation-states but within easy reach of all kinds of actors, Joseph Nye foresaw a fundamental shift of power and potential threat to the international legal order. A dependence on complex cyber systems to support military and economic activities creates new vulnerabilities in major states, which can be exploited by other state and non-state actors. It was evidence of the continued diffusion of power from governments to other types of actors, and was part of one of the great power shifts of this century.[84]

Considering the limited and non-lethal impact of cyberattacks, other analysts doubted whether it was possible to talk about cyberwar in the Clausewitzian sense.[85] Another debate highlighted the specific character of cyberspace, which produces a fundamentally different deterrent dynamic than in the conventional or nuclear domains.[86] Cyberattacks are difficult to counter because (a) so many actors have cyber weapons at their disposal, and (b) the threshold for using them is low, as is the damage caused by them and the chance that retaliation will be enacted. As such, it follows that it will not do much damage. If deterrence is the aim, then this can only be done by being very active in cyberspace and by continuously carrying out exploratory, defensive and also offensive cyber operations.[87] The consequence is that activities in cyberspace can lead to an increase in international instability. It is subsequently no longer a question of whether there can be a cyberwar—it is already a fact.[88]

Another factor featuring in visions of future war was the emergence of new technologies such as hypersonic missiles, artificial intelligence (AI), robotics, and the combination of these last two in the development of Autonomous Weapon Systems (AWS, or in the colloquial 'killer robots'). Swarms of AWS can overwhelm air defences and continue to patrol over an area in search of specific predefined targets such as tanks, artillery systems or mobile missile launchers, as well as specific individuals. AI systems can also offer great advantages in monitoring data streams, video images and in analysing other types of data.

THE FUTURES OF WAR

The proliferation of drones provides an increasing number of state- and non-state actors with the ability to carry out air strikes with relatively simple and affordable means, both on military targets and civilian facilities. In the West, the ethical and legal aspects of these weapon systems are still under investigation, and the requirement remains for a minimum of meaningful human control concerning the decision to use lethal force. Because AWS developments are mainly driven by commercial dual-use technologies (and hence commercial interests), the inevitable proliferation of AWS may result in destabilization and rapid escalation during a crisis. Further dehumanization of warfare lurks in the danger that AWS will be used somewhat casually because the lives of friendly troops are not put at risk. AI will likewise become available to various types of actors, facilitating the production and dissemination of fake news and deep fakes. Furthermore, AI-enhanced intelligence organizations may be able to pinpoint the location of nuclear weapons among rivals in the future, which could result in the erosion of deterrence stability. This risk becomes all the greater when an actor also has access to a new generation of hypersonic, long-range cruise missiles that travel too fast to be intercepted by existing anti-aircraft and anti-missile systems, and so fast that a reaction from the opponent will probably arrive too late.[89]

Five Perspectives on Future War

By the end of 2021, Western analysts had collectively painted a broad and colourful canvas featuring five distinct strategic landscapes. The first one, capturing the many ongoing civil wars in the 'Arc of Instability', predicted that the West must still be prepared for humanitarian crises and corresponding humanitarian operations, and that such 'humanitarian wars' are (and should remain) the primary justification for the use of the military instrument. Kaldor coined this the 'liberal peace security culture', which she associated this with international organizations such as the Organization for Security and Co-operation in Europe, the EU and the UN. Repeating her 1990s argument, Kaldor's normative cosmopolitan vision sees a future in which wars in failing and fragile states must be settled, and

59

violence contained through peacekeeping operations so that human suffering can be alleviated.[90]

A related second vision of the future—sophisticated barbarism—predicts that wars in the future will revolve around the actions of violent non-state actors—an ecosystem of terrorist movements, well-armed criminal organizations, warlords overseeing militias, and insurgents. Ethnicity or religion is occasionally a motive for their struggle, but often this goes hand in hand with economic profit. This view is a continuation of Martin van Creveld's line of argument in his 1989 book *The Transformation of War*, which also echoes Kaldor's 'new wars' thesis, as well as the 'fourth-generation warfare' model from the 1990s and Frank Hoffman's hybrid conflict concept from 2007. Recently, authors such as Kilcullen and McFate predict these groups will increasingly be able to inflict damage on Western countries through cyberattacks and the use of drones.[91] Moreover, they will easily organize themselves into 'smart mobs' via social media and the use 'barbaric' tactics to intimidate populations. Their battlefield is increasingly the city, an environment in which it is difficult for Western soldiers to operate. According to McFate, there will be a 'durable disorder', which repeats Kaplan's 1990s warning on the spread of anarchism in large parts of the world.[92]

The third vision—immaculate warfare—follows on from this and considers what it means for Western armed forces now that 'the future is irregular', as Seth Jones predicts. Western military personnel will mainly be deployed in a COIN fashion across a multitude of protracted conflicts in unstable regions.[93] But unlike before, the West will not employ large formations in a conflict zone, but will rely on the use of special forces, the training of proxy forces, reconnaissance assets capable of observing areas for a long time, and armed drones that can be deployed if necessary to execute a precision attack on a few individuals. The primary aim is not defeat of the insurgents or terrorist movements, but risk management: to contain the risk that these groups will lead to regional destabilization and/or form a direct threat to the West.[94] This is the battle the US is waging in various African countries as well as against ISIS. Andreas Krieg and Jean-Marc Rickli describe this form of warfare as 'surrogate warfare', which is characterized by the West trying to exert influence in conflict areas with minimal physical presence and therefore minimal political risk.[95] In a similar vein, Russia has deployed privatized military companies such as the Wagner Group in various African countries.

The renewed geopolitical rivalry—the return of strategic competition between the US (with its European allies), Russia and China—form the context of the fourth vision: cool war. In addition to the notion of hybrid threats, others have dubbed this vision 'new total warfare', 'political warfare' and 'grey-zone warfare'.[96] These concepts consider strategic competition to be war. Just as during the Cold War, this rivalry involves a wide range of instruments and activities in various military and especially non-military domains, which can affect various sections of Western society. Some perceive this as a 'cool war' or 'soft war', to emphasize the idea that the continuous use of so-called non-kinetic means predominates in attempts to exert influence.[97] Instruments and activities include economic espionage through cyberattacks, economic sanctions and financial warfare, financing the bribery and intimidation of politicians (and liquidation by poisoning if necessary), and financing and even arming militant anti-European political groups in democratic states.[98] Troll armies spreading fake news through social media also feature prominently in the array of influence methods, much like in Singer and Brooking's *LikeWar*.[99] In fact, as Galeotti describes, 'everything has become weaponized'; war and peace merge.[100] Subsequently, defence in cool war requires boosting societal resilience and a whole-of-society approach.[101]

The final vision is that of the return of major war, which reflects shifting power relations. Pace John Mueller, war in the classical sense—as an armed encounter between two large countries—is no longer impossible and less unlikely than, for example, in 1999.[102] The influence of the US (and therefore of the West) is waning relative to China. The liberal world order is eroding.[103] International organizations such as the UN, World Bank and the International Monetary Fund have been under pressure since the financial and economic crisis of 2008. The EU is struggling with the rise of nationalist and populist sentiments and illiberal political movements across the continent. This is not a return to the Cold War, because unlike that period, authoritarian powers are now actively seeking to disrupt stability. Ideology once again plays an important role. Fukuyama has been belied: in authoritarianism, there is indeed a competitor for Western liberalism. And the question is, which international order will win this competition?[104] Authoritarian regimes increasingly pursue aggressive revisionist foreign policies and boost both their offensive capabilities and A2/AD screens to blunt potential Western responses. Limited 'probes'—minor incursions into the airspace or territory of Western countries—are likely; these will probably not justify a large-scale Western response, but may gradually change the status quo, test the Western willingness to resort to serious reprisals and enhance escalation risks.[105]

Such confrontations will likely see the employment of swarms of drones and killer robots. New defences such as electromagnetic pulse systems will be developed, and offensive hypersonic missiles may be used as a new type of threat. AI will be used to analyse large amounts of data from large numbers of networked commercial and military sensors and satellites. AI and quantum computing will provide advice via new cognitive processes on decisions about, for example, the right time for a conventional attack, a cyber offensive, or whether to escalate or whether to launch an anti-satellite weapon.[106] Space will also have increasing military applications, including offensive ones.[107] This warning of war with Russia and/or China, which may well involve a new RMA,[108] inspires Western searches for novel operational concepts such as 'multidomain battle' and 'cross-domain deterrence' to exploit emerging technologies and thus provide an appropriate military response to military challenges such as the A2/AD problem, cyberattacks and nuclear threats.[109]

Visions of Future War Meet the War in Ukraine

Wars are educational moments that serve to gauge the extent to which prior views on future war hold water. A brief excursion into the war in Ukraine will allow preliminary observations about the validity of the five perspectives presented above. At first blush, several predictions seem to have materialized, albeit not in their pure form or with the dramatic impact analysts anticipated.

THE FUTURES OF WAR

When Russia started its invasion in Ukraine—which it dubbed a 'special military operation'—major war, which NATO in 2010 had dismissed as a very small risk, once again tragically returned to Europe. Preceded by 'cool war' methods such as the spreading of disinformation, a rapid victory—hoisting the Russian flag over Kyiv's government buildings and the eradication of Ukraine's identity— seemed assured. Its quantitative superiority and doctrine led to expectations that Russia (the world's ninth-largest economy) would be able to overrun Ukraine (the fifty-sixth-largest economy) easily. Russia had mustered between 150,000 and 190,000 troops along the long Ukrainian border in so-called battalion tactical groups (BTG), divided over four fronts. While insufficient for achieving Putin's maximalist objective (the full occupation of Ukraine), it could have enabled a rapid advance, too fast for Ukraine to mobilize units or for the divided West to generate a timely robust response. As such, ousting the democratically elected government in Kyiv certainly seemed feasible. Russia could count on, if necessary, three times the number of tanks and artillery pieces that Ukraine could mobilize, eight times the number of combat helicopters and ten times the number of combat aircraft. Assuming Russia had embraced the Western RMA capabilities, a campaign not unlike Operation Iraqi Freedom seemed likely, but now with the additional employment of hypersonic missiles and swarms of drones, as well as intimidating nuclear threats.

For the first two or three days of the invasion, this scenario seemed to unfold. Massive cyberattacks attempted to paralyse Ukraine's transport and communications infrastructure. Around 1,000 cruise missiles and stand-off weapons were launched at airfields, military headquarters and air defence positions.[110] Communications and radar systems were disrupted by intensive jamming operations, temporarily neutralizing Ukrainian surface-to-air missile (SAM) systems. Ukrainian fighter jets lost against the qualitatively and quantitatively superior Russian counterparts, who could use airborne early-warning and extended-range air-to-air missiles. Airmobile units landed with helicopters at Hostomel' airfield near Kyiv, waiting to connect with the mechanized columns advancing towards Kyiv from the north and northeast, and ready to

63

receive transport planes transporting hundreds of infantrymen and armoured cars to Hostomel'.

Zelensky, however, won the 'like war' with his savvy use of social media, unifying his nation and creating the moral foundation energizing Western support, which in turn amalgamated into a continuous series of intensifying economic and financial sanctions.[111] His commanders saw their situational awareness benefit from simple cell phone and tablet apps which enabled troops and civilians to spot enemy units and weapon systems, and transmit those locations to headquarters. Those headquarters also exploited the near real-time transmission of drone footage through networks that have been provided and supported by commercial companies such as Starlink communication satellites. Adaptability and the ability to use civilian technologies—drones, commercial communication tools, simple target-location apps, crowd-funding, etc.—have shown their potential. Predictions of the increasing impact of emerging technologies such as AI and killer robots seem validated with the introduction of Switchblade drones.

Immaculate war also seemed evident. First of all, Putin claimed that this was a 'special operation' aimed at regime change, conducted with a limited number of highly trained units in a very short time, and promising quick success with limited risk for Russian casualties. Second, it manifests itself in Putin's use of informal armed groups such as the Wagner Group and Kadyrov's Chechen fighters. Third, it is evident in the prevalent use of stand-off munitions to attack the opponent while keeping their own troops out of range of enemy weapons. Russia employs massive numbers of cruise- and hypersonic missiles, intense barrages of long-range rocket artillery as well as scores of drones, which are now a permanent feature on the ecosystem of the battlefield. Drones combined with artillery significantly improved locating targets, as well as fire accuracy, responsiveness and counter-battery tactics. As a result, artillery caused the most damage to materiel and led to the most casualties. For the infantry, small drones provide cheap intelligence, surveillance and reconnaissance, and can be armed with improvised grenades to provide their own short-range organic air power, often with deadly results against dug-in enemy troops. Swarms of cheap long-range

Iranian Shahed drones reinforce the image that Russia is intent on bludgeoning the opponent from afar and reducing the political risks for the Kremlin regime.

In general, the variety of drones makes it extremely risky for an opponent to mass units and materiel. The same applies to the impact of multiple-launch rocket-like systems, which from the summer of 2022 onward forced Russia to place command centres and ammunition depots at a greater distance from the front, aggravating existing command and logistical challenges. Ukraine's use of high-mobility artillery rocket systems (HIMARS) were also useful in disabling Russia's SAM systems. As a RUSI report summarises, 'There is no sanctuary in modern warfare. The enemy can strike throughout operational depth. Survivability depends on dispersing ammunition stocks, command and control (C2), maintenance areas and aircraft'.[112] The tenets of network-centric warfare seem to have become reality.

There is, however, another potential pointer: the restoration of the Clausewitzian notion of defence being dominant over offense. A cyber version of Pearl Harbor has not materialized despite massive cyberattacks, nor have autonomous weapons systems or hypersonic missiles proven real, strategic-level game-changers offering offensive dominance. The dramatic asymmetry in capabilities between the warring parties that immaculate warfare presupposes also proved absent. After one week, while the Russian advance in the south of Ukraine was going reasonably smoothly, the advance from the north and northeast stalled. Failures in conducting combined arms tactics and logistics, as well as not exploiting its air power advantage to achieve air superiority, conduct air interdiction, strategic attacks and provide responsive close air support, all contributed to the failing of the envisioned ten-day 'special operation'.

These failings were enhanced by Ukraine's surprisingly effective resistance. It managed to quickly bring artillery fire to bear on Hostomel' airfield, shoot down several helicopters, and eliminate the landed Russian units, precluding them from landing with transport aircraft. The Russian advance from the north was bombarded with artillery fire and attacks from small, mobile infantry teams, taking out tanks and armoured vehicles with anti-tank weapons and often making use of drones. Ukraine's mobile SAM systems denied Russia the use of airspace, providing much-needed freedom of manoeuvre for its ground troops and logistics.

When, on 9 April, Putin declared that his troops would retreat from Kyiv and instead focus on the Donbas, that failure was de facto acknowledged. A disconnected, under-resourced, four-front attritional war ensued, including pre-modern siege warfare in which Russia encircled and pulverized Ukrainian cities, resulting in terrible numbers of civilian casualties. The last major cities to fall to Russia after prolonged artillery barrages and costly urban combat were Mariupol, Severodonetsk and Lysychans'k. The defence of these cities cost Ukraine dearly too, but bought it the time it needed to mobilize new units and bring Western artillery, howitzers and later HIMARS launchers to the front. Ukraine re-took the Kharkiv Oblast in September 2022, and the city of Kherson in November, just before winter conditions precluded further manoeuvres.

Well into 2023, barrages of artillery (sometimes Russia fired 30,000 shells a day) and waves of Russian infantry smashed against well-developed defence lines in the terrain, towns and cities along the long, stagnating frontline, daily losing hundreds of soldiers. Russia belatedly launched missiles and drones against Ukraine's logistical infrastructure first, and after the fall of Kherson, also against Ukraine's power plants. Due to Russian shortages of missile stockpiles, the relative inaccuracy of the strikes, increasing intercept rates (aided by supplies of Western air defence systems) and rapid repair capabilities, the frequency and intensity of these strikes failed to have a strategic impact, apart from boosting the West's resolve to support Ukraine. One year into the war, Russia had probably lost half of its deployed tanks and more than 6,000 armoured vehicles, as well as 200,000 soldiers, including between 40,000 and 60,000 dead.[113]

With such losses in armour, the era of tank warfare seems to be over. The faith of Russia's airmobile operations and the heavy losses among attack helicopters also suggest that the future role of aviation needs to be reassessed. The same seems to hold for the dominant role of air power. Here, the effectiveness of large numbers of mobile air defence systems had denied both sides the use of offensive air power over and beyond the frontline. Russia, by default, reverted to the use of cruise missiles, drones and hypersonic missiles to circumvent this problem. Nevertheless, Ukraine's steep learning curve, combined with the introduction of Western air defence systems, improved the interception rate to a stunning 80–90 per cent. The de facto ability to maintain air denial suggests that, once again, the defence is now dominant in air warfare, much as it was during the Cold War. This seems to validate studies of the past decade that warned of the challenges of new A2/AD capabilities whilst undermining concepts such as network-centric warfare that argued the RMA would boost the offensive.

The Future of War Continues

It is unwarranted to draw definitive conclusions concerning the future of war, to use the Russo-Ukrainian war as a touchstone for critically assessing previous visions of future war, or to argue for a radical overhaul of existing defence policies and investment priorities. First, Russia's initial failures have shaped the trajectory of this war. The Kremlin assumed a divided Ukrainian population, a weak regime and weak military resistance. It overestimated its own military capabilities, and the secrecy of its plans meant that (a) the frontline troops received orders far too late—sometimes only hours before the start of the advance; (b) too little coordination had taken place between the BTGs themselves, and between the BTGs and the necessary supporting artillery and Russian air power for close air support; and (c) the logistics were not in order, and the units crossed the border with their tanks and armoured vehicles in non-combat formations. Moreover, the campaign plan was poorly thought out and marred by a weak, corrupt and highly centralized command-and-control system with a culture that stifles lower-level initiative and reliable information. Troop discipline was lacking, in part due to the absence of a well-trained non-commissioned officer cadre. Finally, materiel proved to be poorly maintained and in a bad state.[114] All these elements explain the demonstrated inability to conduct proper combined arms tactics and, at the operational level, to execute joint operations.

Russia's failures and Ukrainian successes remind us that there is a large measure of continuity in this war. The use of drones demonstrates the usual action–reaction dynamic, in which new weapon technology that proves itself on the battlefield quickly results in the development of specific countermeasures in tactics, doctrine and defence systems. As a result, the average lifespan of a drone in this war is five to six sorties. At the tactical and operational level, familiar important factors are reconfirmed, such as quality of training, intelligence (with which the US and UK provide crucial support to Ukraine), logistical organization and capacity, competent leadership, and the importance of troop morale and well-designed defence lines, including minefields and trenches.

Russia's default strategy of attrition harks back to twentieth-century interstate warfare dynamics. The renewed acquaintance with the Russian strategic culture of total war and the realization that the West must be prepared for industrial warfare remind us of the importance of what Michael Howard called the 'forgotten' dimensions of strategy.[115] The quantity of weapon systems, ammunition stocks, industrial capacity, spare parts, redundancy, societal resilience—these are all strategic qualities. Whether the future of war can be gleaned from this clash between two quite similar twentieth-century armed forces is doubtful.

The war in Ukraine holds interesting and worrisome paradoxes. It is postmodern as well as modern (and sometimes even pre-modern). It sees accelerated innovation at the technical and tactical level that seems to validate predictions of future war. It also confirms predictions concerning the return and shape of major war. Moreover, it includes elements of cool war and immaculate war. In Russia's criminal, indiscriminate, horrific and destructive assaults on the identity of the Ukrainian people, we recognize the tenets of pre-modern and modern-style warfare, as well as the tenets of sophisticated barbarism and the brutal strategies long discarded by the West. Mariupol fell after prolonged, almost mediaeval, siege tactics. City bombings and the long battle in Bakhmut show stark similarities to the battle of Stalingrad. In the surrounding countryside, the muddy trenches resemble those of the Somme in the First World War. This war already ranks among the top 10 per cent of the bloodiest wars of the past 100 years, even without counting the civilian casualties. The casual use of nuclear threats by Russian media personalities and senior politicians also echoes a previous era.

Lawrence Freedman was right to conclude that predictions about the future of war and warfare need to be read with a healthy dose of scepticism. Still, while in their pure form, none of the five futures discussed in this chapter present 'the future', they function perhaps like string theory in physics: an esoteric belief that reality is composed of multiple dimensions. Though some of these will probably be wrong, they nevertheless serve to inspire fruitful analysis and experiments.[116] Indeed, as the recent strategic history of the West suggests, Western militaries, in their obligation to prepare for future war, need to study the range of potential futures and understand the specific political, strategic and operational dynamics of each scenario they deem likely to present in the not-so-distant future. To wit, in 2023, another civil war emerged in Sudan; Russia's Wagner Group gained influence in Mali via a proxy warfare fighting style; and Chinese fighter aircraft regularly violated Taiwanese airspace. Each of these indicate that concepts such as sophisticated barbarism, immaculate warfare and major war are relevant notions to make sense of possible futures. Presuming that the future is a singular one that the armed forces can focus on will, as the past three decades have proven, often result in organizational amnesia. Knowledge and expertise concerning other kinds of war are lost. As both Frank Hoffman and Robert Johnson note, they remind us that the future of war is plural.[117]

3

AFTER RUSSIA'S INVASION OF UKRAINE
IS THE DECLINE OF WAR A DELUSION?

Azar Gat

The war in Ukraine has brought back with a vengeance the question of whether the world is becoming more peaceful.[1] This proposition has in any case encountered widespread disbelief. Even before the dramatic events in Ukraine, the United States and its allies were repeatedly involved in messy local wars over the last decades. Furthermore, the relative peacefulness of recent decades has been widely attributable to a transient US hegemony. Have we not been tempted by a resurfacing of old illusions that are again dispelled by a resurgent Russia, by the rise of China to a superpower status, and by vicious wars in South or Central Asia, the Middle East and Africa? With the growing threats of war around Taiwan and in the South China Sea, and with the Russian invasion of Ukraine, 'realists' in international relations theory have been able to raise their heads again and claim that 'we told you so.' Yet, they have missed the main point, comparing cabbages with oranges. There has been an understandable shock about war occurring in the 'middle of Europe.' But the real dividing lines are not a matter of a particular continent or 'race.'

Today's world is divided into two very distinct zones: a 'zone of peace' encompassing all the developed countries (that is, those with GDP per capita higher than US$20,000 from non-oil and gas sources; not indicative of development); and a 'zone of war,' which includes developing or undeveloped countries, as well as failed developers (whose GDP per capita is usually much lower than this threshold).

The 'zone of war'—where war and the threat of war are very much alive—is all too familiar. It extends from the borders of China (around US$10,000 GDP per capita) and North Korea, through South and Central Asia and the Indian subcontinent, to the Caucasus, to Russia's borders (GDP per capita fluctuating with the price of energy, and in any case relatively low and derived from resources), to the Middle East and North Africa (a few thousand US dollars per capita), to sub-Saharan Africa (in many cases, no more than a few hundred US dollars per capita). In all these areas, war—mostly civil wars, but also interstate wars—is either rife or looming.

By contrast, awareness of the 'zone of peace' is distracted by the persistent involvement of the United States and its allies in wars in the less developed parts of the world. Such involvement is prompted by direct threats arising from these parts (e.g. the War on Terror, Afghanistan), but also includes interventions in local conflicts that disrupt the liberal world order (e.g. Iraq), as well as humanitarian interventions (e.g. Somalia, Kosovo, Libya). After all, it can scarcely be argued that the United States has profited from all these wars: on the contrary, it has lost a great deal in resources, life and prestige.

More significantly, whereas the underdeveloped or developing areas are a clear war zone—within themselves and irrespective of foreign involvement—the developed world is an absolute zone of peace. While the United States' futile wars attract all the attention, it is scarcely asked why Canada is not at all concerned about the prospect of being conquered by the United States, which would have increased US wealth and power considerably, even more, one assumes, than a war in Afghanistan. People find it difficult to explain why exactly this is so. Similarly, Holland and Belgium no longer fear a German (or French) invasion in the slightest—an historically unprecedented situation. In East Asia, the most developed countries—Japan, South

Korea and Taiwan—even though there is no love lost between them (for historical reasons relating to Japan's colonial past), do not fear war among themselves, or with any of the other developed countries. At the same time, they are deeply apprehensive of being attacked by less developed neighbors, such as China or North Korea. That Australia and New Zealand do not fear each other seems so obvious that it attracts no special attention. Latin America has been noted as a borderline case: not quite developed and yet relatively peaceful, especially as concerns interstate wars, if not civil wars.[2]

Thus, the probability of war between developed, affluent, liberal democracies has declined to a vanishing point, where they no longer even see the need to prepare for the possibility of a militarized dispute with one another. The 'security dilemma' between neighbors—that ostensibly intrinsic feature of international anarchy—no longer exists among them. Much the same applies to civil wars. Modernized, economically developed democracies have become practically free of civil wars, largely due to their stronger consensual nature, plurality, tolerance and, indeed, a greater legitimacy for peaceful secession.

Figure 3.1: GDP per capita (2023), current prices US dollars

(Source: IMF DataMapper, World Economic Outlook data, April 2023, https://www.imf.org/external/datamapper/PPPPC@WEO/OEMDC/ADVEC/WEOWORLD)

So, is economic modernization the key to the absolute peace that prevails in the developed parts of the world, as is argued here? Or perhaps other theories that have been aired better explain the decline of war in these parts? Such theories include: a nuclear peace; the notion that war has become far too lethal, ruinous and expensive to indulge in, or that it no longer promises rewards; a democratic peace; a capitalist, trade-interdependence peace; peace through international institutions; and US hegemony. To find the answer, we need a broad historical perspective and analysis.

The so-called 'Long Peace' among the great powers (that is, no war between the great powers since 1945) is commonly attributed to the nuclear balance, a decisive factor to be sure.[3] The absence of war between democracies has been equally recognized. However, the decrease in war had been well marked even before the nuclear era, and encompassed both democracies and non-democracies; in fact, it included not only great powers but also relations between developed countries in general. Between 1815 and 1914, wars among industrializing countries declined in frequency to about a third of what they had been in the previous centuries—an unprecedented change. For example, Austria and Prussia, neither of them a democracy, fought about a third to a quarter as many wars after 1815 as they did in the preceding century.[4]

Indeed, the long peace between the great powers since 1945— seventy-seven years to date and counting—was preceded by the second-longest peace ever, without any wars occurring among the great powers between 1871 and 1914 (forty-three years in all), and by the third-longest peace, between 1815 and 1854 (totaling thirty-nine years). The nineteenth century has long been recognized as by far the most peaceful among the European powers until then. Thus, the three longest periods of peace among the great powers in the modern era have all occurred since 1815, with the first two taking place before the nuclear age. No similar long periods of peace occurred before 1815. The widely used Correlates of War database, covering all wars since 1816, has concealed the Long Peace phenomenon from researchers, who are not aware that realities before 1816 tell a very different story.[5]

Figure 3.2: The Long Peace phenomenon between the great powers since 1815

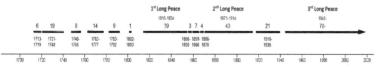

(Source: Gat, *The Causes of War*, 113)

Clearly, one needs to address this entire period of reduced belligerency from 1815 on. At the same time, it is necessary to account for the glaring, Himalaya-size exception to the trend: the First and Second World Wars.

It is tempting to assume that wars have declined in frequency during the past two centuries because they have become too lethal, too destructive and too expensive, meaning fewer but more ruinous wars. This hypothesis barely holds, however, because relative to population and wealth, wars have not become more lethal and costly than they were in earlier times. The wars from 1815 to 1914 were in fact particularly light. Prussia won the wars of German unification in short and decisive campaigns and at a remarkably low price, and yet Germany did not fight again for forty-three years. The world wars, especially the Second World War, were certainly on the upper scale of the range in terms of casualties; yet, contrary to widespread assumptions, they were far from being exceptional in history. As such, we need to look at relative casualties—the percentage of those dying in wars in each society—rather than at the aggregate created by the fact that many states participated in the two world wars.

In the Peloponnesian War (431–04 BCE), for example, Athens is estimated to have lost between a quarter and a third of its population, more than Germany in the two world wars combined.[6] In the first three years of the Second Punic War (218–16 BCE), Rome lost some 50,000 male citizens of military age, out of a total of about 200,000 men.[7] This was roughly the same percentage as the Russian military casualties in the Second World War, and higher than the percentage lost by Germany in that war. Similarly, in the thirteenth century, the Mongol conquests inflicted casualties and destruction on the societies of China and Russia that were among the highest ever suffered during historical times. Even by the lowest estimates, casualties were at least as high as the Soviet Union's horrific rate in the Second World War of about 15 per cent of its population (and in China almost definitely far higher).[8] A final example: during the Thirty Years War (1618–48), population loss in Germany is estimated at between a fifth and a third—either way, it is again higher than the German casualties in the First and Second World Wars combined.[9] People often assume that more developed military technology during modernity must mean greater lethality and destructiveness, but offensive and defensive advances generally rise in tandem and tend to offset each other. As such, developments in military technology also mean greater protective power, as with mechanized armor, mechanized speed and agility, and defensive electronic measures. In addition, it is all too often forgotten that the vast majority of the many millions of non-combatants killed by Germany during the Second World War—Jews, Soviet prisoners of war, Soviet civilians—fell victim to intentional starvation, exposure to the elements, and mass executions rather than to any sophisticated military technology. Instances of genocide in general during the twentieth century, much as earlier in history, were carried out with the simplest of technologies, as the Rwanda genocide horrifically reminded us.

Nor is it true that wars during the past two centuries have become economically more costly than they were earlier in history, again relative to overall wealth. War always involved massive economic exertion and has generally been the single most expensive item of state spending. Both sixteenth- and seventeenth-century Spain and eighteenth-century France, for example, were economically ruined by war and staggering war debts, which in the French case brought about the Revolution. Furthermore, death by starvation in pre-modern wars was widespread (think Yemen and parts of Africa today).

What, then, is the cause of the decline in belligerency? Even before the middle of the nineteenth century, during the first Long Peace, thinkers such as Saint-Simon, Auguste Comte and John Stuart

Mill realized that it was caused by the advent of the industrial–commercial revolution, the most profound transformation of human society since the transition to agriculture some 10,000 years ago. In the first place, it led to explosive growth in per capita wealth, about thirty- to fifty-fold from the onset of the revolution to the present in the countries which have experienced it. Thus, the trap that had plagued pre-modern societies, famously described by Thomas Malthus in 1799, whereby slow growth in wealth was absorbed by more children and more mouths to feed, has been broken. Wealth no longer constitutes a fundamentally finite quantity and a zero-sum game, when the only question is how it is divided, and with force functioning as a major means of attaining a larger share of the pie. The pie has been continuously growing, with wealth now derived predominantly from economic growth and investment at home, from which war tends to be a wasteful distraction.

Secondly, the much-cited significance of trade in the economy has ballooned to entirely new dimensions precisely because of the

Figure 3.3: GDP per capita (1820–2018)

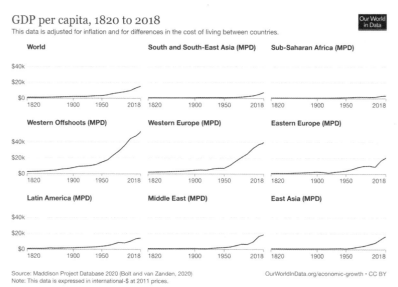

(Source: Our World in Data, https://ourworldindata.org/what-is-economic-growth)

new process of industrial growth. Greater freedom of trade has become all the more attractive in the industrial age for the simple reason that the overwhelming share of fast-growing and diversifying production has now been intended for sale in the marketplace rather than for direct consumption by the peasant producers themselves. Consequently, economies are no longer overwhelmingly autarkic, having become increasingly interconnected by specialization, scale and exchange. Foreign devastation potentially depresses the entire system and is detrimental to a state's own wellbeing. What John Stuart Mill discerned in the abstract in the 1840s was repeated by Norman Angell during the first global age before the First World War, and formed the cornerstone of John Maynard Keynes's criticism of the harsh reparations imposed on Germany after 1918: if the German economy was not allowed to revive, the global economy could not revive either. This was a matter of self-interest for the victors.

Greater economic openness has decreased the likelihood of war by disassociating economic access from the confines of political borders and sovereignty. It is no longer necessary to politically possess a territory in order to benefit from it. Of all these factors, commercial interdependence has attracted most of the attention in the scholarly literature. But the escape from the Malthusian trap with rapid industrial growth has been the underlying cause of both commercial interdependence and free trade—a notion best described as the 'Modernization Peace'. Thus, the greater the yield of competitive economic cooperation, the more counterproductive and less attractive conflict becomes. Rather than war becoming more costly, as is widely believed, it is in fact peace that has been growing more profitable.

If so, why have wars continued to occur during the past two centuries, albeit at a much lower frequency? In the first place, ethnic and nationalist tensions have often overridden the logic of the new economic realities, accounting for most wars in Europe between 1815 and 1945. Consider the wars that separated the first and second Long Peace: with the exception of the Crimean War (1854–6), these were the War of 1859 that led to Italy's unification, and the Wars of German Unification (1864, 1866, 1870–1). Ethnic and nationalist

AFTER RUSSIA'S INVASION OF UKRAINE

tensions continue to be the main cause of war today, especially in the less developed parts of the globe.

Moreover, nationalism was a major contributing factor in the retreat from free trade into national economies during the late nineteenth and early twentieth centuries as the new industrial great powers the USA, Germany, Japan and France—resumed protectionism. This growing protectionism led to the return of imperialism from 1882 on, as the emergent global economy became increasingly partitioned, and each imperial domain became closed to everybody else. A snowball effect ensued as each power hastened to grab what it could, while it could.

The size of a nation makes little difference in an open international economy; the citizens of little Luxembourg are as rich as (or richer than) the citizens of the United States. By contrast, size becomes the key to economic success in a closed, neo-mercantilist international economy because small countries cannot possibly produce everything by themselves. Moreover, in a partitioned global economy, economic power increases national strength, while national strength defends and increases economic power. As such, it again becomes necessary to politically own a territory in order to profit from it.

The heightened tensions between the great powers associated with the imperialist race would eventually lead to the First World War. The change was completed in the 1930s with the Great Depression, as the USA, Britain and France practically closed their territories and empires to imports by imposing high tariffs. For the territorially confined Germany and Japan, the need to break out into an imperial *Lebensraum* or 'East Asian Co-Prosperity Sphere' seemed particularly pressing. Here lay the seeds of the two world wars. Furthermore, the retreat from economic liberalism spurred, and was spurred on by, the rise to power of antiliberal and antidemocratic political ideologies and regimes, incorporating two creeds of violence in the form of communism and fascism.

Since 1945, the decline of major war has deepened further. Nuclear weapons have been a crucial factor in this process, but no less significant have been the institutionalization of free trade and the closely related process of rapid and sustained economic growth. The spread of liberal democracy is another major factor. In today's world, all the developed countries are democratic. Even so, how can we distinguish between democracy and development to determine which of them is the key to peace?

In considering this, the following needs to be borne in mind. First, pre-modern democracies (i.e. ancient Greece and Italy) fought each other viciously; it is only when the entire rationale of wealth acquisition has changed with modernization that modern democracies have ceased to do so. Secondly, non-democracies have fought much less during the industrial age compared with earlier times. Thirdly, the spread of democracy itself has been a result of the process of modernization. Note that no democracy the size of a large country (that is, beyond the city-state or canton size) had ever existed before modernity. Thus, both trade interdependence and democracy—cited as the causes for the increasing peacefulness—are actually a function and consequence of the process of modernization.

That said, although non-liberal and non-democratic states became much less belligerent during the industrial age, it is the liberal democracies that have been the most attuned to its pacifying aspects. Relying on arbitrary coercive force at home, non-democratic countries have found it more natural to use brute force abroad. By contrast, liberal democratic societies are socialized to peaceful, law-mediated relations at home, and their citizens have grown to expect that the same norms be applied internationally. Living in increasingly tolerant societies, they have grown more receptive to the Other's point of view. Promoting freedom, legal equality and political participation domestically, liberal democratic powers—although initially in possession of vast empires—have found it increasingly difficult to justify ruling over foreign peoples without their consent. In sanctifying life, liberty and human rights, they have proven to be failures in forceful repression. Furthermore, with the individual's life and pursuit of happiness elevated above group values, sacrifice of life in war has increasingly lost legitimacy in liberal democratic societies. War retains legitimacy only under steadily narrowing conditions, and is generally viewed as extremely abhorrent and undesirable.

Finally, the much-cited 'Thucydides trap' does not exist among modern economically developed democracies either.[10] It did not occur when the United States replaced Britain as the world's leading power around 1900; nor when the United States dreamt of curtailing the spectacular growth of Japan and Germany after the Second World War; nor in the 1980s and early 1990s, as the Soviet threat was waning and the United States' 'decline' drew much public and scholarly attention. Nor does it occur for the European Union today, despite obvious commercial competition.

Similarly, US hegemony, widely cited as the explanation for peace in Western Europe, is a cardinal factor but far from being the root cause of the phenomenon. Consider that the United States, in its role as the 'world's policeman,' has failed to force peace on large parts of the developing world. The notion that a present-day Germany of 80 million people is constrained from reasserting itself as an armed militant power by US hegemony should be recognized as the preposterous idea that it is. The United States can scarcely enforce much even in Iraq or Afghanistan. Countries in other parts of the developed world (e.g. Western Europe) are simply as interested in peaceful development as the US.

Another element of the modern condition: with the much-increased sexual opportunity within society, young men now are more reluctant to leave behind the pleasures of life for the rigors and chastity of the battlefield. 'Make love, not war' was the slogan of the powerful anti-war youth campaign of the 1960s, which not accidentally coincided with a far-reaching liberalization of sexual norms. Students had to decide whether they wanted to be sent to Vietnam or stay in college and have fun. How difficult was this choice? Note that no such protests had existed only a decade earlier, such as during the Korean War.

Now let us turn to the darker and more dangerous parts of the picture. In 2006, I suggested that the dramatic spread of peace, while very real, is far from being foolproof and free from challenges. One threat which is likely to hang over our heads during the twenty-first century is the prospect of unconventional terrorism. It is difficult to see how it can be prevented from materializing somewhere, someday.

An even greater threat suggested at a time when everybody still celebrated the 'End of History' and the triumph of the liberal world order was the return of a regime type—authoritarian-capitalist—

that had been absent from the international system since the defeat of Germany and Japan in 1945.[11]

Transformed from a backward communist economy to a fast-developing authoritarian-capitalist great power—and communist by name only—China is now recognized to represent the greatest challenge to the global balance of power. As I suggested at the time, Russia, too, had been retreating from its post-communist liberalism, and has assumed an increasingly authoritarian and nationalist character coupled with a more aggressive stance.[12] I argued that even though the new authoritarian-capitalist great powers (most notably China) were deeply integrated into the world economy, there was the prospect of more antagonistic relations with the democratic world, accentuated ideological rivalry, potential and actual conflict, intensified arms races, and new cold wars.

My 2006 article on this subject attracted several rebuttals arguing that China was bound to join the liberal world order, as well as liberalize itself.[13] Since then, China and Russia have become more authoritarian and oppressive domestically; present-day China may even be labeled semi-totalitarian. Simultaneously, they have become far more assertive and aggressive internationally.

Furthermore, since the outbreak of the economic crisis in 2007–8 and the subsequent upheavals in democracies worldwide, the authoritarian great powers have gained in confidence, while the hegemony and prestige of democratic capitalism have suffered a massive blow unparalleled since the 1930s during the rise of fascist and communist totalitarianism. One hopes that the current economic and political malaise in these democracies will not be nearly as catastrophic.

Will China become more assertive and aggressive as its wealth and power increase over the coming decades? Or will growing wealth and affluence make its people and government more liberal and increasingly averse to military action, as is the case throughout the developed world? Given China's size, might and historical traditions, this is, in my opinion, an open question—indeed, the premier political question of the twenty-first century.

The prospect of renewed protectionism fueled by new military tensions increases the likelihood of armed confrontation, as production and trade are again linked to territory and direct rule. The system of free trade has been exploited by China, as in the direct theft of knowledge and the coercion of foreign companies to cede know-how. But the main threat is the use of the wealth derived from the system of peaceful free trade to exert political-military pressure, including the threat and use of war. As Adam Smith, the theorist and champion of free trade, has put it: 'Defence is much more important than opulence.' On the other hand, if protectionism and trade blocs are going to re-emerge, China's incentive to secure its control over vital resources, as in the South China Sea, might grow momentously. There is no easy solution to this bind.

Russia—a country lacking a truly productive sector, whose international trade is limited mostly to the export of energy and other raw materials, and whose power of attraction towards its neighbors/ former imperial domains is low—is left only with the threat and use of arms as its main foreign policy tool vis-à-vis its neighbors trying to escape its embrace. This is how it was with Georgia and, at the time of writing, with Ukraine.

War has been declining and, indeed, has entirely disappeared in all its forms within the developed world. There is no delusion about this. This is an unprecedented and hugely optimistic development. But the deeper causes of peace in the modernization-development sphere, and hence the boundaries of the 'zone of peace,' must be clearly understood so as to calibrate our expectations and prepare us for the rough times ahead. The war in Ukraine—like the wars in Africa, the Middle East, potentially the Indian subcontinent, and around China and North Korea—demonstrate that the world's 'zone of war' is still simmering and occasionally erupting into destructive and lethal armed conflicts, and which might even involve the use of nuclear weapons in Ukraine and elsewhere.

4

THE NEXT WAR WOULD BE A CYBERWAR, RIGHT?
LESSONS FROM THE RUSSO-UKRAINIAN WAR

Paul Ducheine, Peter Pijpers and Kraesten Arnold

As early as 1997, John Arquilla and David Ronfeldt predicted that 'cyberwar is coming'.[1] A discourse was started, culminating in antagonizing contributions from Thomas Rid stating that 'cyberwar will not take place'[2] and John Stone arguing that 'cyberwar will take place'.[3] When assessing the trains of thought, the discourse was not a dichotomy of arguments but rather an analysis based on different perspectives, with a pinch of semantics. It did, however, epitomise the struggle by academics and policymakers over what to make of the disruptive power of operations in or through cyberspace, as well as how to assess, interpret or even frame these in term of international politics and law.[4]

With the Russian invasion of Ukraine in 2022, 'cyberwar' has (finally) come. Or has it?[5] In recent decades, Information and Communication Technology (ICT) has become an integrated part of our societies, penetrating deep into its capillaries. Cyberspace appeared to be well suited for social interaction in general—not only for benign purposes, but also for malevolent goals such as crime, sabotage, manipulation and subversion.[6] In recent years, numerous cyber operations have been executed, including Operation Orchard in 2007,[7] the 2010 Stuxnet attack,[8] the 2015 and 2016 Ukrainian power outages,[9] interfering with the 2016 presidential election in the United States,[10] and the NotPetya ransomware attack in 2017. In the last decade and a half, modern armed forces have embraced cyberspace as a domain for warfare and subsequently developed cyber capabilities (including the formation of Cyber Commands[11]) in order to prepare for a niche within social interaction: war, conflict and competition in cyberspace.[12] After cyberattacks on power grids and nuclear facilities, and after digital meddling with foreign elections, the thought that 'the next war will be fought in cyberspace (too)' was clearly more than imaginary.[13]

The pitfall in the discussion whether the so-called cyberwar will or will not take place is that it is often framed as a dichotomous or even binary approach.[14] Therefore, when Russia invaded Ukraine in February 2022 with brute force—after which both states have been engaged in an armed conflict (or war)—the rationale behind the assumption that the cyberspace would play a decisive part in the next war[15] appeared to be flawed.

But reality is not binary. When states use brute military force on land or at sea, it does not exclude the employment of other instruments of power—whether diplomatic, economic, legal or informational—in other domains, including in cyberspace.[16] The danger is that by following a binary rationale, the Russo-Ukrainian War will primarily be assessed as a kinetic-military confrontation (which is understandable given the immediately noticeable effects)[17] and lessons will be learned accordingly, thus missing out on—and not being prepared for—the new developments of conflict in cyberspace.

Echoing Beecroft, who argues that 'it would be premature to draw definitive conclusions based on eight months of war',[18] some trends that might shape future wars are nonetheless emerging.[19] The main purpose of this chapter is, therefore, to explore what can be learned about military operations in cyberspace from the current war between Russia and Ukraine, including the wider conflict between Russia and Western states, and what we can say about the character of future war.

In order to substantiate the analysis, the chapter starts with a short description of the notion of 'cyberwar'; or to be more precise, the military use of cyberspace in war. The core part of the chapter will be an assessment of the activities in cyberspace that were executed in the run-up to, and waging of, the 2022 Russo-Ukrainian War, specifically both the hard-cyber operations (e.g. hacking ICT infrastructure) and the soft-cyber operations (e.g. digitally influencing the adversary's audience) conducted by the Russians and Ukrainians. The following sections will analyse whether and why cyber capabilities appear to be underutilized, and what the developments in the Russo-Ukrainian War might mean for the future of cyberwarfare.

'Cyberwar'

For the purpose of this treatise, the present authors differentiate between the war—the armed conflict between Russia and Ukraine—and the wider conflict between Russia and other states that support Ukraine in some way. Therefore, warfare and conflict are viewed as distinct terms, each employing typical instruments. In essence, cyberwarfare involves the military use of cyber capabilities in the course of armed conflict to achieve military objectives in or through cyberspace.[20] Cyberwar may be (and often is) part of the wider cyber conflict that may serve non-military objectives, such as espionage, or political or economic disruption, and may also involve other actors and power instruments such as proxy hackers launching ransomware attacks. The domain involved—cyberspace—is an area of engagement similar to the land or sea domain. It encompasses a physical network layer of computers, a logical layer of data and software, and a virtual persona layer. The latter can be regarded as the digital reflections of individual and groups that appear on WhatsApp, Instagram or email accounts. In a similar vein to land and sea, cyberspace is a neutral domain, not a weapon as such.

In cyberspace, numerous instruments of state power can be employed, ranging from diplomatic endeavours and informational activities to military means. The nature of war in cyberspace equals those in other domains such as land or sea, but its characteristics are different. The characteristics of cyberspace are the low cost of entry for state and non-state actors, the heightened opportunities for anonymity, and a seemingly worldwide reach. Cyberspace is boundless, and it enables communication to travel faster and reach a broader audience. The major difference with traditional kinetic warfare, however, concerns the weapons used. Kinetic weapons are tangible and create visible effects, whereas cyber weapons consist of intangible binary code, creating effects which are neither directly perceivable nor tangible.

Similar to earlier wars, present and future armed conflicts will not solely be fought in one domain but in a multidomain environment. Cyberwars are not likely to be executed in isolation but will probably be synchronized with activities on land, air or sea, and may also be aligned with economic (gas), agricultural (grain), geopolitical (access to sea lanes), or informational (narration) levers to persuade, manipulate or coerce target audiences. Once vital interests are at stake, states will forge a strategy to unleash instruments of power for offensive, defensive or deceptive purposes.[21]

Strategies using cyberspace as a domain can encompass three sorts of operations. The first is digital intelligence operations,[22] which can be used to gather data and information without manipulating or destroying them.[23] Second, hard-cyber operations are activities in cyberspace targeting and creating direct effects on the very components of cyberspace itself. Malicious software (malware) and data are the 'weapons' or 'payloads' used during these operations. Examples include attacks on virtual identities (social media accounts, websites), virtual objects (computer programs, software, data) or hardware (servers, routers) that may impact the availability or integrity of services and data. These days, many physical objects and systems depend on computers operating faultlessly; as a result, the effects of cyber operations may subsequently materialize beyond cyberspace, affecting the functioning of physical objects such as (civilian) critical infrastructures or military (weapon) systems. Finally, soft-cyber operations (otherwise known as digital influence operations) may utilize cyberspace as a 'vector' to influence the psyche of targeted audiences. These 'cyber-enabled' psychological warfare, propaganda or information operations use information or content as 'weapons'.[24]

In practice, hard and soft cyberattacks often complement intelligence operations, and vice versa. Consider, for example, a phishing mail to lure individuals into revealing login credentials, after which an attacker may manipulate, destroy or steal data. Sensitive or undisclosed information may be revealed (hack-and-leak) to cause a commotion.[25]

The Conflict between Russia and Ukraine Prior to 24 February 2022

Russian cyber operations against Ukraine have a long lineage. The emergence of cyberspace has invigorated the Russian doctrine of 'active measures',[26] whereby they would seek strategic advantages in the information environment by deception, forgeries,[27] provocation, subversion,[28] and also by the spreading of disinformation.[29] Tensions reached an apex when, in the early months of 2014, Russia invaded and subsequently annexed the Crimean Peninsula,[30] and pro-Russian armed separatists simultaneously took control of parts of the Donbas area in Eastern Ukraine. Although this invasion coincided with fairly straightforward Distributed-Denial-of-Service (DDoS) attacks, these kinetic operations were not accompanied by sophisticated cyberattacks.[31]

Russia subsequently used numerous information tools to influence and tighten their grip on the Donbas and wider Ukrainian population.[32] Historical myths, narratives and symbols—including the Kremlin's 'demilitarization and denazification' narrative—were used via a plethora of media outlets to disorient the targeted audiences,[33] including during Ukraine's May 2014 presidential election, amongst others. Apart from digital influence operations, the pro-Russian hacker group CyberBerkut compromised Ukraine's Central Election Commission's vote-counting software in order to designate ultra-right leader Dmytro Yarosh as winner of the elections, regardless of the actual votes cast. Right after the elections, the very same manipulated results were presented on Russian TV. However, the hack was discovered just in time, and the malicious software was sanitized at the very last moment. In actual fact, the pro-Western politician Petro Poroshenko was elected president with an absolute majority.[34]

The years thereafter, a long chain of Russian-attributed needle-prick cyberattacks targeted a variety of Ukrainian government organizations, critical infrastructure and businesses.[35] In 2015, hackers attacked Ukraine's power grid, causing a power outage in parts of the country. A year later, Ukraine's electricity distribution system was again targeted. The cyberattacks not only temporarily denied service of these Industrial Control Systems (ICS), but the victim's computer monitoring systems were also deliberately corrupted, rendering them inoperable. The Security Service of Ukraine (SBU) attributed both attacks to the Russian state or state-sponsored hackers, which cyber security firm Kaspersky Lab later substantiated.[36] In retrospect, since 2014, Ukraine has appeared to be a 'testing ground' for refining Russian hard-cyberattacks on critical infrastructure.[37]

One example of a hard cyberattack particularly aimed at a (military) weapon system concerns malware (allegedly) implanted on Android devices via the Russian social media platform VKontakte by Russian Military Intelligence (GRU) hackers. The platform's original application enabled Ukrainian artillery forces to more rapidly process targeting data of their Soviet-era artillery units. However, from late 2014 to 2016, the manipulated application was able to retrieve communications and some locational data from infected devices.[38] That intelligence was arguably used to strike against Ukrainian artillery units operating against pro-Russian separatists in Eastern Ukraine.[39]

Of the pre-2022 cyberattacks, the 2017 NotPetya cyberattack—executed on the eve of Ukraine's Constitution Day, which commemorates the country's exit from the Soviet Union—was perhaps the most devastating.[40] Initially targeting Ukraine's financial system, the malware soon spread and infected businesses and critical infrastructures worldwide, resulting in a collective loss of nearly US$10 billion.[41]

In concert, soft cyberattacks were executed targeting both Ukrainian and Western audiences. A number of cyber-enabled influencing operations can be found in the aftermath of the Malaysia Airlines flight MH-17 crash in July 2014, which was shot down while flying over eastern Ukraine near the Ukrainian/Russian border. The international investigation team concluded that the aircraft had been shot down by a Buk surface-to-air missile system, and Russian involvement in the provision of the missile system was demonstrated. Despite the Russian government denying all responsibility and persistently providing alternative possibilities for the crash via social media platforms, web fora and online news sites, the investigation team subsequently excluded other possibilities based on the evidence available.[42]

In the two months preceding the war, pro-Russian hackers executed various cyberattacks. These attacks started with rather straightforward DDoS attacks on Internet service providers, the Ukrainian Ministry of Defence, the armed forces and two national banks.[43] Almost simultaneously, pro-Russian hackers defaced the websites of dozens of Ukrainian government organizations with manipulated political imagery and provocative statements.[44] Spear-phishing mails, ostensibly originating from Ukrainian state bodies and targeting Ukrainian entities, supplemented these cyberattacks, albeit with only modest impact.[45] In mid-January 2022, Microsoft revealed the existence of a destructive malware operation (dubbed 'WhisperGate') targeting multiple organizations in Ukraine.[46] This destructive 'wiperware'-attack, attributed to the GRU,[47] had been designed to inflict permanent damage. Masquerading as mere 'ransomware', this malware not only hijacked data, but also erased data on attacked computers with the intention of rendering these computers useless. In addition, Russia was said to have set up preparations for the broadcasting of a false-flag operation that could have legitimized Russian intervention against Ukraine once it was revealed by Russian media outlets.[48]

Already in the prelude to the war, Ukraine increased its resilience to counter the Russian cyberattacks,[49] cognizant of an upcoming war in cyberspace or even in the physical realm. On invitation, US Cyber Command supported Ukraine for three months from December 2021.[50] On 22 February, the European Union (EU) offered its Cyber Rapid Response Team,[51] but Ukraine did not seize that opportunity. Separately, but related, Belarusian Cyber Partisans hacked their country's railway network to disrupt the transport of Russian troops and equipment in Belarus.[52]

On 23 February 2022, one day prior to the invasion, several cybersecurity firms sounded the alarm after finding other malicious software, again with destructive 'disk-wiping' capabilities, targeting hundreds of machines in Ukraine in the financial, defence, aviation and IT services sectors.[53] This newly discovered HermeticWiper malware showed some similarities with the earlier discovered wiperware, but this time in a more sophisticated guise.

On the eve of the invasion, apart from the expected jamming of communications,[54] one particular occurrence attracted attention: the cyberattack on Viasat, a major satellite communications provider for Ukraine and its military, but also for large parts of Europe. The allegedly Russian cyberattack rendered satellite modems unserviceable and had an immediate effects on the Ukrainian military's battlefield communications, particularly in the Chernihiv region (close to the threatened Kyiv).[55] The loss of satellite Internet communications hampered frontline communications and undermined Ukraine's ability to track Russian troop positions and movements.[56] Notably, the impact was not restricted to Ukraine, as impairing Viasat also affected communication systems in other parts of Europe and Northern Africa.[57] However, after Ukraine's Minister for Digital Transformation Mykhailo Fedorov posted a Tweet asking Elon Musk for help,[58] the latter's Starlink satellite system (which has end-to-end-encryption) restored the disrupted Ukrainian Internet and communications.[59]

Evolution of the War

One notable feature of the cyber dimension of the wider conflict (and sometimes of the war itself) has been the prominent part played by non-state actors.[60] The night of the invasion, the notorious hacker group Anonymous declared a cyberwar against the Russian government.[61] Activists launched numerous DDoS attacks and website defacements, the effects of which were often difficult to verify and value.[62]

In addition to strengthening its own cyber security ecosystem,[63] the Ukrainian government was supported by numerous IT and cyber security firms such as Microsoft, Mandiant and ESET.[64] For instance, shortly after the Russian invasion, Vodafone cut communication in large parts of the Donbas region, leaving the pro-Russian separatists digitally blind.[65] In addition, both the US and the UK acknowledged having supported Ukraine's defence *in situ* up to the invasion, and later remotely.[66] According to the Cyber Group Tracker, 117 non-state entities with a pro-Ukrainian inclination joined the conflict,[67] amongst them the Anonymous hacker group, the IT-Army of Ukraine and the North Atlantic Fella Organization (NAFO).[68]

Already on the first day of the invasion, Fedorov—using social media platforms Telegram, Facebook and Twitter—called on tech-savvy volunteers worldwide to join the newly established IT-Army of Ukraine to protect his country from Russian digital attacks.[69] Although the IT-Army executed offensive cyber operations, such as DDoS attacks against Russian government and company websites,[70] they were also used by the Ukrainian government as a symbol of Ukrainian cyber-resilience and to boost morale.

The IT-Army goes beyond a random group of worldwide sympathizers, and is allegedly mixed with in-house expertise from the Ukrainian defence and intelligence services. According to Stefan Soesanto, the IT-Army is 'a hybrid construct that is neither civilian nor military, neither public nor private, neither local nor international, and neither lawful nor unlawful'.[71] This mixed group thus runs the risk of crossing boundaries of legal frameworks regarding norms and rules for state behaviour in cyberspace, which in turn may wreak havoc on the future stability of cyberspace.

Not all 'cyber patriots' sided with Ukraine, however, with many hacker groups choosing to fight for Russia. Some seventy-four entities are active, including agents of the governmental intelligence units,[72] such as the GRU's units 74455 (aka SandWorm) and 26165 (aka APT [Advanced Persistent Threat] 28/Fancy Bear).[73] Pro-Russian hacker groups also include criminal groups such as Stormous Ransomware, Ghostwriter, Conti and UNC1151.[74]

In June 2022, Microsoft published a report on their lessons from the first four months of the 'cyberwar' between Russia and Ukraine.[75] Although the document was likely written from a commercial perspective, the report provides insights from a first line of cyber defence. The company saw that Russia launched multiple waves of sophisticated destructive cyberattacks against dozens of distinct Ukrainian agencies and enterprises, eventually destroying thousands of computers. Microsoft argued that these cyberattacks were conducted largely in concert with kinetic operations.[76]

As the first Russian troops crossed the Ukrainian border, cyber security company ESET discovered two more destructive wiperware families—HermeticWiper and IsaacWiper—targeting Ukrainian organizations.[77] Similarly, Symantec revealed yet another form of disk-wiping malware—Trojan.Killdisk—launched shortly before the Russian invasion, targeting the Ukrainian financial, defence, aviation and IT services sector.[78]

In March 2022, Ukraine's railways and transportation systems transferred weapon systems and military supplies from Western allies to the east of the country, while large numbers of internally displaced persons and refugees used the same means to flee in the opposite direction. These railway and transportation systems were targeted with a variety of weapons.[79] Destructive cyberattacks on, for example, a key logistics centre for the movement of military and humanitarian aid in Lviv were executed (allegedly) in concert with missile attacks on railway substations.[80]

In April 2022, ESET reported a hard cyberattack with so-called 'Industroyer2' ICS malware on the Ukrainian power grid. The attacker had modified pieces of malware known as Industroyer, which had been used in 2016 to attack the power grid and cause power outages. This time, the ICS-malware was accompanied by yet more sets of destructive wiper malware.[81]

On 30 June, Ukraine's State Service of Special Communications and Information Protection reported that since the invasion, Russia had launched some 800 cyberattacks, mainly targeting the government and local authorities, security and defence, and the financial, energy, transport and telecoms sectors. Over time, the intensity of cyberattacks had not decreased, yet their quality has been declining.[82]

Next to the aforementioned hard cyberattacks, Russian state and supportive non-state actors conducted cyber-enabled influence operations: disruptive propaganda and disinformation campaigns (both mainstream and on social media) to influence target audiences.[83] Russia not only targeted the Ukrainian government and domestic Russian audiences, but also Western audiences. In Russia, television is still the most used mainstream media outlet; domestic audiences are influenced via talk shows, news reports and educational programs. Social media channels, including VKontakte and Telegram, are also extensively used to disseminate framed images, memes and deepfakes.[84]

The main purpose of Russian influence operations is to demoralize the Ukrainian population, to drive a wedge between Ukraine and its Western allies, and to enthuse the domestic audiences of Russia and their foreign allies. Narratives used in this context are the fear of Russophobia—a sensitive topic to Russian diaspora or ethnic Russians living in Ukraine, framing Ukraine as a Nazi-state conducting genocide against ethnic Russians[85]—or the endemic corruption within the Ukrainian government.[86] Another narrative depicts the West as applying double standards and imposing liberal values to try and keep Russia small and maintain its own global dominance.[87] A conspiracy theory that received considerable attention in the Russian discourse dealt with US bioweapons made in Ukraine.[88]

Ukraine's President Volodymyr Zelensky recognized and fully utilized the potential of social media,[89] and this has affected the cognitive dimension of both friend and foe.[90] Zelensky (and his ministers) not only remained in Kyiv from the very start of the war, but he also addressed his own population, military forces and the international community online. Zelensky's quote 'I need ammunition, not a ride',[91] after he had been offered transportation to safer areas, went viral. Several other impassioned speeches to foreign parliaments, given on-screen via the Internet, have resulted in much-needed international sympathy and support, especially in terms of the supply of funds, weapons and ammunition.

Civilians and soldiers alike use social media to show the online world the actual situation on the ground. Russian troop locations, strengths and movements are constantly being reported, whilst such information about Ukrainian positions is carefully kept offline. Social media platforms are used to mobilize the population and share information and instructions for non-violent and armed resistance; this varies from passing on information about Russian weapons and troop movements,[92] to instructions on how to attack enemy forces with Molotov cocktails.

On social media, the story of a Ukrainian fighter pilot with the nickname 'the Ghost of Kyiv' became very popular early on in the war. The pilot had allegedly shot down six Russian aircraft on the first day of the invasion. The identity of the pilot could not be confirmed, but a heroic legend was born. Later, the Air Force Command of the Ukrainian Forces admitted that the Ghost of Kyiv did not exist in reality, at the same time warning 'not to neglect the basic rules of information hygiene [and] to check the sources of information before disseminating it'.[93]

Another occurrence concerned the bold response of Ukrainian troops defending Zmiinyi Island (also known as 'Snake Island') after Russia's Black Sea Fleet flagship *Moskva* demanded their surrender. The explicit refusal 'Russian warship, go home'[94] became a popular phrase online, and was frequently used in Internet memes, as was the subsequently introduced postage stamp to honour the heroes. On 13 April 2022, Ukrainian officials reported that the *Moskva* had been hit by two Ukrainian anti-ship missiles,[95] and its sinking a day later[96] was used as theme for another stamp and online memes. The narrative of failing Russian activities has been amplified by online footage of Ukrainian farmers towing away abandoned Russian combat vehicles. It demonstrates the swift exploitation via the virtual dimension of physical battlefield successes to the cognitive dimension.

In March 2022, the SBU announced that it had discovered and shut down automated networks of computers ('botnets') mimicking over 100,000 social media accounts, which had been used to spread disinformation and create panic among the Ukrainian population, and to destabilize the socio-political situation in the country.[97] One may assume that Russia was well prepared to start influence activities parallel to a kinetic ground war on Ukrainian territory. Nevertheless, Russia's efforts seem less successful than anticipated. Moreover, unlike Russia, it appeared that Ukraine has a good hand in exploiting the online media capabilities: from showing the heroism of their own soldiers and the determination of their people, to disclosing war crimes of Russian fighters, revealing the mismanagement of Russian troops, and the exposure of poor equipment and destroyed Russian weapons systems.

Analysing the Differences in Cyberwar Expectations and Reality

Several observations relating to hard-cyber operations, soft-cyber operations, activists, and IT service and infrastructure companies should be made. The general observation is that the Russo-Ukrainian War is not the decisive cyberwar some analysts were expecting, but neither is the war void of extensive cyber operations. There have been numerous cyberattacks in the prelude to it and during the armed conflict itself. Up until now, these cyber operations persisted, but they are different from what was to be expected, both in number and appearances.

Firstly, hard-cyber operations appear to have played a smaller role than expected. Hacking or digital sabotage operations directed against critical infrastructure and military systems were relatively sparse. Though hard-cyber operations were executed in the prelude to the war, they appeared to play a less prominent role during the actual war, despite Russia having executed numerous operations in the years leading up to it.[98] In that sense, the dogs of war failed to bark loudly.[99]

Secondly, influence operations that make use of the Internet and social media have been significant. However, no new forms of operations were witnessed. Often, existing techniques were reused—for example, hack-and-leak operations against Russian organizations by actors such as Anonymous; the 2015 malware-attack by APT SandWorm against the Ukrainian power grid;[100] or the deepfakes of President Zelensky[101] and Putin.[102] Strategically though, these information campaigns are essential for this war's centre of gravity: namely, international support. For Ukraine, this represents its lifeline, as this is what guarantees the supply of weapons and ammunition. However, at the same time, it is an Achilles heel, as Russia tries to undermine this support and to divide Ukraine's allies.

Thirdly, the cyber operations, in the prelude to and especially during the war, have not solely been executed by the Russian and Ukrainian armed forces. Most prominent is the unexpected involvement of independent non-state actors[103] on both sides,[104] who have targeted both military targets as well as civilian critical infrastructure, media and financial systems. The actual impact of these groups' involvement, however, remains a matter of further research.[105]

Finally, a completely different set of non-state involvement involves the crucial role played by commercial parties, especially ICT-related firms like Microsoft,[106] Amazon,[107] Vodafone,[108] Facebook, Twitter[109] and Starlink.[110] Although the role of these companies is evident, they are often overlooked in many analyses of the war, despite playing important roles in such crucial areas as attacking or protecting supply chains. Likewise, the threat intelligence shared by Microsoft, the hardware delivered by Starlink, even the neutral position of ICANN and RIPE, demonstrate their fundamental role in the effective functioning of the Internet and cyberspace.[111]

Despite all these cyber operations, the question is why there are not more. When parsing the evidence as depicted in the previous sections, two explanations can be put forward.

There Are Attacks, but We Do Not See Them

One of the reasons Russian attacks may not be noticed by spectators like the present authors is that the Ukrainian cyber defence is well organized[112] and attacks have been intercepted before substantial damage was caused.[113] Over the years,[114] Ukrainian cyber infrastructure has been hardened, with state (US, UK) and commercial (Microsoft) support cushioning the impact of an attack (though this coordinated defence should not be overstated).[115] Moreover, ongoing cyberattacks since the annexation of Crimea have exposed Russia's *modus operandi* and preferred targets; for their part, Ukraine probably also withholds information. Attacks may have affected targets, but the impact has not been made public, thereby hindering the opponent's conduct to make a battle damage assessment or to avoid disheartening one's own forces. Another possibility is that hard cyberattacks definitely took place and caused damage, but were not recognized or detected as such.[116]

The Number of Attacks is Limited—Deliberately or Due to Failure

It cannot be ruled out that as Russia was initially expecting a *blitzkrieg*-like victory over Ukraine, traditional military capabilities and cyber capacities were only geared for a short operation. The means of cyber used were to shape the battlefield and create the conditions for a successful military intervention. Following this rationale, it must not come as a surprise that in the period after the invasion, where fighting lingered on, cyber means were not instantly available to follow-up. It is common knowledge that more sophisticated cyber operations take time to prepare and are quite complicated to synchronise with other (i.e. kinetic) lines of operations.

A second reason might be that cyberattacks have been planned or prepared as a back-up scenario awaiting to be executed. Logic bombs might have been placed, lying dormant until they are remotely activated or when particular conditions are met to trigger their payload. They also might have been overrun by hastily drawn-up campaign designs later on in the conflict.

Thirdly, Russia might be able to conduct more hard-cyber operations against Ukraine but is reticent to do so because revealing these capabilities would put them at a disadvantage relative to North Atlantic Treaty Organization (NATO) and EU member states.[117] While there is no absolute need for stealthy operations during the war with Ukraine, cyberattacks against other states or entities benefit from covertness in order to hide their *modus operandi*, or to preserve scarce 'zero-day vulnerabilities and exploits'.[118] Moreover, the Ukrainian ICT infrastructure might be of use for Russia either currently or in the postwar phase.[119] Fourthly, cyber weapons' inherent characteristics of reach, time, speed and versatility are instrumental in the element of surprise,[120] especially when used in a supporting role in a wider effort to achieve an attacker's objectives. While cyberattacks can be conducive to subversion—as seen during the prelude to the invasion—Russia may realize that cyberattacks have less strategic utility to contribute to the vital interests of the state compared with kinetic attacks, especially in a war-like

BEYOND UKRAINE

scenario.[121] The efficacy of cyber weapons pales into insignificance when it comes to kinetic effects during armed conflict. The absence of operational and tactical cyber weapons is therefore perhaps not that unexpected.

Fifthly, Russian cyber operations need to cover at least four audiences: domestic, former Soviet republics, Ukraine, and NATO/EU member states.[122] As such, they may simply be overstretched to meaningfully concentrate their resources. Moreover, the agents working in the Russian cyber entities, as in many states, have a division of labour, and each hacker cannot easily master all techniques.

Sixthly, Russia might want to avoid escalation. Russia is willing and able to unleash devastating cyberattacks; however, the risk of unintended and unexpected second- and third-order effects beyond Ukraine's cyber borders could be too great. The aforementioned 2017 NotPetya cyberattack initially targeted Ukraine but, by virtue of the malware's design,[123] rapidly spread worldwide. Striking the US or other NATO member states unintentionally with disruptive and indiscriminate cyberattacks may elicit unintended responses.[124]

A seventh reason might be that Russia does not have 'cyber weapons' it can use in wartime. Recent armed conflicts (e.g. Georgia, Syria, Ukraine) show that Russia prefers conventional kinetic weapons rather than using destructive hard-cyber operations.[125] Russian doctrine values the use of plausibly deniable stealthy cyberattacks in peacetime, but once an all-out war commences, their usefulness diminishes.

Finally, apart from failures in campaign design, planning and synchronization of physical military operations on land, sea and in the air, designing and developing cyber weapons that create surgical effects is a time-consuming endeavour. Gathering actionable intelligence about an actual target, its computer and network infrastructure, hardware and software, determining the desired effects to be created, the methods and techniques to penetrate a system—or rather, successfully completing the entire 'cyber kill chain'—often requires months or even years of preparation. Cyber weapons with strategic effects can be prepared in peacetime, but after the failed invasion, the dedicated state actors involved in cyber activities may have lacked the required preparation time to produce effective hard-cyber weapons. Though some ill-prepared efforts were made targeting the energy and finance sectors, and other critical infrastructure (e.g. the Ukrainian power grid attack), these operations had an ad hoc character and were not synchronized with other instruments of power. As Russian military commanders were obliged to increasingly focus on prosecuting the conventional war and on avoiding a military disaster, cyber operations may have been deprioritized in the overall Russian campaign.[126] This could explain the lack of synchronization with other instruments of power (including hard-cyber operations) in the months after the invasion commenced.

Conclusions

The intent of this chapter was to explore what could be learned from the first twelve months of war between Russia and Ukraine (from February 2022 to February 2023) concerning military operations in cyberspace, and what this implies for the character of future war. We observed that cyber operations were used both in the prelude to and during the war, though in a different manner than initially expected. Prior to the invasion, several state actors executed hard-cyber operations to undermine, sabotage or merely destroy Ukrainian systems. As far as we know, after the invasion, the impact and quality of hard cyberattacks declined. More than expected, soft cyberattacks have been engaged persistently both in the prelude to the 2022 invasion and during the subsequent war. Moreover, the sheer size of the engagement of hacktivists worldwide, along with active involvement by commercial ICT services, was not yet part of the rulebook of cyber operations.

What does this mean for the future of war? Antony Beevor recently argued that 'Putin doesn't realize how much warfare has changed',[127] as he is trapped in an obsession with the Second World War. Though this is a fair point, it might similarly be valid for Western policymakers and academics. Several pundits of military thought saw the Russian invasion as a resurrection of the Cold War template, discarding the cumbersome new ways of warfare as reflected in buzzwords such as 'influence', 'information', 'post-truth', 'hybrid', 'cyber' or 'cognitive'. Subsequently, they assessed the Russo-Ukrainian War exclusively through the lens of armed conflict, conveniently neglecting or minimizing the use of other instruments of power via other domains of engagement in the wider conflict arena. However, that the next war did not turn out to be a cyberwar in the way many had expected should not imply a relapse into preceding conceptual frames.

The authors argue that the Russo-Ukrainian War underscores several elements that might impact the future of cyber operations. Firstly, a crucial lesson to be learnt is that competition, conflict or war is not fought in one domain (e.g. on land or at sea) with one instrument of power (e.g. the military). Contemporary conflicts are multidomain, feature in all dimensions of the information environment (physical, virtual and cognitive), and make use of all instruments of power available. In this sense, modern warfare can be considered hybrid warfare. In the future, malign actors will not solely shift between the employment of different military assets (adjusted to the air, land, sea or cyberspace domain), but will also alternate or create synergy between different instruments of power. To be effective, the quintessential attribute in future wars will be the ability to coordinate and synchronise all of these elements.[128] This is easier said than done, as became clear after the first three days of the Russo-Ukrainian War.

Though the use of hacker groups, proxies and even commercial actors should not be overestimated,[129] they are an enduring reality in modern-day conflict. A proliferation of cyber activities and cyber actors can be witnessed, potentially leading to democratization of the use of force. These actors, ranging from malign cyber criminals[130] to bona fide hacktivists or multinational enterprises, have interests that fluctuate amongst themselves and differ from the states they endorse. The result could be that states, hackers or commercial parties may form ad hoc coalitions, generating effects that appear to be aligned in a specific time and place. Once financial, physical or ideological interests fan out, so will the role of these non-state actors. It is no longer possible to unambiguously attribute these non-state actors to a state. While the APT SandWorm or the St. Petersburg-based Internet Research Agency (aka Glavset) can be linked to a state (Russia in this case), this does not apply to Anonymous, Conti or Microsoft.

The development of cyber weapons aimed at having strategic effects requires ample preparation. An effective manner of undermining adverse military power is to affect weapon systems, command centres, radars or logistics systems. However, to pave the way for the development of operational or even tactical weapons, extensive knowledge about these systems is necessary, which may require the involvement of these systems' manufacturers (with or without their knowledge). Therefore, to be effective in wartime, a proactive cyberspace posture in peacetime might be necessary. No advantage is permanent, however, since new vulnerabilities and opportunities frequently rise, targets change, and offensive or defensive capabilities lose effectiveness over time. To increase cyber resilience, 'persistent engagement'[131] arguably allows for greater freedom to manoeuvre in cyberspace, even during peacetime. Within the present international legal order, this may yet be a bridge too far.

Finally, while the dominant lens of reference in war might be armed conflict, this does not and should not preclude covert subversive actions in cyberspace, digital espionage or cybercrime in the context of a wider strategic competition away from warfare. In fact, in future wars, all domains and conventions might overlap. In the Russo-Ukrainian War, despite there only being two warring states, numerous other actors are involved. Commercial ICT firms and hacker groups will take a forward-leaning stance, siding with one of the warring entities while avoiding becoming a belligerent party. For Western states, this may imply that, though the differences between war and peace, and between internal and external security might exist in theory, in reality this distinction is obsolete. This urges a revisit of how the West prepares its armed forces (or rather, Western societies) against malign aggressors and attacks in the cyber-related information environment.

While the jury is still out on whether cyber operations have strategic value, the ongoing war makes clear that despite the apparent limited strategic leverage of hard-cyber operations, the extensive use of framed information and narratives via social media seems to underline the strategic centre of gravity of this war: that is, international support for Ukraine. This proves to be both the lifeline and the Achilles heel for Ukraine: on the one hand, guaranteeing the flow of weapons and munitions, and on the other, encouraging Russia to do all it can to degrade Western cohesion.

PART II

LANDSCAPES OF FUTURE WAR

5

FOUR FACES OF WAR

Frank Hoffman

The future of war is inherently opaque under the best of circumstances, but the risks involved can be reduced by drawing upon the lessons of contemporary conflict. Events from Ukraine suggest that war will reflect many enduring continuities, but that war also demonstrates how change can rush upon us. The war offers signposts about the future of warfare for those willing to objectively evaluate the evidence, dissect the unique circumstances of its geography and terrain, and understand the interaction of the antagonists. Learning from ongoing wars is not an easy exercise, however, as institutions have historically struggled to capture clear lessons beyond existing preferences.[1] Nevertheless, learning from recent wars is an important source of data for policymakers and their national security advisors as they assess risk over the horizon.

To future-proof their security preparations, policymakers need foresight into the potential principal challenges and their implications.[2] This requires a continuous scan of social, political, economic, and technological developments, and an open and curious mind as to their possible impacts. As this book shows, there are numerous trends and factors driving the changing character of warfare in the twenty-first century, including shifts in the frequency, length, and lethality of war, altered geopolitical conditions and new technological advances.

There are numerous other questions that defy easy predictions. Who will fight for their country, and what will the cause of the next war be? Where will the next war be fought—Asia or the Middle East? What will be the impact of new domains like space and cyber? How does urbanization and environmental degradation impact the frequency or character of tomorrow's wars? How will our adversaries fight? Will the character of the conflict be unconventional or fought principally with high-tech systems like hypervelocity missiles and quantum computing that outclass our own?[3] Which technologies present the most risk or opportunity, and what organizational changes are needed to leverage them properly?

Failure to anticipate and adapt to the questions posed by these emerging realities will generate a significantly higher risk of violence and instability, and a lower level of readiness to respond. History suggests that the price of this risk could be quite high.[4] The only certainties are that the future will bring surprises and that the cost of complacency is rising.

The Four Faces Framework

This chapter focuses on key trends in conflict that suggest four distinctive types of warfare. This suite of potential modes of modern warfare can be depicted by a simple two-by-two matrix, as shown in Figure 5.1. The horizontal axis represents a continuum of indirect/ non-kinetic versus direct, kinetic or physical trends; the vertical axis of the matrix reflects a similar range of state and non-state or proxy actors. The four boxes represent four schools of thought about the most salient forms (or 'faces') of future warfare. Each 'face' has numerous proponents and supporting arguments for its rising salience. This model is not without limitations (particularly for intrastate conflict), but what it does do is raise relevant issues in our understanding of future conflicts. It is also expected to be useful to assess the risks associated with a broader conception of threats to democratic societies, as well as how they allocate resources and adapt institutions to address these risks. After reviewing issues related to each of these 'faces,' this chapter concludes with a discussion about their relative risks.

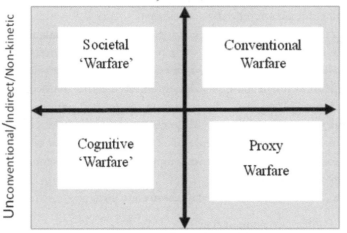

Figure 5.1: The Four Faces framework

(Source: Frank Hoffman)

Societal Warfare

The initial 'face' of the future captures the increased use of non-kinetic means to directly attack a society, as well as its means of economic production and day-to-day functioning. This mode of warfare seeks to disrupt or degrade the population's sense of invulnerability and to generate popular discontent to change the government's policy. It can also aim to undercut the popular support and legitimacy that democratic governments are based upon. This openness is, of course, a source of resilience and adaptability as well. In today's highly networked and interconnected social and economic systems, there are numerous opportunities for states or state-sponsored actors—and even some non-state cyber saboteurs—to wreak havoc in the information-intensive operations of modern society.

Societal warfare seeks to exploit that opportunity by penetrating critical but weakly defended economic systems: the 'hardware' of society. As a working definition of societal warfare, I offer the following: the application of direct or indirectly violent methods against critical infrastructure to dislocate and paralyze the civil society and economic activity of a target state in order to extract desired political outcomes by coercion.

Certainly, there is little new in the idea that combatant states should strike directly at key economic nodes to undercut the morale and productivity of the adversary. In the interwar years, the concept of strategic bombing was predicated upon the notion of circumventing long wars of attrition on the ground with precision bombing of industrial nodes.[5] The novelty emerges in the growing recognition over the last couple of decades of how vulnerable the open societies of the West are to computer attacks directed on their civilian operating systems.[6]

The target system for societal warfare includes:

- Public/government services;
- Economic infrastructure (roads, tunnels, ports, bridges);
- News and communication services;
- Food and resource security;[7]
- Energy distribution networks (nuclear, oil and electricity);[8]
- Information networks and connectivity (including servers and cloud systems);
- Healthcare system (hospitals and databases);[9] and
- Transportation systems.

Hostile states have already attempted to exploit these systems. For example, Russia stands accused of engaging in a broad array of cyber activities aimed at undercutting the stability and effectiveness of many of these systems.[10] Recent advisories published by the US Cybersecurity and Infrastructure Security Agency (CISA) reveal that entities sponsored by the Russian state are targeting the following sectors and organizations in the West: Covid-19 research, governments, election organizations, healthcare and pharmaceuticals, defense, energy, nuclear, commercial facilities, water, aviation, and critical manufacturing.[11] The same reporting associated Russian actors with a range of high-profile malicious cyber activity, including the 2016 leaks of documents stolen from the US Democratic National Committee, the 2017 NotPetya ransomware attack, the 2018 targeting of US industrial control system infrastructure,[12] the 2020 compromise of the SolarWinds software supply chain, and the 2020 targeting of US companies developing Covid-19 vaccines.

The SolarWinds hack, attributed to Russian intelligence, was highlighted by CISA as a demonstration of how easily US critical infrastructure and networks can be compromised.[13] European banks and energy facilities face similar challenges. Likewise, the NotPetya attack perpetrated by Russia illustrates the changing character of cyber threats.[14] The chaos started in battle-torn Ukraine in June 2017, when Russian cyber operators launched destructive malware adapted from a series of widespread vulnerabilities common to unpatched aspects of the Windows operating system. The attack quickly spread from targeted Ukrainian banks, payment systems and federal agencies to power plants, hospitals and other life-critical systems worldwide. While its intended targets were in Ukraine, it quickly diffused, eventually infecting commercial firms beyond that state, including the shipping company Maersk and global pharmaceutical producer Merck, resulting in losses estimated near US$10 billion.[15]

The pace of these attacks has not been alleviated by their detection or attribution. The US intelligence community concludes that Russia finds cyberattacks an acceptable option to deter adversaries, control escalation and prosecute conflicts. In an annual threat assessment, the Director of National Intelligence in Washington reported that:

> Russia continues to target critical infrastructure, including underwater cables and industrial control systems, in the United States and in allied and partner countries, as compromising such infrastructure improves—and in some cases can demonstrate— its ability to damage infrastructure during a crisis.[16]

The ongoing Russo-Ukrainian conflict displays the relevance of societal warfare, and although Russia is now targeting critical infrastructure with more kinetic means, its cyber-based means, while less visible, are still in play.[17] While many think the latest Russian invasion only began on 24 February 2022, Moscow actually launched a cyberattack on 14 January 2022, targeting nearly two dozen government institutions in Ukraine in an attempt to paralyze the country's leadership and disrupt its ability to respond to the pending attack.[18] However, it failed to significantly disable Ukraine's cyber infrastructure.[19]

Google says Russian cyberattacks aimed at North Atlantic Treaty Organization (NATO) members have risen by more than 300 per cent since the invasion began, while attacks against Ukrainian targets have risen by 250 per cent compared with the year before February 2022. A surge in phishing attacks was used to gain access for aggressive attacks, with 'wipers' used to destroy Kyiv's databases. The cyber offensive placed 'a strong focus on critical infrastructure, utilities and public services, and the media and information space,' Google's Threat Analysis Group (TAG) reported, 'but with limited impact.' Even so, according to TAG, 'It is clear cyber will continue to play an integral role in future armed conflict, supplementing traditional forms of warfare.'[20]

Most countries are aware of the risks in this domain and have built the necessary mechanisms to monitor and deflect malicious cyber activity. But the danger of strategic reconnaissance for future offensive computer network attacks cannot be discounted. Officials are now raising alarm in national security terms about the scale and severity of efforts by China.[21] Until recently, Chinese cyber efforts were focused on commercial espionage. In 2021, Microsoft's threat intelligence arm identified state-level activity from China consisting of sophisticated efforts to penetrate information networks.[22] The US intelligence community assess that China would almost certainly undertake aggressive cyber operations against critical infrastructure in a crisis, specifically naming oil and gas pipelines as targets, as well as rail systems.[23] Going forward, the security community must understand that some competitors view this domain as being key to achieving strategic outcomes without resorting to war.[24]

Cognitive Warfare

The second 'face' of warfare is an ancient contest with new tools and technology. Today it is most often labelled 'cognitive warfare.'

Societal warfare targets the main operating systems of the modern state's economy or its hardware, disrupting key systems that serve society, and works indirectly to change or coerce governments and their publics. Cognitive warfare, in contrast, goes after a much softer target: the human mind.

This form of warfare is marked by a combination of advanced cyber techniques and social media communications, along with the manipulation aspects of psychological operations. The manipulative element uses a biased or distorted depiction of ongoing events, usually digitally altered or manipulated to favor the sending state's interests. New communication tools now offer nearly infinite possibilities of digital distortion, opening the way to the achievement of one's desired objectives. Key competitors from autocratic states are not content to merely control their own population, instead, they have weaponized social media with 'algorithmic amplification.'[25]

While information and narratives have shaped wars in the past, present-day and forthcoming tools will accentuate this part of the battlespace; in doing so, the fight for the minds of noncombatants expands both the battlespace and our conception of what constitutes war.[26]

Theorists have been studying information warfare for some time,[27] but most of their focus has been on the content of messages being sent as much as the hardware and infrastructure used. However, in the future, this focus will shift to reframing beliefs and fighting for dominance of attention via narratives. The emergent definition of cognitive warfare is the weaponization of public opinion by an external entity for the purpose of (1) influencing public and governmental policy, and (2) destabilizing public institutions.

In cognitive warfare, the message is the munition, and the target is the mind of either specific individuals (elites, influencers, policymakers) or the collective population of a democratic state. By tailored and repetitive messaging, these mental munitions can successfully dissuade or dissolve the cohesion and social consensus that underpins the way our societies operate.[28]

Cognitive warfare can be defined best with a theory of victory focused on coercion, and with some specificity on targets. With this in mind, my definition of cognitive warfare is as follows:

The application of targeted and tailored messages and nonviolent methods used against decision-makers or the general population to influence, dislocate or paralyze the government and civil society of a target state in order to gain a positional advantage in the cognitive domain or gain desired political and informational outcomes.

Russia has long employed a robust and versatile toolkit of so-called 'active measures,' including cyberattacks and disinformation campaigns.[29] Russian intelligence services, moreover, are well versed in designing and conducting campaigns to utilize these tools.[30] The latest innovation in this longstanding practice is the exploitation of social media, which is seen as an extension of the battlespace.[31] Election interference in Europe and the United States is just one aspect of this playbook.[32]

Vladislav Surkov, an advisor to President Putin, unabashedly bragged about Moscow's influence campaigns: 'Foreign politicians talk about Russia's interference in elections and referendums around the world,' he quipped, but 'the matter is even more serious: Russia interferes in your brains, we change your conscience, and there is nothing you can do about it.'[33] Experts warn that a general understanding of the impact of Europe's susceptibility to Russia's malign forms of influence is completely lacking,[34] with one study warning:

A disunited, politically paralyzed and antidemocratic Europe would erode the ability of NATO to defend and uphold transatlantic norms, values and institutions, seriously undermining and ultimately questioning the future of the alliance. The stakes are enormous.[35]

China has also recently studied this aspect of modern conflict, and its government has supported research into how to possibly develop the concept into an advantage for the People's Liberation Army (PLA). They describe cognitive warfare 'as using public opinion, psychological, and legal means to achieve victory.'[36] This comports with the famous maxim of Sun Tzu that asserts winning without fighting as the highest form of warfare, which some scholars believe remains a key feature of Chinese strategic culture. Evidence of this as an operative concept can be seen in the establishment and activities of the PLA's United Front Work Department and the promulgation of its 'three battles' concept.[37] These 'three battles' include (1) public opinion warfare to influence domestic and international audiences; (2) psychological warfare to demoralize enemy soldiers and civilians; and (3) legal warfare to gain international support through both international and domestic law.[38]

The US intelligence community is monitoring increased PLA interest in a holistic approach to generating influence via cognitive domain operations.[39] These operations are touted by Chinese authors as the next evolution in warfare, moving from the physical domains—land, sea, air, and electromagnetic—into the human mind. The goal of cognitive domain operations is defined as 'mind superiority' (*zhinaoquan*), using psychological techniques to shape or even control the target society's cognitive thinking and decision-making.[40] Researchers at RAND find that China is increasingly seeking to exploit the manipulation of social media to support online disinformation campaigns.[41]

Further evidence of applied cognitive domain operations is found in an internal TikTok investigation that identified over 1,700 accounts promoting a concerted disinformation campaign to European audiences on its platform.[42] Additionally, new technologies are being applied to increase the potency of these psychological techniques. In late 2022, a pro-Chinese influence operation posted video content that included footage of computer-generated fictitious 'people' acting as newscasters, entirely created using artificial intelligence (AI). Numerous actors will continue to experiment with AI technologies and can be expected to produce increasingly convincing media artifacts that are harder to detect and verify.[43]

As the technological advances in AI and machine learning become more operational, they can create a perfect storm for a new era of mass disruption, using low-cost but possibly impactful forms of influence. In the very near term, it will become more difficult, if not impossible, to distinguish between real and manufactured products. As authoritarian states such as Russia and China exploit these new technologies, the global competition in influence operations or political warfare is going to intensify. Moreover, as the digital technologies expand, democracies will increasingly have to defend against cognitive assaults from hostile states.[44] As such, the proliferation of deep fakes is likely to constitute the virtual deep strikes of the coming age.[45] Democracies will be increasingly challenged, as their citizens have to withstand the 'instantaneity and immensity' of social media assault.[46]

Proxy Wars

Proxy wars involve the indirect sponsorship of actors by an external state in a violent conflict to influence its outcome for its own strategic purposes. This definition captures the desire by a state actor (the 'Principal') to avoid direct action while supporting actors (state or local militia or contractors) to obtain desired political goals on their behalf. Some scholars define proxies as solely non-state actors, but this does not capture the historical range of sponsorship employed by great powers in the midst of strategic rivalry.[47]

There is a tendency to equate great power conflict with conventional methods of warfare. Yet, great power competition does not always manifest itself in direct, high-intensity and potentially decisive wars. Instead, they often employ proxy forces to pursue their interests, and may extend their substantial history behind more covert forms of conflict.[48] Covert or deniable confrontation pose less risk, including that of potential escalation against nuclear opponents. As one scholar stressed: 'Proxy wars are not merely relics of Cold War superpower competition. Indeed, they are likely to be an increasingly used facet in the rivalry between today's existing and rising superpowers.'[49]

Students of this mode of war have reinforced this argument, with Stein and Fishel observing, 'The U.S. military will need to resist the urge to conflate direct, head-to-head conflict with great-power competition. Napoleonic, linear conceptions of war may be less relevant between large, nuclear-armed states in the twenty-first century.'[50] Given this, proxy wars represent an indirect and non-Westphalian mode of conflict that is increasingly relevant for future conflict.[51]

Both China and Russia have good reason to avoid competing with the United States through overt, direct military clashes, but their histories both suggest an inclination towards indirect methods. China is modernizing its conventional forces for modern forms of warfare, particularly investing in anti-power projection systems to be able to thwart an outside power like the United States from intervention in the Western Pacific. But it also has a history of supporting proxies in North Korea and Vietnam, and a recent history of operating short of the threshold of armed conflict in its efforts to shape its region and coerce states taking positions contrary to Beijing's wishes. It has also increased its outsourcing of security, with twenty international private military companies (PMCs) employing over 3,000 personnel in Iraq, Sudan, and Pakistan.[52]

Russia has also significantly increased its reliance on PMCs in the last few years.[53] Readers will by now be familiar with the Wagner Group, a Russian PMC that operates almost as a subsidiary of the Kremlin.[54] There is nothing novel with this, as Russian PMCs have been employed several times in the past.[55] Historically, Russia has extensive experience with indirect strategies, including the support of separatists and mercenary forces.[56]

Iran, too, has a longstanding history of training and using proxies as a key element of its international statecraft,[57] with its Quds Force considered a master trainer for proxy forces, most notably Hezbollah.[58] Such indirect methods are key to Tehran's way of warfare.[59]

The use of proxy forces has inherent limitations. One of the tensions and paradoxes of proxy sponsorship is that the stronger and more effective a group becomes, the greater its ability to stay independent of, or operate against, the Principal's interests.[60] The allure of proxies for policymakers of an indirect, 'limited' war is reduced by these tensions, yet the use of proxies remains attractive in great power competitions so as to avoid the risks of direct military intervention.

In sum, an era of great power competition can be expected to generate conflicts where one or both sides use proxy forces and surrogates. These surrogates may be PMCs with high-tech capabilities, or even maritime militias swarming like pirates at sea. China's use of PMCs and its various maritime militias can be seen as an effort to undermine the rules-based order at sea, and undercut US credibility and alliance architecture in the Indo-Pacific region. As Andrew Mumford observes, China is likely to emphasize 'the use of indirect mechanisms in an attempt to alter the balance between the two countries; and this is increasingly likely to involve some form of proxy, largely because of the high levels of Sino-US economic interdependence.'[61] This assessment may ultimately be altered by China's significant military modernization, compounded by its economic decoupling from the West. Some scholars, however, anticipate proxy wars as more likely to occur in the future.[62] Stein and Fishel, meanwhile, concluded, 'The United States is more likely to face a Syria-like scenario in the near future, where large powers seek to shape narratives and garner leverage, than to fight a repeat of the Gulf War in the Baltic, or in a Taiwan-type event.'[63] This suggests the study of this mode of conflict warrants more attention than is recognized today. Both great powers and assertive regional powers will prefer intervening indirectly so that they 'can outsource the problem to others at a fraction of the human, economic and political costs.'[64]

Conventional Warfare

The final face—the upper-right quadrant of the framework—deals with conventional wars. Due to the potentially high consequences of this form, conventional warfare between states is the dominant professional focus and competency of modern military institutions. Interstate conflict dominates military histories and is the focus of much of the defense planning of major powers, making up 90 per cent of their defense budgets and a major share of the training and educational program. It is characterized by the integration of infantry, armor, artillery, and ground- and air-launched missiles to gain territory and attrit or outmaneuver an adversary's armed forces. Protagonists apply kinetic force directly against the armed forces of the opponent to defeat them or reduce their war-making capacity. Since the publication of the 2018 US National Defense Strategy, US strategists have focused on the potential scale and duration of such a conflict with China or Russia.[65]

118

Future conventional wars are likely to exhibit three attributes in the coming decades: kinetics, connectivity and the synthetic. Each of which will impact future wars depending upon the particular circumstances, including the competing parties.[66] The first two factors are extensions of the preceding firepower and information-age revolutions, albeit magnified by greater range and precision. The latter element reflects the growing impact of robotics and the dawning of the age of AI.[67] Each of these factors will influence the conduct of conventional conflict, most notably in how combined arms are executed and in terms of system destruction warfare (that is, efforts to attack the connectivity that such operations require to function).

Cross-Domain Combined Arms

Future conventional warfare is increasingly focused on applying combined arms in a more expanded conception, which includes a greater utilization of the cyber and space domains. Known as 'cross-domain synergy,' it was identified almost a decade ago as a source of advantage in future warfare, acting as an extension that goes beyond traditional combined arms.[68] This concept appears likely to animate conceptual development and key capabilities across NATO for the near future, with Western military developers stressing the synergistic application of force across domains via the use of traditional symmetric operations by single-domain applications. This is a longstanding evolution at the operational level of war going back to Germany's fusion of combined arms in the 1930s.

Emerging US Army doctrine is built around multidomain operations (MDO), which seeks to continuously and rapidly integrate capabilities from and across domains to gain what it calls 'cross-domain overmatch at decisive spaces.'[69] The synergy is gained by making combined arms maneuver more effectively by better incorporating space and cyber contributions. By combining these capabilities, the US Army projects that it can incapacitate the opponent's understanding of its intent and axis of action, which is hypothesized to complicate the opponent's situational awareness and reaction time. The theory of victory in this doctrine comes from that convergence, using the complementary or synergistic combat power against key vulnerabilities to disrupt, degrade, destroy, or disintegrate enemy combat systems. This should efficiently produce an effect greater than the sum of the individual components.

NATO is also basing its future concepts around MDO. This would be enabled by the building of cognitive superiority and developing insightful cross-domain command capacity. The one distinction NATO makes from the US is the generation of simultaneous effects across the virtual, physical, and cognitive dimensions. They believe that it is this orchestration, rather than overmatch in any particular capability or domain, that produces decisive advantage.[70] NATO has also extended the concept from offensive operations to 'Integrated Multi-domain Defence,' which is now central to the NATO Warfighting Capstone Concept.[71] NATO seeks to provide Allied commanders with the capability to assess and understand the fluid operating environment, formulate responses in a prompt manner, and take effective action across domains to create the greatest effect. This is a demanding form of command capacity as it involves orchestrating activities across a coalition of many partners, from more than one domain, to generate simultaneous effects across the battlespace.[72]

The Chinese are stressing joint integrated operations along similar lines, and have published concepts and doctrine to promote their equivalent approach to modern conventional warfare.[73] While it is modernizing at a fairly impressive pace, China understands that it remains behind Western powers in terms of its command-and-control systems and its ability to successfully confront the United States in the near term. The Chinese military seeks cross-domain synergy in warfare, which they describe as 'systems of systems warfare.'[74]

Russian theorists have also stressed the employment of advanced military technology in the air, at sea, on the ground, and in the information domain. Russian military modernization has been significant, but as their experience in Ukraine highlights, systemic shortfalls are evident in leadership, doctrine, and other non-material challenges. Space is also emphasized as a critically salient domain, but cross-domain integration has not been explicitly promoted.[75]

Systems Disruption

The Chinese are the originator of another approach to conventional warfare designed to attack a perceived vulnerability in Western operating concepts. This emergent approach attacks the connectivity that ties together the various operating systems of a modern military force. The PLA calls this concept 'system destruction warfare.' This approach, sometimes called 'systems confrontation,' seeks to paralyze and even destroy the critical functions of an enemy's operational system. According to this concept, the enemy 'loses the will and ability to resist' once its operational system cannot effectively function.[76] This theory is a key element in China's most recent Defense White Paper that stated the PLA's 'integrated combat forces [are] employed to prevail in system-vs-system operations featuring information dominance, precision strikes, and joint operations.'[77]

Reportedly, the Chinese conceive of five types of operational systems for targeting, including command systems, reconnaissance intelligence systems, firepower strike systems, support systems, and what is known as 'information confrontation systems.' This latter element extends the PLA's concept of 'intelligentized warfare' with an offensive character oriented against the networks or connectivity of its adversary.[78] (The term 'intelligentized' is their term to describe a conception of future warfare that leverages disruptive technologies like AI; this includes AI-enabled systems capable of generating intelligent swarms and conducting cross-domain mobile warfare at a high operational tempo, and operations across an extended battlespace that includes space.[79]) The PLA theory of victory is that the use of autonomous systems, algorithms, machine learning and human–machine teaming can disrupt the links between various operating systems and make the opponent disoriented and ineffective.[80] This focus on disrupting the connectivity of the opposing force targets the essence of cross-domain combined arms being developed by Western planners. The PLA has in essence absorbed and reframed net-centric warfare concepts from the Pentagon.[81]

Other competitors envision similar approaches to offset Western technological superiority. While attrition via firepower remains a principal component of the Russian way of war, authors suggest that in a war against NATO, its opening moves will include targeting C4ISR (command, control, communications, computers intelligence, surveillance and reconnaissance)—including assets based in space—and critical infrastructure in order to 'disorganize' the alliance's command and control systems.[82]

As Western militaries move ahead with incorporating cross-domain applications in warfighting, each of these three approaches is likely to be utilized to some degree in order to generate combat power and enhance military effectiveness. Different states will adapt these capabilities to suit their strategic culture and their particular operational needs. Certainly, the adaptive use of uncrewed systems will proliferate, and AI-enabled cyberattacks on command-and-control nodes will be employed by many states.

Assessing Risk

The Four Faces model presents a set of potential challenges for national security leaders to consider. The four categories do not, however, represent equally weighted problems or have the same level of risk. Risk is a function of both probability and consequences. Given the openness and connectivity of modern democratic societies, their exposure to societal and cognitive warfare is fairly high. Many states and non-state actors can generate attacks against social hardware and public opinion. However, as shown in Ukraine, modern societies are also resilient, and non-kinetic means have not thus far demonstrated the ability to generate significant costs.

Conventional interstate wars present far more destruction potential, yet have a much lower likelihood of occurring based on the last few decades of history. For Europe, Russia's limited performance and high materiel degradation in protracted high-intensity operations will buy some time. However, the relative chances of future interstate conflict should be measured by looking forward into an era of anticipated strategic competition and eventual Russian reconstitution. A number of rivalries and 'dueling dyads' suggest that violence between states remains an enduring element of international relations.[83]

122

Overall, the greatest risk is projected to be from proxy and surrogate warfare, as it is violent, destructive, and can be a persistent method for great powers to contest their interests forcefully whilst avoiding direct confrontation and the related escalatory risk. Deterrence against conventional challenges may remain the mainstay investment for the armed forces of the West, but dealing with well-armed and nimble surrogates is projected to be the greatest security risk in the aggregate.[84]

As the Russo-Ukrainian War unfolds, we can see that combinations of different methods are evident. It appears that Gerasimov's disputed 'doctrine' is being operationalized, to some degree at least, against Ukraine.[85] What he meant when describing the 'reduction of the military-economic potential of the state by the destruction of critically important facilities of his military and civilian infrastructure' is now clearer. Likewise, the expectation that Russia will seek 'simultaneous effects on line units and enemy facilities throughout the entire depth of his territories' is also in evidence.[86]

Regardless of whether it is called 'full-spectrum conflict,' 'new-generation' warfare or 'hybrid' warfare, a principal characteristic of future conflicts is the amalgamation of military tools with non-military aspects within an integrated operational design, and which operates within an expanded battlespace that impacts civilian society.[87] Looking at the ongoing conflict in Ukraine, one can observe all elements of the Four Faces model being applied. Instead of distinctive and decisive modes, the framework may depict the menu that tomorrow's aggressor states select from and apply in combination depending on the circumstances.

Conclusion

The future remains opaque, but war—including large-scale interstate conflict—is clearly not an historical relic. We have to assess the future with some humility, but it is clear that war in some form will remain a part of our future. The End of History has not arrived, nor have we eclipsed the darker angels of human nature. The character of warfare will continue to evolve, but we should anticipate future battlefields that blend traditional kinetic means with more nuanced and indirect 'weapons' and modes. This convergence is both expanding and altering the battlespace beyond what Western militaries have historically focused on. The ultimate question is whether states and their policymakers will be prepared for the new complexity and endurance that various combinations of warfare will require. To future-proof our societies and meet future security challenges, all four faces of warfare in some combination should be anticipated.

6

PEOPLE'S WAR VS PROFESSIONAL WAR
WHICH HAS THE FUTURE IN EUROPE?

Jan Willem Honig

At the beginning of 1812, the Prussian staff officer and military thinker Carl von Clausewitz wrote several memoranda in which he argued that his country should begin a *Volkskrieg* ('people's war') to drive the French occupiers from his fatherland.[1] The survival of the Prussian state, he claimed, could not be assured in any other way. As the smallest of the major powers in Europe, Prussia was unable to rely on the might of its own regular standing army alone to defend itself successfully. Either it needed allies—which in early 1812 were in short supply—or, as a last resort, it could appeal to 'the people' to join the fight. A popular insurrection could make the occupation so costly for the French that the small, indigenous regular army would stand a chance of fully liberating the country. Such a war, he predicted, would be a gruesome affair. The menfolk from the countryside and town must attack the French, literally, with pitchforks and cudgels. They would suffer horrendous casualties, while their women, children, and the old and infirm could be expected to fall victim to increasingly pitiless and indiscriminate reprisals. The Prussian king rejected advice of this ilk out of hand.

He feared that mobilizing the people would endanger his regime by exposing him to popular pressures for democratic constitutional reform. But moral scruples and questions about the people's aptitude for war reinforced his reticence. Exposing a people untrained for war to its horrors and ravages was an immoral act for a Christian ruler tasked with protecting the welfare of his people. What the king eventually accepted was a carefully controlled system of limited conscription and voluntarism of men who were embedded in the regular army.[2]

Over two centuries later, the question of war becoming 'once again a matter for the people', as Clausewitz wrote, is re-emerging with the war in Ukraine.[3] Without allies willing to intervene directly, the ability of the Ukrainian government to garner domestic popular support and participation in the war was a central factor in fighting the initial Russian invasion to a standstill. Sustaining that support will indubitably also be key to an eventual victory. It may, however, no longer be immediately apparent how contested, yet continually debated and considered, the issue of the involvement of 'the people' in war has been in Europe for the last two centuries (that is, until the end of the Cold War). The collapse of the Soviet Union presented what seemed to be a decisive end to the debate. Since then, the trend across Europe was in the direction of full professionalization. Reliance on compulsory military service all but disappeared in most countries, with remarkably little debate.[4] In their wars and military interventions, European powers (including Russia) have relied on professionals. Yet, the unexpected failure of Russian regulars to overcome Ukrainian popular resistance, first quickly and then slowly, and the re-appearance in a developed country of the people on the field of battle breaks a trend that seemed decisive. Looking back, one can also wonder whether the defeats of Western regular forces in Iraq and Afghanistan against popular insurrections were earlier signs that a reliance on regulars was a mistake. Does Europe's future hold a need for the people to prepare once again to become part of war?

It is useful to start an analysis with Clausewitz and the period of the French Revolution, since this is the period when 'the people' first received recognition as a potentially legitimate actor in war. That is the subject of the first section. What emerges from this analysis is that the key unresolved issue was whether the people should be entrusted with involvement in war. Should war not be considered a special sphere of activity that was best left to a specially trained group of professionals? Or should the emerging political constitutions of European states be followed, which were founded on the principle that ultimately war was the business of the people? Informed by political preference (lowing either from social prejudice or humanitarian goodwill, this tension between regarding war as an essentially technocratic or a deeply political phenomenon was already present in Clausewitz. The second section sets out the ways in which European states in the nineteenth century attempted to limit and control popular involvement. As the third section explains, these ways faltered in the twentieth century, but the powerful political and (professional) military distrust of the people never completely disappeared, and neither did the liberal humanitarian wish to excuse the people from the horrors of war. The latter two especially asserted themselves after the end of the Cold War, with wars undertaken by Western powers following a professional expeditionary model, only to be called openly into question again with the war in Ukraine. The chapter concludes with the claim that given the constitutional make-up of European states since the French Revolution, victory in war, through the concession of defeat, remains in the gift of the people. That they have not consistently chosen to exercise this right, and that it is extremely hard to predict with confidence when they will (as the Putin regime and many others before found out), should not blind us to the people's immense immanent power in future war.

The Emergence of 'the People' as a Legitimate Actor in War

When linked with war, the concept of 'the people' nowadays takes on several meanings that tend to lead to some confusion. Contrary to the general meaning of 'the people' as an inclusive term, in the context of war they are generally seen as a group of human beings that is distinct from the military—the entity that fights war— but also the government—the entity that directs war. The classic formulation of this distinction is the practical world example Clausewitz gives of his famous trinity.[5] However, there arises an ambiguity as to what distinctive quality the people possess in relation to war. For Clausewitz, the people were first a resource; a reservoir for manpower on which the state called to man its regular, standing army. This view implied that the transition from 'people' to 'soldiers' was recognisable through a change of attire into military uniform and an ability physically to fight (the latter being the most essential distinction for Clausewitz). To modern readers, this implies that those members of the people who have not transitioned to the military can be regarded as civilians who, as non-combatants, should be kept out of war and not suffer from its violence.

But Clausewitz also regarded the people as a powerful force which, in a less tangible way, influenced the quality of war the state fought. That is illustrated by the primary trinity through which he understood war as a composite of three elemental drivers: political instrumentality; a play of chance and probability; and hatred and enmity. He presented the last-named emotions (which are often rendered more innocuously in English as 'passions') as practically embodied in a socio-political context by 'the people'. (The other two are embodied by government and armed forces respectively). However, the association of the people 'with the original violence of [war's] element, with hatred and with enmity, which are to be regarded as a *blind natural instinct*', still leaves a major issue unclear.[6] Must the influence of the people manifest itself only by direct physical participation and open use of violence? Or does their animus more indirectly influence the level of violence and increase the intensity of the conflict, for example, by pushing the government to raise the political stakes, by influencing their fighting menfolk to brutalize the battlefield, or perhaps by provoking their opponent into an escalatory response through their peaceful, passive resistance?

Clausewitz's take on 'the people' reflects a deeply suspicious, politically conservative view of them. For the late eighteenth-century noble officer class, seeing the people as an irrational, uncalculating social group represents an old prejudice, going well back to the Middle Ages. The ruling nobility had for centuries declared the common people unfit for war, and therefore endeavoured to keep them away from direct involvement. If commoners were increasingly called upon to man the noble-run regiments of absolutist monarchs,

PEOPLE'S WAR VS PROFESSIONAL WAR

the soldiers were socialized tightly into their units by means of usually life-long service and draconian discipline. If the people resorted independently to violent, insurrectionary action, as they did regularly in peasant and city revolts, they were generally dealt with pitilessly in a show that put on public view that the people were socio-politically inferior and had no legitimate place in war.[7] In Clausewitz's trinity, one can see a traditional view which believes that permitting the people (read: the mob) to enter into war leads to a loss of control over violence and escalation.

The unresolved ambiguity he leaves in the people's role also points to two important elements of novelty in his attitude. The first is that throughout *On War*, and in line with modern Enlightenment ideals, he accords a consequential role to public opinion. Although more a middle-class phenomenon than a mob one, public opinion influences and shapes the quality of the war. As such, it reinforces the second novel element: under the influence of the French Revolution, Clausewitz comes to see that the blind instinct of the people can have an instrumental value. French military successes were not only the result of a morale-boosting injection of popular hatred and enmity into war, but also represented a belief in a just political cause. This combination of fury and cause helped them overwhelm the small professional armies of the monarchies. If the subjects of the Prussian king could be made to see that their hatred and enmity was underpinned by a just cause, and could be directed at the French, then their innate sensitivity to rabble-rousing could become militarily and politically productive for the Prussian state. The question then became: why should the royal subjects that inhabited the lands of the king of Prussia hate the French and rise up, as one, against them?

The Prussian king could not use the script the French pioneered. The Revolution of 1789 had brought a heady cocktail of radically novel political concepts to prominence that were fundamentally antithetical to monarchical rule. Implementation of the ideas of democracy, equality, and liberty meant that the constitutional order was no longer centred on royal and noble privilege policed by mercenary standing armies, but on the will of the people. The latter transformed from subjects into citizens and so became responsible, individually and collectively, for defending the new order against its enemies. In the early years of the French Revolution, this led to a surge of tens of thousands of volunteers from all classes who self-organized into military units.[8] At first, they protected the revolution at home, but as Europe's monarchies made ready to invade France to restore Louis XVI's regime, this also turned to contend with foreign enemies. As the invasion was defeated, the volunteers took the next logical step: the eradication of the counter-revolutionary threat by the overthrowing of the European monarchical order.

In fighting off this existential threat, the kings of the *ancien régimes* were forced to consider mobilizing their own peoples without offering them the full political package of the French Republic.[9] What they arrived at was a compromise solution.[10] First, they offered the lower and middle classes greater equality before the law and more rights as citizens, but not democracy or liberty. The regime then no longer needed to rely on state coercion to man its armies, but, as in democracies, could appeal to their citizens' duty to protect their newly won rights. However, the citizenry remained marginalized as regards to the setting of state policy and strategy. Secondly, the royal regimes, especially in Prussia, played up the element of an onerous, exploitative occupation by 'foreigners', who went by the collective national noun of 'the French' and who meddled heavy-handedly in the long-established local political order. The regimes thus turned the war outward into an incipient 'national' war of liberation from the French rather than a war for political freedom.[11]

The experience of the French Revolutionary Wars shows that the constitution of armies and the wars they fight are deeply political matters. Mobilizing the people and involving them in war is not simply a matter necessitated by improving the technical odds of winning; it is a product of a political recognition that they have a legitimate place in war. It also goes together with a belief that the cause for which they are called to fight is vitally important, and generally involves either a defence of the existing political order or the establishment of a new and better one. These high stakes mean that popular participation in war depends significantly on volunteers who present a double-edged sword. While they offer a highly motivated source of manpower, they do not necessarily fight

PEOPLE'S WAR VS PROFESSIONAL WAR

for the same objectives as their regimes. These differences may not manifest themselves clearly during the war, but they are bound to emerge after the primary objective that united them is achieved. To reduce the dangers of deferred revolution, regime types of all stripes must try to make them a more dependable instrument in their hands by formalizing recruitment and organization. This is not simply to make the citizen-soldiery militarily skilled, but also to embed them within structures of regime loyalty. Nonetheless, when the people become involved in war, the distinction between foreign and internal war can blur. The function of the armed forces readily becomes one of protection against foreign and domestic enemies. Both French Revolutionary and Monarchist armies constituted a bulwark against enemies of fundamentally opposed political ideas that were in principle non-territorial, and thus did not break down easily into a domestic-foreign distinction. Hence also the blurring of a distinction between foreign and civil war. Mobilization of the people therefore introduces a critical element of potential political instability in regimes of all colours. We do not know yet how this will pan out in the Ukrainian (and perhaps the Iraqi) case, but in Afghanistan, the victors fought not only to be rid of foreign occupation, but also to free themselves from an ungodly domestic constitutional order.

A further observation—which we do well to remember today—is that the selective recourse to old and new ideas in the political rhetoric of legitimization matters greatly. A judicious combination allows for initial social traction across the political spectrum, while their interaction generates change. Over time, two major political discourses emerged regarding the role of the people in war: one nationalist and the other democratic. For example, Prussia applied the name *Landwehr* (or defence of the land) for the popular volunteer contingents. In the Germanic languages, *Landwehr* was a familiar term, going back to the Middle Ages.[12] As invading armies had laid waste to the regions they passed through for centuries, forceful requisition in enemy territory (contrary to modern views of eighteenth-century 'limited warfare') had remained standard practice.[13] An idea of local defence by all inhabitants in territories that suffered recurring invasions therefore also survived. The French revolutionary mobilizational appeals to *Patrie* (fatherland) and *citoyens*

131

(citizens) likewise appealed first of all to long-established sentiments of local community and city residence, rather than primarily the modern ideas of nation and national community. As such, the truly revolutionary does not appear instantly in full force if it is to have broad political appeal and immediate effect.

Finally, where does the idea of the people as civilians feature in all this? The short answer is: it doesn't. A French state that famously decreed the following in 1793, does not recognise civilians as a category of people excused from war:

> Henceforth, until the enemies have been chased from the territory of the Republic, the French people are in permanent requisition for army service. The young men shall go to fight; the married men shall forge arms and transport provisions; the women shall make tents, clothes, and serve in the hospitals; the children shall tear old linen into shreds [for wound dressings]; the old men shall carry themselves to public places, to arouse the courage of the warriors and preach hatred of the kings and the unity of the Republic.[14]

For the monarchies, the distinction between the armed and unarmed was not one of material consequence either. Like the revolutionaries, they recognized that war encompassed different categories of people, but active involvement in the war effort was not an immediately relevant criterion. Rather, those who were presumed to be politically loyal to an enemy ruler were in principle considered part of war. They may not have been supposed to fight, but that did not give them blanket immunity. Hence, the brutal treatment of enemy subjects (whom we now consider civilians) was not only permitted (as in plundering peasants or massacring town dwellers), but considered right and even mandatory.[15] Still, as we will consider in the next section, both the growing democratic and conservative nationalist discourses were attracted by the possibility of excusing the non-combatant civilian population from war. And it was an increasingly professionalizing, technocratic military which provided them with an answer.

PEOPLE'S WAR VS PROFESSIONAL WAR

The People as a Limited Resource for Technocratic War

In 1814, with the conclusion of the Congress of Vienna, it appeared that the conservative tradition of monarchical rule had re-emerged victorious. However, pressures for democratic constitutional reform did not abate for long. The years 1830 and 1848 saw major revolutions sweep across Europe.[16] Despite extensive political and military interventions by coalitions of the old monarchies (Clausewitz succumbed to cholera, for example, when preparing a Prussian army to aid the Russians in putting down a revolt in Poland), many regimes were ultimately forced to seek constitutional compromise with their peoples. In many, if not most, countries, however, the growing appeal of nationalist discourses succeeded in taking the sting out of the revolutionary pressures for democratic reform. It seemed that a sense of national commonality could override the demand for democratic representation if it was believed that the regime did not act out of elite, class interest, but out of shared, national interest. Whereas class antagonism fed domestic division and constant struggle, nationalism presumed a unity more fundamental than unhappiness when it came to political and economic inequality.

A secondary effect was that by diminishing the antagonism over the internal state constitution, nationalism sharpened the idea of a naturally existing enmity between nations. Since nationalism is an exclusive ideology, it assumes fundamental differences to exist between nations; these can easily include perceptions of superiority (and lead to claims of special 'responsibilities' over other nations) or inferiority (and a right to correct inequalities in resources or power, for example). The pursuit of selfish 'vital national interests' (by force if necessary) can easily become the legitimate, overriding goal of 'foreign' policy. As a consequence, in the second half of the nineteenth century, the era of counter-revolutionary wars drew to a close in Europe, to be replaced by international wars.

Nationalism as the glue that held regime and people together also seemed to solve the conundrum of who should defend the state. As in a democracy, the true nation-state must call upon all its citizens to serve. Thus, the idea of mass mobilization spread across Europe irrespective of regime type. Such mass armies could in principle

133

be considered more loyal (read: better) to their regime than the armies of the old monarchies, which had consisted of the dregs of society and included many individuals who would be considered foreigners.[17] There was, however, a difference between political ideas and prudent practice. Few regimes felt completely at ease with the reality of creating mass armies of their own voters or even of their own fellow nationals. European societies remained too politically divided for their regimes to feel confident that they could entrust all their citizens with weapons and training. The explosive growth of the industrial proletariat only reinforced their nervousness. Still, few states chose to continue to rely on long-serving professional forces.

Two cases exemplify the fundamental political dynamics at work during this time. The first of these was Britain, who avoided conscription until the early twentieth century. Although conscription was consistently presented as unnecessary for an island power protected by the Royal Navy (as well as costly), it was also subject to a large dose of socio-political prejudice. In repeated government attempts to reform recruitment and increase manpower, any element that smacked of compulsory service was consistently defeated in Parliament as 'an infringement on political rights'.[18] It was judged better to get the desperate 'loafers and idlers' to enlist than the productively employed, who could easily fall prey to 'an ambitious and aggressive spirit', as one leading Liberal politician put it.[19] As such, the political reliability of the British people only seemed assured once conscription was introduced in 1915, after the call for volunteers at the outbreak of the First World War led to such a massive—and surprising—influx of young men joining the colours.

Democratic France displayed a similarly ambivalent attitude to national service, with left-wing popular governments favouring general conscription and right-wing ones preferring caution.[20] Despite pursuing an ambitious foreign policy, Napoleon III hesitated to expand the army in the 1860s, with around 100,000 men over the age of twenty-one drafted annually by lot. Of these, only a minority were made to serve for five years, with the rest assigned to a largely untrained reserve (these conscripts were mostly poor and thus could not afford to buy off their service). These conscripts joined regiments located away from their hometowns, thereby reducing potentially

PEOPLE'S WAR VS PROFESSIONAL WAR

dangerous domestic cohesion, but turning wartime mobilization into a prolonged mess.[21] When the Franco-Prussian War broke out in 1870, Napoleon's 'regime-safe' army was quickly defeated by a larger Prusso-German army.

This victorious army was also not a true *Volksheer*. As Gerhard Ritter, the prominent historian of German militarism, put it:

> It was no people's army in the modern [Second World War] sense ... but a tightly disciplined royal army, commanded by a predominantly noble officer corps with a singularly strong *esprit de corps* .... A well-trained pick of the militarily fittest and the young from the people took to the field.[22]

Critically, between 1858 and 1961, Germany drafted relatively more men than their international competitors: 40 per cent of all eligible males per year group (by comparison, the French only managed 30 per cent for five years).[23]

Denuded of his army, Napoleon III surrendered and ceased playing a political role. The predominantly republican 'government of national defence' which came to power in his stead attempted to replicate the revolutionary fervour that had gripped the nation eighty years before.[24] This time, the volunteer numbers did not initially keep pace with need. The regime therefore resorted to mobilizing levies of conscripts, whose lack of training made them no match for the Prusso-German forces, and they were repeatedly defeated whenever they sought to confront invaders in regular battles. Nonetheless, the regime did not fold: they kept raising new armies and, most worryingly, called upon irregulars (or *francs-tireurs*) to join the fight. As the war dragged on—partly because of the underdog's refusal to surrender—German political will to defeat the French and unilaterally impose peace weakened. Ultimately, an uneasy compromise permitted the Germans to leave France. The Republican government then faced the challenge to assert itself over the politically divergent groups of mobilized fighters, the most notorious example of which was the bloody suppression of the working-class-supported Paris Commune.

The Franco-Prussian War is a watershed moment. For the next century or so, it settled the question of whether conscription was necessary in Europe. Nevertheless, the unreliability of popular sentiment as seen in France would lead to attempts to restrict and control the involvement of the people in war, which can be discerned in the following four examples. First, the officer corps—which remained largely noble in composition—internalized, and promoted a technocratic conception of war which elevated the fight between regular, uniformed armies to the norm, and which claimed that the battlefield decision between such armies is final.

Second, whether nationalist or democratic in nature, all regimes would propound discourses that emphasized military service as a duty for all able-bodied male citizens. With the aid of their professional caste of officers, the soldiery was trained and drilled according to carefully laid down programmes of tactics to suit regular warfare, but which also promoted the state's preferred ideology. Conscription periods would tend to last several years so that when war arrived, it would take the form for which the available forces have trained. In doing so, it was hoped they would be able to force an outcome so quickly and conclusively that the issue of further popular involvement need not arise.

Third, the increasingly technical issues of organization, preparation and conduct of regular mass warfare now required officers to dedicate their lives to the technocratic endeavour. It was believed that doing so would slowly separate the officer corps from the political class and thus help elevate international war—now usually called 'great war', *grande guerre* and *großer Krieg*—to a unique activity which requires a dedicated, specialist, professional workforce. In this conception of war, and in the way the workforce is structured and trained, there is ostensibly no need for the *franc-tireur* (the private citizen who elects to self-mobilize and join the war).

Finally, in the body of international law that is formalized at the end of the nineteenth century, one can see an attempt to reserve war for professionals and to deter the people from joining. Within an elaborate construct of occupation law that demands and expects obedience from a population whose regular forces have been defeated, it specifically legalizes brutal reprisals against populations that spawn irregular forces. The practice of the random taking and execution of hostages, for which the German armed forces became

136

PEOPLE'S WAR VS PROFESSIONAL WAR

notorious in the two world wars (particularly during the latter of these), was a legally permissible tool and only outlawed after 1945.[25]

By the beginning of the twentieth century, it was unclear whether the nationalist–monarchical and democratic regimes and their professional officer corps had built a strong enough dam against the uncontrolled involvement of the people in war. Much depended on the people's loyalty to the regime or their passivity. Yet if they took an active interest in war, either out of revulsion for governmental policies or war in general, then the odds were off. And if the professional recipes for quick, decisive international war failed, sustaining the war effort would leave governments and the military with little choice but to call upon them, with results that were hard to predict.

Will the Dam Break? The Pressures for People, Military and Government to Become One

Limiting the involvement of the nation in war to carefully trained and disciplined units of conscripts under a professional body of officers came under intense pressure as the First World War progressed. Nonetheless, the trinitarian separation of actors was maintained. Towards the end of the war, however, Germany would prove to be an exception. With its regular forces facing collapse in October 1918, its new left-of-centre parliamentary majority government— which styled itself a 'Ministry of National Defence'—considered a *levée en masse*.[26] The military Supreme Command strongly resisted, mainly because it would imply admitting its leadership and way of war had failed. Realizing that the horrors it was certain to unleash unto the German people would not likely improve terms much, the civilian government cooled to the idea—just as the military warmed to it.[27] Resorting to the German vocabulary of *Volkskrieg* and *Kampf bis zum Äußersten*, the Supreme Army Command now clamoured for fighting to the last man to save Germany's national honour. The government dismissed the demand, but contravening policy, the Naval High Command ordered their fleet to sea in a suicide mission on 26 October 1918, which set off a mutiny that proved a spark for revolution across the country.

137

The German military leadership may not have been entirely sincere in their support for *Volkskrieg* if one considers the lateness at which they called for it. In fact, given the growing popular resistance against continuing the war abroad, the whole idea may have been unrealistic. Furthermore, the war actually seemed to confirm the validity of the professional vision of warfare, as Germany was eventually decisively defeated by the regular forces of the Allies on the battlefield. Nevertheless, the war's overall experience put the idea of *Volkskrieg* firmly on the interwar military and political agenda. While a last-ditch people's war was clearly undesirable, the inordinate and sustained effort required to fight modern war mandated far-reaching popular involvement. Despite this, the people's reliability remained in question. The challenge of the interwar period became the search to assure the early and reliable integration of the people in any future war effort.

In 1918, a new term to define modern war emerges: total war. Léon Daudet, editor of the far-right periodical *L'action française*, defined it as:

> ... the extension of the struggle ... into the domains of politics, economics, commerce, industry, thought, law and finance. Not only armies fight, but so do traditions, institutions, customs, codes, spirits and, above all, the banks. Germany has mobilised on all levels, in all these sectors .... It has constantly pursued, beyond the military frontline, the material and moral disorganisation of the people it attacked.[28]

Daudet vastly exaggerated German intent and success—and the role of banks (read: Jewish-owned banks)—but his words retain a prophetic ring. His description would not look misplaced in current debates about 'hybrid war', which is also presented as a form of warfare encompassing all of Daudet's spheres and which enemies usually fear the other has mastered.[29] What blurs in this understanding of war is the distinction between the public and private spheres, between military and civilian, and between war and peace. It raises immediate questions of the powers of the state over its citizens, the position of the military and their concerns in state and society, and of war losing its own specialist, technocratic

138

PEOPLE'S WAR VS PROFESSIONAL WAR

sphere, and becoming an all-encompassing domain. It is perhaps no coincidence that precisely at the moment when the state explicitly becomes regarded as an organism that fully merges people, military and government, 'civilians' become a recognized, protected category in the laws governing war.[30]

In the period between the world wars, it appeared that, as the concept of total war gained ground, the state and the military were steadily extending their control over society. Whether now fascist or communist in regime type (both quickly dubbed 'totalitarian') or democratic, the preparation for war increasingly intruded into civil society.[31] A key concern was the nurturing and protecting of what a key member of the German Supreme Command in 1918 and a convert to total war, General Erich Ludendorff, termed the '*seelische Geschlossenheit*' ('spiritual unity') of the nation.[32] Achieving such unity required a radical transformation in the constitutional make-up of states.[33] Turning nation-states into total war-fighting machines mandated a redefinition not only of political life (so as to make it wholly subservient to the needs of war), but also of the military profession. No longer would war be narrowly focused on the management of the application of violence against other regular armies. Instead, in peacetime, the military needed to take on a role in militarizing all social activities—from education and journalism to infrastructure and even recreation. In wartime, at one with the people, they would be ready to fight enemy peoples who, like their own, would all count as combatants.

When it came to the real crunch, however, all states held back on fully mobilizing their peoples. Despite the self-professed totalitarian Nazi state and its armed forces going further than all others in extending their uncompromising enmity to whole peoples in the Second World War, even it hesitated when it came to total mobilization. To give one example: when Nazi propaganda minister Joseph Goebbels asked his audience whether it wanted total war in his famous February 1943 Sportpalast speech, their enthusiastic affirmative response is generally seen as typical of the total grip the Nazis held over German society. However, the event was a carefully staged trial balloon to test public support for the full mobilization the military were demanding in response to the unfolding disaster at Stalingrad. When reports came back that the German people were lukewarm, the Nazi political leadership desisted and continued to try to insulate the home front as much as possible from the war. As a result, Germany ended the war with an inefficiently exploited nation.[34]

The Nazis were not unique in worrying about public support. All major warring states kept a close and wary eye on public opinion. Again, however, the lessons from the war were ambiguous. Across a diversity of regime types, their peoples remained loyal to their national government. They did not rise up or give up until, in the Axis case, their state's regular armed forces were destroyed in battle. Although Germany and Japan had toyed with the idea of a last-minute *Volkskrieg*, neither bit the bullet. And although armed popular resistance to Axis occupation and partisan warfare was much more widespread than in earlier wars, they never escalated to the full popular insurrections that the communist left hoped for, or the fascist right feared.

Later, when the enmity between fascism and the liberal democratic–communist alliance was replaced by that between communism and the liberal democracies, the question of total societal mobilization remained on the agenda. Nevertheless, the response in the liberal democratic camp was half-hearted. Conscription was more than ever seen as an infringement on political rights rather than a basic civic duty, and war was regarded as unacceptably costly in human and economic terms, as well as wrong in principle. Most North Atlantic Treaty Organization (NATO) powers retained conscription, but in a form that drafted ever fewer men for ever shorter periods. The invention of nuclear weapons and their deterrent effect meant that the preparation for a Third World War could take the limited form that it did. If the people served a role in war, it was as hostages. Deterrence through mutual assured destruction was better served by leaving them defenceless. Civil defence initiatives were begun in the 1950s but abandoned when this paradox sank in. The maintenance of a conventional defence force remained important to the credibility of deterrence, but planning for war settled on what was in effect a three-phase war. Since NATO members judged taking the offensive and attacking the

PEOPLE'S WAR VS PROFESSIONAL WAR

Soviet bloc with nuclear weapons amoral, and with conventional means impractical and unwise, conventional forces were tasked to defend as close to the eastern border as possible for as long as they could. By thwarting the Soviet plan for a quick conventional victory, it was hoped common sense would prevail and the war would end. If this failed, the use of nuclear weapons would be threatened. As a last resort, if NATO political resolve to actually use nuclear weapons dissolved, the Alliance would give way to occupation and undertake guerrilla warfare. To that end, a network of weapons caches and operatives was prepared in peacetime.

Conventional defence was increasingly seen as a tripwire offering just enough time for the imminent threat of nuclear war to focus minds. In any case, a conventional war could not be sustained for long; this was not so much because of a shortage of manpower, but more one of equipment. Warfare had become so equipment-intensive, and equipment so expensive, that preparing a capability in peacetime to sustain a long period of conventional war was prohibitively expensive (as the collapse of the USSR seemed to confirm). Moreover, after war broke out, there would be no time to scale up production. Less obviously, however, investment in a short war also fed a perception that fighting in the end was better undertaken by long-serving professionally trained technicians. Conscript armies, after all, had not only failed to save the British, Dutch and French empires, but they had even faltered in defending the cause of freedom and democracy against a communist uprising in Vietnam. It seemed better, once again, to keep the people out of war. When the end of the Cold War came, it was therefore no surprise that NATO's armies put conscription on ice and began to rely on regular professional armed forces.

Developments appeared to have come full circle. Technology and an environment of diminished international threat assured professional forces with a new, undisputed lease of life. In liberal democracies, populations could be excused from bearing the financial and personal cost of preparing for war, and less trusting regimes could prize loyalty over mass in their security forces. Whether the military-professional way of war still worked was unclear. The setbacks suffered in a long series of post-Cold War military interventions could be blamed on other factors, like insufficient political commitment and hazy government objectives. Then the war in Ukraine happened.

Conclusion: Implications for Now and the Future

In the first instalment of the Russo-Ukrainian War in 2014, the lightning success of Russian regulars in taking the Crimea without firing a shot, and the utter failure of the Ukrainian army to contest the attack, obscured two Russian miscalculations: the first was the expectation that the largely Russian population in Donbas would come out in support; the second was the success of Ukrainian nationalist volunteer forces to stabilize the frontline there.[35] Looking back, these events signalled a critical asymmetry between the parties: whereas the people chose to come out actively on the Ukrainian side, they did not do so on the Russian side. In the war's second instalment in 2022, the miscalculations of Russian regular forces became clear much sooner. The twin expectation that a mere demonstration of regular force would make the Ukrainian regular army quickly give way, and that the Ukrainian people would submit as Russian internal security forces took over the state infrastructure, proved wrong within days. The Putin regime underestimated the stake the Ukrainian people had developed in their state's freedom and independence. As the war dragged on, the Putin regime also feared and mistrusted its own people more than the Ukrainian government did theirs. The war thus harshly revealed the fundamental problem of technocratic war: it is unwise to approach and conduct war as an apolitical, professional activity. It is, as Clausewitz had concluded at the end of his life, at all times permeated by politics. As the historical examples surveyed in this chapter indicate, even in the cases where regular warfare ended and seemingly decided war, this only happened by the grace of the people. In each instance, the people retained the option to confirm the result, or to resort openly to arms and to join or rekindle the fight at a time of their choosing.

In retrospect, the same issue bedevilled the Western interventions in Iraq and Afghanistan. The stunning, but purely military, successes of Western regular forces denuded the Saddam Hussein and Taliban

PEOPLE'S WAR VS PROFESSIONAL WAR

regimes of their immediate means of protection, and so quickly toppled them from state power. But these technical military exercises did not, in the end, succeed in commanding the consent of the people. It is they—to recall the words used in October 1918 by the then German vice chancellor Friedrich von Payer when rejecting army demands for a *Volkskrieg*—who ultimately 'make a lost war lost'.[36] Instead, the Iraqi and Afghan peoples decided to deny the attackers their victory in numbers that were sufficient to prove largely immune to years of persistent, professional, apolitical warfare. Moreover, what these examples show is the extent to which the ideas of the French Revolution have spread, as have the expectation of peoples the world over to insist on reserving a say in the way they are governed. In other words, a people's war is no longer a European phenomenon.

What the chapter has also illustrated is the difficulty in predicting when the people will come out on the streets and into the fields to resist and fight. It is often assumed that humanity's aversion to war is so strong that the people are hard to rouse to violence in the first instance and, even if that succeeds, their support for war will weaken as route to victory extends and casualties mount. Democracies are often believed to be especially vulnerable to protracted wars as they do not possess the tools for popular coercion held by autocratic and totalitarian states. The only antidote is a strong and clear just cause. The moral clarity of the Ukrainian war thus seems to offer a textbook case (as the Second World War also did). Nonetheless, nervousness about the strength of their people's support is palpable among all sides at the time of writing and leads, as so many earlier wars did, to talk about the need to find a way to keep the war short and decisive. Both Putin and Zelensky are hesitating about enforcing conscription. Again, the historical record suggests that this fear, as well as the need to find a speedy end, can be misplaced. One only needs to recall the debates about what the First World War was about: despite a lack of clarity regarding cause, all regime types on both sides managed to carry on for over four years at the cost of unprecedented casualties. What makes the robustness of popular participation in war even more puzzling is that undemocratic regimes (such as Nazi Germany and even Soviet Russia and communist North Vietnam) succeeded in fighting long wars without relying on mere coercion. The successful appeal of 'alternative facts' and 'false narratives' in democracies should make one cautious about the exact reasons people become animated to fight. One may conclude that given the unpredictability and disagreeable variability of causes, it is preferable to professionalize war and keep the people at arm's length. As has been explained, however, modern regimes do so at their peril.

The predictive failures of Western and Russian governments in the twenty-first century regarding popular involvement are not without precedent. There is no reason to suppose that this difficulty will diminish radically in the future. Those who predict that technology will make resistance easier may be right in that it facilitates organization and recruitment to some degree,[37] but it is less clear whether an improved and more democratized ability to organize resistance will more easily translate into a higher willingness to risk one's life. What is clear is the continuing importance of what was once called political rhetoric, and which is now more usually termed political discourse. Rousing the people into action is not new to politicians, but it is not usually seen as part of the military brief. Nevertheless, a sensitivity to language and political effect is an essential skill in future warfare. One should be wary, however, that the discourses that unify people and induce them to fight will remain constant. In the West, it is often assumed that the most potent idioms are either nationalism—which the French Revolution supposedly encapsulated in the phrase '*la Patrie en danger*'—or democracy, since that is the political system which gives everyone a naturally appreciated stake in their polity. Even so, there are other possibilities, such as discourses grounded in religion or other political belief systems, such as at one time (incredible though it may now sound) communism. The latter example underlines that legitimating, unifying ideologies are subject to invention and change. Current examples are the adventurous Chinese attempts to root the regime in some combination of capitalism and communism, and Putin's venture to invent a legitimating nationalist ideology rooted in a mythical past and solidify this through an active foreign policy.

Nevertheless, keeping the people away from an involvement in war is an uncertain project. Again, as some argue, it is not so much

PEOPLE'S WAR VS PROFESSIONAL WAR

technological development which blurs the line between civilian and soldier,[38] but the constitution of most polities in today's world. Regardless of how much these claim to represent their peoples, they often vary in their veracity. That ultimately gives the people not only an immediate stake in politics and war, but a deciding vote. As pointed out repeatedly, they may not choose to exercise that vote actively, but whether they withhold it or not, their vote implicates them in the decisions for war or peace. A further implication is that the modern construct of equating the people with civilians, and thereby according them special immunity from war, may be a laudable sentiment that rightly attempts to reduce the horrors of war, but it will often constitute a pipe dream. The protections civilians are accorded in what we now call international humanitarian law are not simply a tenuous good, but something that is in contradiction with the modern popular politics of war. With the French Revolution, war became (and will likely remain) the business of the people— with all the ill effects that implies.

7

URBICIDE AND THE FUTURE OF CIVIL WAR

David Betz

At the time of writing, scholars and pundits across a range of disciplines have concluded that the United States is again on the verge of civil war.[1] While alarming, such claims ought not to be surprising, as for years other scholars have been tracking the West's civilizational decline generally and warning of its increasingly imminent demise.[2] In addition, one must add the arguments of further scholars and pundits, who have argued that the condition of life in several major Western states is already one of low-grade civil war. One ought also to consider that what heretofore has rather metaphorically been called the 'culture wars' is increasingly taking on the character of actual war in terms of its levels of mutual demonization and loathing.[3]

The future of war literature is nearly always focused upon hypothesized interstate conflicts and typically employs extrapolations of technological trends when it attempts to describe how such wars will be fought. That is irrelevant. The character of the war which will really matter to the people of the West—probably soon—is not going to be dominated by hypersonic missiles and directed-energy weapons, next-generation tanks and airplanes, big ships and submarines, let alone robotics and cyberattacks. After twenty fruitless years of the War on Terror, followed by extensive materiel support to Ukraine, the West's armed forces will be exhausted and demoralized, their arsenals drained, and their economies too weakened to rebuild them. What remains of their unused bespoke weaponry that has not already been cannibalized for parts will also falter as delicate logistics chains and maintenance regimes crumble.[4] As a result, any potential war will be fought by starving and angry civilians hastily formed into militias, stabbing, shooting and bombing each other in the burning rubble of Western civilization that has reached the end of its rope economically, culturally and morally.

It is impossible to predict how exactly the future civil wars will transpire nor how long they will last. However, it is possible to assay some educated guesses. The current best estimate in the literature is that the chances of civil war breaking out in a country in which the conditions for it are present is 4 per cent per annum.[5] All things being equal, then, we may conclude that the chances are about one in three of it breaking out in a decade. All things, however, are not equal because the rate at which the status quo is destabilizing is accelerating. Governments are doing nothing useful to ameliorate the principal grievances and mutual antagonisms in society that are driving it towards the precipice, as I shall describe below. Economically, their actions since the 2008 financial crisis amounts to no more than kicking the can down the road, while foreign policy would seem to consist of launching one doomed and pointless foreign war after another.

In other words, one chance in two is optimistic. When the war comes, it will impact directly upon your life. It will be very difficult to know what is happening, let alone what to do about it. When the power starts going out for more hours in a day than it is on, when your local magistrates court and police station begin to look something like the Alamo, and violence is occurring all around, the following is intended as a general guide to what is probably happening to you.

There are two objects of this chapter. The first is to explain briefly why this will occur. The second is to describe the character of the war which is coming, including its main actors, strategies and direction of operations. I shall not conclude with many thoughts on what might be done because there are few to none. We are in for a wild ride, and that is that.

URBICIDE AND THE FUTURE OF CIVIL WAR

Before proceeding, it is useful to say a few words about the method employed here, which is straightforward 'futurology' extrapolating from current trends in society and global political economy, rather than technology per se. In this, I take inspiration from the method suggested by H.G. Wells in his famous *Anticipations* of focusing on broad issues such as economics, demography and social structures, and when having determined 'something of the nature' of the times, proceed to specculate.[6] All data are open-source and verifiable except in two categories.

First, I refer to some examples of the guardedness of civil infrastructure, which I have observed directly. Second, somewhat more complicated is the basis of my assessment of the mood and amorphous 'plans' of incipient revolutionaries; for this, I have relied upon over a decade of consistent lurking on the anonymous image boards 4Chan and 8Chan /pol, a range of Twitter and Telegram channels and websites, and (surprisingly useful) the user review sections of controversial books on Amazon and Goodreads book sites.[7] While admittedly unsystematic, my impressions are broadly in accord with those of other researchers of extremist groups, notably the Southern Poverty Law Center and the International Centre for the Study of Radicalisation, several publications of which are cited further on.

Things Fall Apart

Since the end of the Cold War, scholars have argued convincingly that the trend for war is toward more intrastate and less interstate conflict.[8] The West is not free from this development. The fundamental reason that civil war is coming is that Western society is configured in a way that is fractionated and inevitably fratricidal. The violence of already existing conflicts within it remains mostly latent because of the buffering effect of high rates of material consumption buoyed by debt and cheap entertainment. 'Bread and circuses' are what the Romans used to call this system of mollifying the incipient mob. When the bread runs out and the circuses cease to entertain, conflict will cease to be latent.

The Edelman Trust Barometer has tracked a global collapse in levels of trust in society across a range of institutions. 'Distrust', it

149

concludes, 'is now society's default emotion.'[9] Related research by Pew has shown that the drop in trust in Western societies, especially the US, has been especially acute. As of 2019, before the contested Biden election and the Covid epidemic, already 68 per cent agreed it was urgently necessary to repair public levels of confidence in government and in each other as citizens, with half holding that fading trust represented a 'cultural sickness'.[10] Barbara Walters described US institutions as being in a state of decrepitude in which the esteem of politicians could hardly fall any lower.[11]

Such findings are in accord with that of the American sociologist Robert Putnam, who has long tracked a collapse in social capital.[12] The work of his which is most pertinent now, however, is that in which he connected this decline with multiculturalism.[13] The implications from a policy perspective are so sobering and so vindicative of far-right views that they are often politely ignored or vaguely explained away, including to an extent by Putnam himself. The trouble is that there is nothing wrong with the scholarship, which is quite obviously valid. It has been more than a decade since Angela Merkel declared multiculturalism in Germany to have 'utterly failed', an idea echoed by then British Prime Minister David Cameron, who elaborated that 'It ghettoises people into minority and majority groups with no common identity'.[14]

Additionally, the intercommunal antagonism normally referred to as 'political polarization' has been powerfully enhanced, ironically, by the twin forces of social media and identity politics further undermining trust and the existing social order.[15] Digital connectivity tends to drive societies towards greater depth and frequency of feelings of isolation in more tightly drawn affinity groups, guarded by carefully constructed membranes of ideological disbelief (often referred to as 'filter bubbles'), beyond which lies nought but an increasingly contemptible and alien 'other'.[16]

Moreover, the key factor is economic. A 2019 study of the US fiscal outlook up to 2030 by the then Australian Consul General in New York (a respected banker) concluded that the US Social Security fund would be insolvent by 2031.[17] So alarmed by his findings which projected the US debt–GDP ratio reaching 144 per cent by 2050, the author asked experts in the New York Federal Reserve whether

URBICIDE AND THE FUTURE OF CIVIL WAR

he was being too pessimistic. The answer: not at all—perhaps he was too optimistic.[18]

As it happened, America's debt–GDP ratio as of 2022 has already reached 129 per cent on a ballistic trajectory that is set, conservatively, to reach 225 per cent by 2050.[19] For a point of comparison, the historic high in 1946, bearing all the costs of the Second World War, was 106 per cent. The truth, though, is that the numbers will never rise to that height because the system will collapse first. As the author of the above study put it, 'it can't happen as investors won't accept it. This sort of thing normally ends in hyperinflation, conflict, and loss of empire!'[20]

In the above, I have focused upon the United States because it is the lynchpin of the existing Western-led global order. Nonetheless, the fact is that no other Western economy is in much better structural order, and several are in significantly worse. In terms of economic sustainability, further debt issuance and consumption, the collective West has reached the end of the line. For Europe, stark rises in energy and freight costs, combined with recessions that have reduced domestic demand for industrial products on top of an already higher cost base than other advanced economies, make energy-intensive firms unprofitable. The homeland of the Industrial Revolution is being de-industrialized at a rapid pace.[21] The currently unfolding economic downturn—a long overdue recurrence of the 2008 financial crisis—is now combining with a progressive de-dollarization of global trade. This raises the inevitability of a gigantic gap in expectations of individual, group and societal well-being. If there is one thing that the literature on revolution agrees upon, it is that expectation gaps are very dangerous.[22]

To this mix must be added one final element: the availability of weapons. Normally, this is a thing remarked upon in the American context where for well-known constitutional reasons, there is a high degree of availability of small arms. Since February 2022, however, the West has injected tens of thousands of unaccountable high-grade military weapons, including man-portable air defence and anti-tank missiles, into Ukraine, a country from which Europe is 'separated' by a border considered one of the world's most permeable to smuggling. The current most optimistic guess is that such weapons

151

have not made their way on to European black markets in large numbers—yet.[23]

In sum, civil war is imminently and eminently plausible because there is no shortage of dry tinder either literally or metaphorically.

Urbicide

The strategy that anti-status quo groups would seem to generally comprehend is simple, direct and by no means a secret: they intend to collapse the major cities, causing cascading crises leading to systemic failure and a period of mass chaos that they will wait out from the relative security of the rural provinces.[24] What they imagine might be rebuilt in the aftermath is highly underdeveloped. 'Let it all burn' is the main element of their thought process. Like the strategies of many terrorist groups, whose ultimate aims are far-fetched to say the least, it is the chaos produced that matters most.

Consider this passage from a 1974 booklet by Murray Bookchin on *The Limits of the City*:

> Just as there is a point beyond which a village becomes a city, so there is a point beyond which a city negates itself, churning up a human condition that is more atomizing—and culturally or socially more desiccated—than anything attributed to rural life ... Either the limits imposed on the city by modern social life will be overcome, or forms of city life may arise that are congruent with the barbarism in store for humanity if people of this age should fail to resolve their social problems.[25]

Bookchin, an American Jewish social theorist, Trotskyist, influential urbanist and ecologist, is hardly a man of the far-right—though his identification of the problems of society as being atomization and degeneracy (a fair way to describe 'cultural desiccation') are both far-right beliefs. That is why it is interesting that I first found this quote at the beginning of a post on 4Chan's /pol discussion group, which comprised a short essay on tactics for a coming civil war. I will get to those in a moment, but for now consider that the essential premise of the strategy as described is based on a thesis promulgated by a man of the left nearly fifty years ago in an obscure

152

academic text which states that there is an intrinsic wrongness of the 'human condition' centred upon the configuration of the modern city.

The most important thing we might conclude from this is that the author of the comment in which this passage is quoted is not an ill-educated dummy. They might be wrong to believe that contemporary megalopolises like London or the whole north-eastern seaboard of the United States are highly vulnerable to disruption and potential ruination. It is not, however, crazy to suspect that the seeds of urban destruction are already embedded in the social and physical architecture of such places, and that what is needed to realise their demise is a push in the 'right' direction by determined groups.

'Urbicide' is a relatively recent neologism and an obvious portmanteau of 'urban', 'urbanism' or 'urbanity' with 'homicide' or indeed 'genocide'. A few years ago, David Kilcullen, in his description of American urban counterinsurgency operations in Baghdad, defined it colloquially as 'killing a city'.[26] A more formal definition describes it as the destruction of the possibility of a particular condition of urban life (or 'urbanity' in the jargon) through the targeting of structures. As a result of such attacks, a condition of urban 'agonistic heterogeneity' is transformed into one of 'antagonistic enclaves of homogeneity'.[27] In layman's terms, that means destroying the condition of intercommunal comity in ethnically or otherwise mixed urban environments, separating them into warring neighbourhoods.

Originally employed as a way of explaining the tactics of the Yugoslav civil war, the idea resonates with contemporary examples much closer to home, and has merit looking ahead to the civil war which I anticipate. In September 2022, for example, the city of Leicester in Britain, the eleventh-largest in the country, witnessed serious intercommunal violence between the local Hindu and Muslim populations. A Hindu mob marched through the Muslim part of town chanting 'Death to Pakistan' (bear in mind that both populations are nominally British).[28] That would seem to be a reasonable example of agonistic heterogeneity transforming into antagonism. Similar examples from France and Spain, often even more violent, might easily be cited.[29]

The most frequently talked about tactics amongst anti-status quo groups are quintessentially asymmetric. They do not rely upon main force, but rather work on a kind of judo logic of finding points of extreme unbalance in a system and striking there with the intent of causing it to collapse under its own weight. The ultimate vulnerability here is social, specifically the configuration of identities that are increasingly antagonistic, and a parallel, decades-long draining of reserves of social capital. These are intangible, somewhat abstract 'targets', though they are targetable nonetheless.

The intended method of achieving this is to strike physical targets, in particular the electrical energy grid as well as gas networks, on the logic that disruption of essential utilities will rapidly lead to social breakdown. This is an entirely plausible line of thinking because a great deal of critical infrastructure is practically unguarded (likely un-guardable), its location is perfectly obvious public information, it is relatively easy to damage, and the knock-on effects are potentially very severe.

Electrical pylons for long-distance, high-voltage transmission can be brought down with small amounts of simple explosives; transformer stations can be shot up or just as easily set alight. Likewise, gas facilities are vulnerable. It is a matter of a few minutes on the Internet, for example, to find high-quality maps of the UK's National Transmission System—the network of gas pipelines and pumping facilities supplying forty power stations, big industrial users, and liquefied natural gas terminals located on the coast.[30] The pipes conveying dangerous substances like gas are commonly referred to as Major Accident Hazard Pipelines (MAHP); the clue as to their vulnerability is in the name.

The typical MAHP is up to a metre in diameter and is usually buried at a depth of 1 to 1.5 metres. The locations of these pipelines are widely known to a high degree of precision, which makes sense in order to avoid any potential accidents caused by normal digging during construction. In July 2004, in Ghislenghien, Belgium, for example, twenty-five people were killed and 150 seriously injured when one was damaged by construction work. The ensuing explosion produced a fireball of 150–200 metres in height with a core temperature of 3,000 degrees Celsius. An 11-metre section of pipe weighing one tonne was thrown 150 metres, marking the radius of an area in which dozens of buildings and cars were set alight.[31]

The pressure in a large gas system is maintained by a network of compression stations, also known as 'transmission relay stations', of which there are twenty-four in the UK (two main ones serving London), and all of which are in semi-rural environs. Interestingly, those in the UK are not labelled on Google Maps, but they are easily discovered by postcode search. One of the largest and most important, located near Cambridge just south of the Royal Air Force Museum at Duxford, is no more guarded than any of the other nearby light-industrial facilities.

This is in sharp contrast with the fortification efforts that have been extensive inside cities. In April 1992, a 1-tonne bomb concealed in a Ford Transit van was exploded outside the Baltic Exchange Building in London, killing three people and wounding ninety-one others. A year later, an even more powerful truck bomb, equivalent to 1,200 kg of TNT, was detonated at Bishopsgate in the heart of the financial district. As it was a weekend, there were mercifully few casualties—one killed and forty-four injured—but the damage caused was enormous. The economic cost of the two bombings was estimated at about £800 million and £1.45 billion, respectively.[32] As a result of these and other terror attacks, the London landscape is now densely fortified with hostile-vehicle mitigation barriers and highly monitored by a range of sophisticated and expensive surveillance measures. It would be harder to conduct such attacks now. Attacking the average gas compression station, by contrast, requires no more than being able to plough through a chain link fence.

It is not the primary effect of attack, however, which is the big problem, but the secondary effects. Think, for example, of the city of Toronto, Canada, which has a greater-metropolitan population of over 6 million people. In the winter months, heating is a matter of life and death in that climate. Currently, 44 per cent of the population of Toronto are apartment dwellers. When the power goes out, how long will it take people in electrically heated apartment buildings to start trying to warm their frigid homes and cook with jury-rigged fires? How long after that will it take for apartment buildings to start going up in flames? Answer: not long.

Transportation infrastructure is also a likely target. It is well known that American infrastructure is already severely run down, even without active efforts to disrupt it. Hundreds of road and rail bridges, overpasses, logistical nodes and transportation hubs are held together with jury-rigged repairs. Moreover, many major cities—New York being a prime example—are accessed via bridge or tunnel—natural bottlenecks and easily attacked. If nothing else, the recent Covid lockdown made obvious the precarious dependence of social order on the smooth running of civil logistics.

The fact is that the average modern urbanite has on hand no more than a few days of food, while the cities they live in possess typically no more than a few days' worth in warehouses and on store shelves. Britain's food security, for example, depends heavily upon just-in-time logistics for much of its supply. The government considers the existing system to be resilient under normal conditions while at the same time acknowledging the significant risk of disruption arising from the agri-food sector's reliance upon energy and transport. Attacks on such systems would put food supply chain continuity at risk.[33]

In the event of serious efforts to attack transportation and logistics infrastructure, the authorities would be rapidly faced by a gigantic challenge on two fronts. First, they would need to guard a vast and distributed system which is generally unguarded and extremely open to native attackers who know perfectly well the vulnerable chokepoints. Second, they would have to do so while at the same time trying to maintain social control of cities full of hungry, cold, angry, frightened and socially atomized people who have literally and metaphorically been suddenly thrust into the dark.

In other words, urban riots are practically inevitable and likely to be compounded by simultaneous outflows of people from the cities to perceived safety outside. To get a sense of the potential danger, we might consider that for a week in August 2011, London and other British cities were wracked by widespread rioting. Nationally, just over 3,000 arrests were made. It has been estimated, though, that on any given day in London during the crisis there were not more than a couple of thousand rioters, and perhaps only a couple of hundred of those were seriously violent, out of a metropolitan population of almost 10 million.[34] Even so, the police struggled massively to restore order. How would they have fared if 100,000 people, just 1 per cent of that population, revolted? What about 10 or 25 per cent? Answer: badly. The security services would be overwhelmed.

We should assume, moreover, that some fraction of the military and security services—possibly a large fraction given our understanding of the tenor of political conviction, which is generally more right-leaning—would support or remain neutral towards an uprising. This possibility is strongly suggested by the events of early December 2022, when twenty-five members of a German far-right group were arrested for attempting a coup.[35] In fact, there is probably no more important factor in the outcome of revolution than the security forces' willingness (or lack thereof) to crush it.[36] Quite clearly, the security services are prepared to behave highly politically, perhaps by not acting at all.

Would Western armed forces and police react to domestic revolution in the way that Iran's has to domestic upheaval, China's did during the Tiananmen revolt in China, or Arab armies did during the 2010–11 Arab Spring? The question is not easily answered; in fact, it has hardly been asked openly, though one might surmise that the oft-stated apprehension of the Pentagon over the last two years about political 'extremism' in the ranks of the armed forces is a good sign that at some level it is seen as a relevant concern.[37] The armed forces may actively rebel, (i.e. attempt a coup d'état), support the government against the people in revolt, remain neutral, or it may fracture and attempt all of these things simultaneously.[38]

It is well understood that in any war, the primary objective is to defeat the enemy's will to resist. The intimacy of civil war, its political intensity and its fundamentally social quality, along with the acute accessibility to attack on all sides of everyone's weak points, can make them particularly savage and miasmic. Deliberately calculated violent attacks intended to derange the collective psyche of opponents and to provoke them into savagery in kind are to be expected. If we call these 'informational attacks', the intent is not to gloss over their gruesome ugliness but to be precise about the way that they are intended to shape the mental landscape of the conflict.

BEYOND UKRAINE

Firstly, there will be attacks on the information infrastructure directly, with the intent of disrupting communications through the normal media, enhancing the apprehension of disconnectedness, confusion and fear in the population, as well as hampering government efforts to coordinate responses to multiple and cascading crises. As with transportation and energy, the difficulty of doing this is not great because the infrastructure is lightly guarded, widespread and well understood.

A few examples suffice to illustrate this: one of the main UK transatlantic fibre-optic cables, carrying a significant bulk of data traffic between North America and beyond, is located near a popular beach in Cornwall. The cable cuts its way beneath a nearby car park, where it can readily be accessed simply by lifting an unlocked manhole cover. This is widely known because it featured in a Google promotional video in 2021, which was subsequently reported on quite widely.[39] Likewise, the handful of routing stations and data exchanges that underpin Britain's telephone network are basically unguarded (in fact, they are often unmanned). The facilities are usually discreet, indeed deliberately seriously non-descript, but neither are they secret.

It is not commonly known that the core of Britain's government communication system, known as 'Backbone', was designed in the 1950s and 1960s to survive a nuclear attack, which is the reason that all its most critical nodes are located outside of major urban areas.[40] At its core is a series of about twenty 'radio relay stations', now in use also as microwave and cellular phone towers, located at key points across the country. A good example is the 90-metre-high concrete tower located near Stokenchurch alongside the M40 motorway—a well-known if generally un-regarded landmark in southern England. Again, it is only guarded by a chain link fence.

Britain is not atypical with respect to the configuration of its infrastructure—quite the contrary. The fact is that for all the talk and effort over 'critical infrastructure protection' in recent decades, it is nearly always predicated on the idea of external (often cyber) attack, against which there are possibly adequate defences. What it is not predicated on is the possibility of local, domestic and insider attack, against which there is essentially no credible defence. In a normally functioning society, where the default condition of civil life

158

is a widespread consent to be governed, there is no need to heavily defend these public goods: there would be no point, and doing so, moreover, would come at great expense. Nevertheless, the key words here are 'normally functioning' society.

Finally, there will be deliberate attacks designed to take existing social divisions and fractures in society and tear them into wide unbridgeable chasms. At this point in the history of conflict, it hardly seems necessary to explain the technique of propaganda by deed; after all, the employment of terrorism, assassination, kidnapping and torture of selected symbolic and/or otherwise 'important' figures—as well as just random collections of individuals in targeted groups—is simply open knowledge.

The fact is, as superbly exemplified by post-Saddam Iraq and post-Qaddafi Libya, it is practically impossible to maintain a peaceful integrated multivalent society once neighbours start kidnapping each other's children and murdering them with hand drills, blowing up each other's festivals, slaying each other's teachers and religious leaders, and tearing down their cultural symbols. It is soberingly worth noting, moreover, that instances of some of those things have occurred already in many Western countries, and all of them have occurred in France in the last five years.

What, then, is the answer to what future civil war might look like in the case of a very serious breakdown of government, and in a context where all belligerents were of a mind to escalate punishment and provocation of their domestic opponents to the maximum of their capability and ingenuity? The simple answer is spectacularly awful. More specifically, with history as a guide, the least we can look forward to is compelled population movements on the scale of the partition of the Raj after Britain's departure from India, combined with highly urbanized intercommunal violence rather like the Yugoslav civil wars but on a continental scale, and likely something of the genocidal horridness of the Congolese civil war.

What is to Be Done?

The short answer to the above is: not much, because it is already too late.

BEYOND UKRAINE

Civil war is 'political war' *par excellence*. John Paul Vann famously observed that the best weapons in a political war were the gun and the knife, not the airplanes and massed artillery of a conventional war.[41] Admittedly, Vann was a counterinsurgent, not a revolutionary, but it seems to me that the logic holds from either perspective. The coming internal conflicts will combine the knife and the gun, the bolt cutter and the sledgehammer, the improvised explosive device and any other means available, all applied with the 'utmost discretion' not to limit casualties, but in a manner aimed precisely at well-known and widespread points of vulnerability that will lead to mass derangement.

The tactics likely to be employed are not especially complex or difficult to perform. The means are widely available, and the important ones are essentially the tools of civilian life that are just lying around. They are merely the continually more potent appliances physical science offers the citizen in revolt. There is nothing important about these tactics that has not already been worked out in dozens of examples of civil wars outside the Western world.

Similarly, the strategy that is likely to be employed is based upon observations made by completely mainstream social scientists—generally left-leaning—going back more than half a century. Every strategy is a gamble against chance because that is war's nature. The one at hand here, in my opinion, looks like a safe bet. There are no implausible assumptions or yet-to-be-discovered technologies required for it to occur; on the contrary, it would require a departure from basically understood historical norms and social processes for it not to happen.

In the preceding pages, I have focused quite a lot on the attitudes evinced in far-right 'extremism', howsoever we might define that term. I might as well, however, have drawn on far-left extremism. Indeed, if left and right agree on anything it is that the existing system is beyond saving, and that some sort of revolution, perhaps involving violence, is necessary. Neither believe that 'democracy' as currently constituted means much more than cynically managed oligarchism.[42]

Likewise, to judge from such people as Pentti Linkola, formerly a leading figure in the deep ecology movement, there is no shortage of contempt from that quarter for the supposedly democratic status

160

URBICIDE AND THE FUTURE OF CIVIL WAR

quo. He argues that, in reality, a tiny elite minority possessing 'the power and cogency of a shaman, the potency of a fanatic, the mysterious, irrational, and persuading strength of an idiot' actually rules.[43] These are not the feelings of someone who thinks the problems of society will be solved by 'normal' politics. Indeed, Linkola's conclusions on Lenin's famous question, 'what is to be done?', are strikingly millenarian: he believes that about four-fifths of the human population needs to die quickly in order to preserve the life of the planet.

But, realistically, is there anything that could be done? The problem is that governments do not possess the wherewithal to defuse the bomb that they have themselves constructed at the heart of their own societies. We are in for a wild ride, and that is that.

Perhaps, though, certain steps might limit the scale and duration of damage and speed up the return to a more peaceful status quo. There is nothing unprecedented about such steps in principle. Civil defence, including thinking about the preservation of knowledge and expertise after a nuclear disaster, was taken seriously during the Cold War. Indeed, as one famous study of Britain's Cold War-era civil defence observed, there was a high degree of overlap between the needs of a 'revolution-proof' seat of government and a nuclear stronghold. Each required defensible and physically remote installations, were much improved by being split into small parts (the better to prevent collusion between separate subordinate institutions), and required communications systems that were diverse and redundant.[44] Such measures may sound extreme from a perspective of relative stability, but in the event societies do go off the metaphorical cliff, they will not seem extreme; instead, they will be viewed as the sort of sensible forward thinking that should have occurred but regrettably did not. By comparison with nuclear apocalypse or a major interstate war, the disaster which I have described is merely terrible. Yet in the past it was considered a proper function of government to take seriously the problem of how to reconstitute a nation from the ashes of war. We should try thinking about it again now.

161

8

LIVING WITH DENIAL?
GREAT POWER COMPETITION OVER ECONOMIC AND MILITARY ACCESS

Paul van Hooft

Russia's invasion of Ukraine in February 2022 failed to accomplish its objectives. Despite the assumption among many observers before the invasion that Russia had successfully modernized its armed forces and restored them from their decline in the aftermath of the Cold War,[1] Ukrainian forces successfully held off Russian advances in the early stages of the invasion, raising the costs of entry, and then pushed them back, with Russia unable to even establish clear air superiority. Notably, this occurred prior to the large-scale delivery of US and European weapons. This suggests that, even for major powers, enforcing access in the face of denial by weaker actors has become a serious challenge.[2]

Much had been made before the war of the increasing possibilities to impede or constrain access to and deployment of an adversary's forces into a given theater of operations, thereby limiting their freedom of action, as a consequence of improvements in precision weapons and their widespread availability.[3] Some are calling this phenomenon an 'age of denial.'[4] Here, denial is distinct from defense as it does not assume that an actor can be stopped, but rather that their life will be made exceedingly difficult. As a result, the stronger power may be constrained from fully employing their military potential, as was seen in the Russian invasion. Given this, US planners fear that the United States' 'command of the commons'[5] is increasingly challenged by combinations of long-range stand-off weapons and integrated air defense systems, which provide adversaries such as China, Russia and Iran with so-called 'Anti-Access/Area Denial' (A2/AD) capabilities. Moreover, other scholars argue that three decades of globalization have made the military offshoots of dual-use technologies, such as cyberweapons, unmanned vehicles, autonomous systems and artificial intelligence (AI), widely available. Even before Ukraine's widespread use of unmanned aerial vehicles (UAV), Azerbaijani forces had used UAVs to great effect against its unprepared adversary, Armenia.[6] Consequently, cutting-edge capabilities are now within reach of non-state actors and minor powers. This development has been coined the 'democratization of war' by Audrey Kurth Cronin, in which, as T.X. Hammes argues both in this volume and elsewhere, small powers and non-state actors will have clear advantages in pursuing denialist approaches.[7]

In contrast to these claims, this chapter argues that the vision of the future as being denial-centric, and benefiting small and middle powers, is too simplistic. Though caution must be exercised when drawing lessons from the Russo-Ukrainian War about the possibilities for denial strategies, when taking into account the events since February 2022, as well as longer-term trends, a more nuanced appraisal should take into account the various levels of competition during peace and war.[8] In future war, the effects of denialist technologies will have different effects at the tactical, operational and strategic levels of war (see Table 8.1). At the tactical and operational levels, denial-enabling technologies, including dual-use technologies, are likely to have deep effects; however, at the strategic level, access to highly advanced technologies that enable access-enforcing technologies, such as stealth, quieted submarines and specific AI applications, is likely to remain highly uneven.[9] In fact, current trends may well suggest that great powers, if they choose to do so, can exploit and deepen existing structural inequalities to an even greater extent because the economic and military forms of denial compound each other. If so, current trends also suggest that the United States continues to be well positioned to make use of these structural inequalities, and, to a large degree, so will China. However, Russia—and arguably other major powers—are likely to face an uphill struggle. As Russia's war in Ukraine shows, it is difficult to overcome denial capabilities, and difficult to maintain access to advanced technologies, even in an age of globalization.

The rest of the chapter builds on this argument. Throughout the chapter, I use denial beyond its narrower use in the military domain, where it highlights the ability to impede the movement of military forces across physical space on land, sea, or in the air. I also explicitly include access to technology, whether knowledge, components or material inputs. The chapter first discusses how the constraints of distance have changed, as many globalization scholars have noted, but also that the manner in which states are constrained is highly uneven. Second, it considers how these constraints unevenly affect which states can deny or enforce access at the strategic level, even if the proliferation of denial capabilities at the operational level among non-state actors and small powers is taking place. Third, the chapter briefly looks at how these dynamics affect the options of states at multiple stages of production and deployment, and considers how they relate to Russia's war in Ukraine. Finally, the chapter discusses the implications of these developments for great power competition in the coming decades of the twenty-first century.

Increased Access Due to Globalization, Increased Options for Military and Economic Denial

Both the current economic and military forms of denial developed in parallel over the past half-century have created an ironic counterpoint to the more common narrative of increased openness and access as a consequence of globalization.[10]

The decline in maritime costs (as well as land and air transport costs), and expansion of instantaneous communication and options for information collection and analysis, have allowed both civilian-commercial actors and states to operate more effectively over far greater distances. The sum value of exports and imports amounts to more than 50 per cent of the value of total global output,[11] with the vast majority of these goods transported over the world's seas.[12] The decline in transport costs has also facilitated specialization by economic actors, leading to increasingly complex supply chains.[13]

At the same time, the United States has been better positioned than any potential rival to reap the benefits of globalization. The development and manufacturing of semiconductors is the example that is most often used to illustrate the extensive degree of specialization and complexity in global supply chains and the mutual dependencies that have emerged. The United States still plays a disproportionate role in the research and design of semiconductors for commercial and military applications.[14] Focusing in particular on military applications, while the production of weapons has internationalized, the United States occupies a unique position among its allies and partners at the core of the supply chains, with US defense firms specializing in being 'system aggregators.'[15] Proliferation of technologies is also uneven, as the complexity of most advanced weapons, and their increasing costs, actually makes it more difficult for certain knowledge to diffuse.

The deepening of the US military's advantages preceded the collapse of Soviet Union, with the former making use of its advances in the emerging information and computing technologies to gain a qualitative conventional advantage over Soviet and Warsaw Pact forces. One example of this was the development of long-range stand-off precision weapons that led to the AirLand Battle doctrine; these weapons could target Soviet and Warsaw Pact follow-on forces far behind the frontlines, thus offsetting their conventional preponderance.[16] Yet, while a precision strike is a capability that has now become more widely available, it is important to note that the same does not hold true for technologies to avoid detection. In this case, the United States retains advantages in stealth and other concealment capabilities.

These capabilities carried over to the post-Cold War era to give the US a significant lead in both advanced conventional technologies that could deny adversaries access to the theater, as well as enforce the US's access into the theater itself. These military capabilities allow the US to not only threaten the chokepoints through which military forces must pass, but also to militarily limit the access of adversaries to key economic inputs such as critical material resources, energy, components, and so on. Other capabilities cannot do this; for example, despite the much-touted large-scale introduction of UAVs, adversaries that have and know how to use integrated air defense networks have been able to limit their effectiveness in practice.[17]

The globalization of the world's production has also created more economic chokepoints that can be exploited, even without the use of military force. The decline in the costs of transport and communication has increased technological complexity and the interdependence of actors over global supply chains, which can then be weaponized.[18] This unprecedented degree of interdependence provides multiple opportunities to deny access; however, once these approaches have been used more often, actors are likely to work to insulate themselves as much as possible, making the economic-denial approach less successful. Weaker actors are therefore less well positioned to insulate themselves.

In a global economy dependent on information collection, processing and analysis, access to semiconductors (specifically advanced logic chips) has come to present another chokepoint.[19] Semiconductors are not only key components of computers and smartphones, but of cars, washing machines, and so on. They are also essential for the information-dependent military domain, being crucial to precision guidance, fly-by-wire, and specifically the emerging field of AI. Most of the horror stories surrounding AI envision killer robots run amok, but it arguably offers the greatest potential in terms of data fusion and analysis, and jamming applications.[20] China has invested heavily in AI as a technology, hoping to bypass US advantages in the other domains.[21] Specifically, China hopes to use AI to dramatically enhance their ability both to hide from adversaries and to locate them. If they succeed in doing so, the US ability to project power will be severely hampered, and it may not have sufficient combat power to come to the aid of its allies within the timeframe needed to stop Chinese advances (for example, in an escalation over Taiwan).

BEYOND UKRAINE

Deepening Inequalities to Enforce and Deny Military and Economic Access

The possibilities for increasing and combining technological-economic and military denial cannot be separated. However, the trends above have varied effects because the ability to possess and master technology at the highest level is highly complex and unevenly distributed. Advanced military technologies are proliferating, and this is driven by increasingly dual-use capabilities; yet the effects vary greatly depending on the level of competition and conflict. This complexity and uneven distribution have consequences for both economic and military denial.

Denial or access enforcement work entirely differently at the strategic level, where states can be prevented from accessing and maneuvering into a theater, as well as at the operational level, where armed forces are in closer contact with each other. The United States was and remains the one state that is able to not only project military power across the globe, but also prevent others from doing the same. The United States spent most of the post-Cold War era (with the partial exception of the 1991 Gulf War) involving itself in conflicts with non-state actors and minor powers. Arguably, the US experiences in the so-called Global War on Terror (GWOT) skewed the perception of the abilities of these actors. The struggles of American armed forces in the alleys of Baghdad or hills of Kandahar do not take away from the fact that all great powers struggle in the situations where the full extent of their power cannot be applied. The fact that the United States encountered great difficulties in winning against these much weaker actors when it fought them up close, however, should not lead us to overlook a more obvious trend: namely, these actors were absolutely helpless in stopping the US from moving armed forces and their logistics support across vast regions. The problem is not that the democratization of war postulated by conflict scholars is entirely wrong; it is rather that they are looking at the lower, more operational level of conflict and competition, rather than the strategic level that supersedes theaters or campaigns. For example, Iran has developed the ability to raise the costs of access for the US Navy to operate in its littoral waters,[22] and developed

LIVING WITH DENIAL?

a relatively sophisticated air defense network, including both Russian S-300 systems alongside indigenous Mersad and Bavar-373 systems.[23] Iran's UAV production has been vaunted, and its arm sales to Russia have been held up as a sign of its growing influence. Be that as it may, the effectiveness of its Shahed 131 and 136 drones in the Russo-Ukrainian War has been limited. Iran's armed forces cannot effectively deny the US armed forces anywhere outside of its air and maritime littorals; moreover, they have not been able to stop the attacks on their defense industry. Ukrainian armed forces were able to skillfully deny Russian victory in the earlier stages of the invasion, but even before the delivery of long-range precision weapons, much of the Ukrainian successes depended on US space-based intelligence to move their mobile air defenses and to target Russian assets.[24] Likewise, the initial successes Ukraine had with their UAVs—specifically the Turkish Bayraktar—declined once the Russian forces adapted their defenses.[25]

In short, dual-use technologies and denial capabilities are proliferating among non-state actors and small states, and they can create new challenges for major powers; these actors, however, do not have the ability to deny major or great powers the access to resources, components or movement of their military forces outside of their territories, littoral waters or air spaces. So far, non-state actors and smaller powers cannot block and deny access to key technological-economic and military chokepoints. In contrast, that the United States could (and arguably still can) project power at the strategic level is the more striking feature.

Improvements in sensing options, data collection and analysis have increased the visibility of military (and civilian) surface vehicles and vessels, submarines and aircraft;[26] nevertheless, so has the ability to avoid detection or defend against attacks whether through more advanced stealth aircraft or quieted submarines. These technologies are not new, but fully developing and integrating them with networked sensors across multiple platforms means that, in the case of aircraft, the transition to fifth-generation fighters (let alone sixth-generation) has only been fully accomplished by the United States.[27] Similarly, the competition between seeking and hiding also takes place in the subsurface domain. In the face of the oceans

169

becoming increasingly transparent, submarine technology has been directed more recently towards quieting.[28] Conversely, other states have been more successful at developing missile technology—a more conspicuous form of warfare—and the past decades have seen horizontal proliferation of ballistic and specifically cruise missiles by many international actors.[29] The diffusion of military power throughout the international system should therefore be understood as differentiated at different levels. While long-range stand-off weapons have spread through the international system, quieted submarines have not, nor have the United States' adversaries so far managed to emulate a technology at the same rate of success that the US had already fielded at least three and a half decades ago.[30] China has encountered these difficulties, but it is arguably coming the closest, and may leapfrog the US with regards to specific technologies such as AI in the future. However, this means that the United States and China are in a league of their own as only they can militarily target the movement of the armed forces of other states in the global commons.

Combining Military and Economic Denial

The increased possibilities of technological-economic and military chokepoints interacting, to create possibilities to deny or enforce access, are part of longer-running trends. These possibilities to target chokepoints and deny access can be conceptualized as consisting of four distinct steps: (1) research and development; (2) pre-production; (3) production; and (4) pre-deployment. The section below discusses these steps and provides some recent illustrations before underlining how the war in Ukraine showed Russian weaknesses in terms of denial and countering denial.

First, research and development, whether for commercial or military purposes, requires increasingly specialized knowledge. Technologies tend to be stacked upon one another, and each of these requires significant skill sets that are not easy to transfer.[31] Yet, any form of hyper-specialization creates vulnerabilities that can be exploited. Moreover, cyber espionage and sabotage are increasingly possible; for example, scientists suspected of involvement in

Iran's potential nuclear program have been assassinated,[32] while its nuclear enrichment facilities were sabotaged by the infamous Stuxnet cyberattack.[33] The financial resources for comprehensive research and development are more difficult to acquire, however. Due to these tremendous costs, informational technology has been primarily driven by commercial applications.[34]

Second, during the pre-production process, access to key economic inputs is uncertain, as supply chains are complex and concentrated in specific locations. This applies to many critical material or energy resources that are mined or extracted in one location and refined elsewhere. Ownership of both mining and refining can be controlled by small numbers of companies, with specific states acquiring ownership. Key components are manufactured across vast and complex supply chains. Consequently, there are both technological-economic and military-logistical chokepoints. Technological-economic chokepoints can be weaponized through export controls or price manipulations; even if they are not fully effective in stopping access entirely (due to black market networks or patrons of great powers providing the technologies), the targeted states will pay more to get less and less quality. Military-logistical chokepoints are more straightforward. Most goods—including critical material resources and electronic goods, not to mention energy supplies—pass through maritime chokepoints like the Strait of Hormuz, the Taiwan Strait, the Malacca Strait, and so on.[35] Without inputs, timely and cost-effective production becomes more and more difficult (if not impossible) once possible stockpiles are exhausted during a conflict.

Third, production itself is vulnerable depending on how the facilities are spread out in specialized complex supply chains. If production facilities are spread out, then the routes between them are vulnerable; if facilities are concentrated, they are vulnerable to increasingly precise, stealthy and destructive weapons, including those that produce stealth bombers and submarines. During the twentieth century, targeting production facilities was key. As Philips O'Brien shows, Allied air and naval powers were highly successful in destroying German and Japanese military production of aircraft, vessels and other weapon systems.[36] Since then, the possibilities for doing so have only increased; for example, Iran's air defense network was incapable of stopping drone attacks on its UAV factories.[37] The possibilities for economic denial are already apparent as well: in response to the Covid-era disruptions, major powers, including the United States, China and EU member states, have started renationalizing or 'friendshoring' key parts of their supply chains.

Fourth, pre-deployment of weapon systems to their strategic uses takes place over the same physical lines of communication. States that are not defending and deterring attacks on their home territory and its approaches will thus be vulnerable as they attempt to project power. Likewise, as the experience of the Second World War shows, movement of military equipment constitutes a key vulnerability, with Axis weapon systems being destroyed by Anglo-American air and naval powers long before they were brought to bear in battle.[38]

The longer-term trends of increasing possibilities for denial across these four phases were also visible in Russia's war in Ukraine. Before its invasion, Russia was believed to have recovered its military potential and modernized its armed forces. In the decade before the invasion, Russia invested heavily in advanced air and missile defense systems to counter US aerospace power, as well as ballistic missiles, cruise missiles and its much-advertized hypersonic missiles. However, beyond mistakes in the early stages of the war, including a lack of planning and of logistics, other limits to Russia's ability to produce and maintain denial have also become apparent during the war.

Russia's research and development is still effective at both creating innovative new designs and improvising. Yet in the wake of the war, Russia is likely to face real issues with maintaining a robust research base. Russian researchers have left the country in large numbers, and access to foreign direct investment has drastically declined.[39]

Russia's defense industry relies heavily on foreign components; from advanced components like semi-conductors to more basic production machinery and ball bearings.[40] This means it has become increasingly vulnerable in the pre-production phase. Many advanced (as well as less advanced) components are produced by the US, EU member states, Japan, South Korea and other US allies that are participating in a comprehensive set of sanctions and export controls.[41] As such, Russia has struggled to maintain an inflow of semiconductors as its indigenous production capabilities are highly limited. Before the start of the war, Russia built up a stockpile, and it has still been able to acquire some supplies through illicit means or by repurposing semiconductors from household appliances and gaming computers. Chinese military assistance has been slow in coming, though this may be changing. Of course, Russia's own attempt at leveraging weaponized interdependence against Europe through the use of its energy was moderately successful in the years preceding 2022,[42] but this has failed as Europe cut its imports after the invasion. Simply put, Russia does not have the same pressure on Europe's access to pre-production inputs.

Russia's production has not been targeted to the same degree, outside of the odd act of sabotage. However, Russia's pre-deployment of equipment to the battlefield has been repeatedly struck by Ukraine's UAVs,[43] as well as precision strikes on key storage facilities. These weaknesses underline how limited the possibilities for autonomy are in the current era even for ostensibly great powers.

Implications: The Enduring Preponderance of Great Power

Combining the foreseeable options for economic and military denial underlines that most actors will not be able to master denial at the grand strategic level to enforce military access to multiple regions or within regions. Nor will they be able to ensure access to research and development, resources and components. Most minor states face chokepoints at the research and development, pre-production and production stages that major powers can exploit. Nor can these weaker actors hope to overcome these chokepoints through import substitution. As a result, they are left highly dependent on global markets, from which they can be cut off, or on patron–client relationships with major powers, thereby enhancing their vulnerabilities in other ways. Nor can minor states mount any significant defenses against the power projection of major powers up to their littoral air or maritime spaces; this means that their physical access can be cut and major powers can target production and military targets far inside their territories during wartime, as well as threaten such attacks during peacetime. Military capabilities for the close-in fight on land might be more accessible, but advanced weapons that can evade detection and in turn detect, find and strike targets from great distance will remain out of reach (specifically aerospace and (sub)marine capabilities). The strategic level of denial has become underappreciated and taken for granted: essentially, if actors are denied at the strategic level, they will lack the ability to deny at the theater and operational level. Table 8.1 summarizes the relationships between great power, middle powers and small power/non-state actors with regards to military and economic denial across the various levels.

The argument regarding differential effects has three major implications. First, as technologies that facilitate military denial at the tactical level likely diffuse, the costs of invading and occupying territory are likely to further increase, continuing a trend that has been present for over a century.[44] The war in Ukraine is an illustration of that, as is the war in Nagorno-Karabakh. Innovation at the operational level can accelerate, and the versatility of emerging technologies may actually allow smaller organizations (i.e. non-state actors, and small and middle powers) to adapt faster. Furthermore, denial at the operational level is going to benefit small and middle powers to a large extent, suggesting that mutual denial will be the more likely outcome in the future.[45] Whether in Eastern Europe or in the Western Pacific (specifically Taiwan), states facing the threat of revisionist great powers may be able to mount significant cost-raising defenses to dissuade them. In doing so, their survival becomes less immediately dependent on a more unreliable US superpower. This also implies that the United States can lean away from a more escalatory access-enforcing approach such as the Joint Concept for Access and Maneuver in the Global Commons (previously known as AirSea Battle), a highly offensive and destabilizing approach that would look to destroy Chinese denial capabilities in the initial phases of a conflict.[46] However, as the Ukrainian case shows to the extreme, and as will be likely in the case of middle powers and even some major powers for some time to come, there is a heavy dependence on the US for advanced intelligence, surveillance and reconnaissance (ISR) capabilities. The lack of sufficient ISR capabilities is certainly a serious constraint for any European attempts to focus on deterrence-by-denial as part of an approach that would strengthen their strategic autonomy.

Second, widening inequalities to execute denial at the strategic and theater level means that close-in fighting is increasingly avoidable for great powers when facing weaker states. If they choose to, great powers can thus continue to rely on stand-off capabilities and coerce without direct confrontations. Most states—and certainly non-state actors and minor states—will lack the financial, intellectual, organizational and other capital to master the capabilities that are needed for evading detection, and for detecting and destroying targets. Access to highly advanced, access-enforcing technologies is likely to remain limited. For example, the United States has successfully pressured allies like the Netherlands, Japan and South Korea to sever Chinese access to advanced semiconductors, as well as the technology needed to produce these.[47] It has also used the Inflation Reduction Act and the Chips Act to impede Chinese production. If these policies are successful, China will not be able to fully substitute imports for the most advanced semiconductors,[48] which it needs to use AI on a large scale.

Third, systemic distribution of power will be largely concentrated in the US and China. China has invested heavily in an ability to deny US power projection. Its medium-range ballistic missiles can reach the stepping stones the United States has in the region, whether ports or airbases located on the Japanese islands or Guam;[49] it may also be able to target US aircraft carriers.[50] China, meanwhile, has constructed a sophisticated integrated air defense system.[51] It has an independent research and development base, has ensured access to critical material resources, and maintains the ability to refine, produce and complete key products. This applies to both its defense industry, where it experiments, builds and then discards generations of weapons, as well as dual-use products such as semiconductors. Outside of the United States and China, other major powers will have difficulty attaining and maintaining capabilities that can challenge their access outside of the littorals, or develop advanced weapons without these two nations' cooperation. This development is likely to push a great power like Russia into further decline, and to limit the possibilities for any European state or collective of European states to secure their own access outside of US support. That said, economic denial is a weapon that had its greatest potential at the height of globalization; if great powers continue to pursue long-term denial strategies, the outcome will be de-coupling in most high-value-added sectors and the re-nationalization (or at least re-regionalization) of production. According to Azar Gat in this volume, precisely because this includes many dual-use capabilities, prosperity in turn will likely decline, thereby removing a potential dampener on any conflict that exists. The conventional capabilities both great powers have developed would be immensely destructive, and because they have perceived first-mover advantages, there is a much greater risk that both United States and China will escalate during a crisis.[52]

In short, as a consequence of shrinking distances, technological complexity and interconnectedness, we have entered a new era of industrial competition and warfare where the great powers— foremost the United States and then China—are in fact much better positioned than middle powers, small powers or non-state actors to exploit the technological-economic and physical chokepoints, both during peacetime and during wartime.[53] What this means for the future of war is that if the two sides are evenly matched, the opening stages of a confrontation will be expensive and painful as each side tries to enforce access across domains. This will have the potential for highly decisive outcomes early on if one of the sides manages to disable whatever strategic-level means of denial the other side has, which will put it a prime position to exploit that weakness. This is likely to particularly hold if one or both sides lack strategic depth. However, if there is no decisive outcome at the strategic level in the early stages, then back-and-forth attritional warfare at the operational and tactical levels becomes more likely.

While we have seen some of these trends in the war in Ukraine, it is important to note that the Ukrainians had access to strategic enablers—foremost US intelligence and North Atlantic Treaty Organization (NATO) airborne radar and space-based sensors— that they themselves did not possess, but which were also off the table for Russian counterattacks (with the exception of jamming).

176

Table 8.1: Impact of military and economic denial on interactions between great powers, middle powers, and small powers/non-state actors

	Great power		Middle power		Small power/non-state actor	
	Military denial	*Economic denial*	*Military denial*	*Economic denial*	*Military denial*	*Economic denial*
Great power	(I) Great powers engage in a competitive struggle between access-enforcement and access-denial technologies across all domains and at strategic levels; mutual denial is likely, though possibly uneven over the long term	(II) Great powers engage in competitive struggle over access to (1) R&D; (2) pre-production; (3) production; (4) pre-deployment; mutual denial is likely, though possibly uneven over the long term	(III) Great power advantage on access-enforcement at the strategic level, but difficulties in maritime and air littorals at the operational level; close-in tactical fight on land is very costly	(IV) Great power advantage on R&D and pre-deployment; mixed on pre-production and production, depending on black market/great power–patron relationships	(V) Great power enforces strategic and operational access, small power/non-state actor can mount costly challenge for close-in tactical fight on land	(VI) Great power can deny access, small power/non-state actor is dependent on black market/great power–patron relationships

		Great power		Middle power		Small power/non-state actor	
		Military denial	*Economic denial*	*Military denial*	*Economic denial*	*Military denial*	*Economic denial*
Middle power		(VII) see cell III	(VIII) see cell IV	(IX) Competitive on strategic level and costly to enforce access on operational level, depending on first-mover advantage	(X) Depending on black market/great power–patron relationship	(XI) Middle power has advantage in strategic access-enforcement, but difficulties at operational level in maritime and air littorals; highly costly tactical fight on land	(XII) Middle power has limited ability to deny access, depending on black market/great power–patron relationships of both
Small power		(XIII) see cell V	(XIV) see cell VI	(XV) see cell XI	(XVI) see cell XII	(XVII) Neither small power/non-state actor has ability to deny access at strategic level outside of littorals; highly competitive and costly operational and tactical struggle over territory	(XVIII) Neither small power/non-state actor can deny access, depending on black market/great power–patron relationships of both

(Source: Paul van Hooft)

LIVING WITH DENIAL?

Moreover, Ukraine could not execute economic denial on Russia, which also depended on the United States, Europe and key East Asian partners. With the theoretical Russian advantage on the strategic level less effective than it seemed on paper, Ukrainians were able to deny (though not always stop) Russian advances into Ukraine even though Russia had clear qualitative and quantitative advantages (notwithstanding failures in Russian operational art and planning). For the future of war, the war in Ukraine and long-term trends suggest that long-lasting, grinding conflicts—with sharp and destructive opening salvos punctuated by intermittent bursts of extreme violence—will be a fixture of wars in the coming decades.

PART III

MILITARY INNOVATION AND THE FUTURE OF WAR

9

MILITARY-TECHNOLOGICAL INNOVATION IN THE DIGITAL AGE

Audrey Cronin

The academic work that shapes our thinking on military innovation comes primarily from the twentieth century.[1] It supposes that military power and effectiveness will follow if states can get these processes right. This work structures how we discuss technological change in military organizations, focusing, for example, on peacetime versus wartime dynamics; intra-service and inter-service competition; top-down and bottom-up practices; adaptation under fire; the effects of culture on military organizations; civil-military dynamics in decision-making; and the critical role of theories of victory.[2] Military innovation studies would be nothing without the pioneering work of academics and practitioners such as Samuel Huntington,[3] Williamson Murray,[4] Allan Millet,[5] William Owens,[6] Stephen Rosen,[7] Barry Posen,[8] and Robert Work[9]—all brilliant pioneers whose writings set the debate and determined the vocabulary of the field. And they achieved an enviable track record. In the last decades of the twentieth century, military-technological innovation yielded lopsided victories with low casualties on the side of advanced military power—from Israel's victory over Syrian air defenses in 1982, through the Iraq War of 1991, and the Bosnia and Kosovo wars of 1995 and 1999.[10]

But things changed in the twenty-first century. Technological innovation did not ensure US victories in Afghanistan and Iraq. It's not helping the Saudis and the Emiratis defeat Iranian-backed forces in Yemen. And it's not favoring Russia, the more advanced military power, in Ukraine. We need to think harder about how the fast-changing social-technological context of the digital age is changing the role of digital technology, its adaptation to military use, and the future of war. In this chapter, I will look at seven crucial blind spots in the connection between digital technologies and military force.

Looking to the Wrong Historical Precedents

Firstly, military-technological innovation is too mired in the lessons of the Cold War. The formative years between the aftermath of the Second World War and the breakup of the USSR in 1991 was a period of concentrated state power that featured different technologies, communications, political dynamics, innovation processes and global threats. But it may have been anomalous. As Winston Churchill said in 1944: 'The longer you can look back, the farther you can look forward.'[11]

By looking farther back, we put our current era of military-technological innovation into a broader context, better differentiating what is old from what is new. We can widen our focus beyond the classic case studies that built military innovation studies and still shape its arguments; case studies such as the interwar period of fast tanks and heavy bombers in the 1920s and 1930s,[12] submarine warfare in the early years of the Second World War,[13] the US Navy's submarine-launched Polaris ballistic missile system between 1956 and 1960,[14] or Under Secretary of Defense William Perry's acceleration of precision-guided technologies and battle networks in the 1970s.[15] With security clearances and high levels of expertise, military or scientific elites of the last century could limit the availability of technologies like nuclear weapons, satellites, guided munitions or the global positioning system, all proprietary systems requiring long-term government programs with deep capital investment over time. They were remarkably successful and, some believe, tipped

184

MILITARY-TECHNOLOGICAL INNOVATION IN THE DIGITAL AGE

the balance in the US's favor during its military contest against the Soviet Union.

But the past two or three decades have been different. Unlike that time of concentrated state power during the Cold War, our current era of military-technological innovation echoes the open innovation patterns of the nineteenth century, specifically between about 1850 and 1914, when the Second Industrial Revolution matured and technological innovation was driven by more diffuse commercial processes.

As is the case today, the late nineteenth century was marked by an atmosphere of giddy techno-optimism and no clear dividing line between amateur and professional scientific communities. During the late nineteenth century, amateurs bought cheap wiring kits, chemicals and high explosives, following instructions in brand new journals like *Scientific American*, which was founded to guide the amateur inventor. Today, tech-savvy civilians tinker with off-the-shelf drones and robotics, learning how to use them on YouTube.

A remarkable range of innovations emerged from the work of tinkerers and hobbyists in the late nineteenth century. Italian electrician Guglielmo Marconi invented the radio in the attic of his home using homemade equipment. Bicycle manufacturers Orville and Wilbur Wright made the first powered, sustained and controlled flight in 1903, flying a rickety craft built in a garage in Dayton, Ohio. Alfred Nobel tinkered with explosives in a shed in his backyard and invented dynamite, which set off the first wave of modern terrorism.

In short, today's military innovation resembles the commercially driven, open experimentation that happened in the nineteenth century more than the state bureaucracy-driven weapons-acquisition processes of the twentieth century. Broader patterns of change in general-purpose technologies, such as the invention of steel, the development of fine machine tooling, or the transition from coal to petroleum, offered a maturing technological context for burgeoning commercial innovation of specific military applications at the end of the century. We are at a similar point now with digital technologies, especially the maturing of the computer and the Internet. The evolution of today's drones, robotics and additive manufacturing looks more like radios or biplanes or mercury fulminate blasting

185

caps—all of which began on the workbenches of tinkerers at the end of the nineteenth century, before being drawn into military purposes and doctrine, and becoming vital in the First and Second World Wars.

Regardless of where inventions originate, human beings will find ways to use them to kill each other. For example, open technological innovation with unmanned aerial vehicles has surged during this century, not just because the US Central Intelligence Agency ramped up the secret use of armed Predators, but also because hobbyists learned how to use a smartphone to fly cheap quadcopters carrying a payload. Nonstate groups like Islamic State in Syria (ISIS) picked up on this in 2013, using fixed-wing small drones and quadcopters for propaganda, reconnaissance, attack management, and delivering small payloads to hurt and intimidate Iraqi civilians. Now, jury-rigged drone tech is changing the war in Ukraine. These are examples of what I called 'Power to the People' in my last book.[16]

Today's potentially lethal technologies are emerging in an open technological revolution. Many have more in common with accessible, lethal open technologies like dynamite or improvised explosive devices than with twentieth-century military technologies like nuclear weapons, aircraft carriers and tanks. Functions that for centuries required a well-funded and well-trained army, such as reconnaissance and precision targeting, are now accessible to private actors and individuals; not always at the same level of competency, but good enough to kill and to have widespread political impact.

Of course, as is true of any era, the digital age has elements of old and new; fresh features that affect the future of war. The Second Industrial Revolution created a class of highly wealthy entrepreneurs such as Andrew Carnegie, John D. Rockefeller and Cornelius Vanderbilt; today we have robber barons like Elon Musk, Mark Zuckerberg and Jeff Bezos. Whilst the Second Industrial Revolution saw a sharp increase in the size of the middle class, now the middle class is being gutted. This resulting tension increases conflict. The nineteenth century's mega-millionaires enriched themselves by building monopolies on railroads, oil, steel, high explosives and electrical utilities—all businesses well suited to military innovation and adaptation. Today, venture capitalists' short-term leveraged financing has powered the digital revolution—a key reason as to why

MILITARY-TECHNOLOGICAL INNOVATION IN THE DIGITAL AGE

commercial innovation is leaving the US Department of Defense (DoD) behind.[17] Finally, unlike the Second Industrial Revolution, the use of data is critical to today's military-technological innovation. The power of data grows not with monopoly but with scale and scope; moreover, unlike steel or oil, most data resides in the open world.

Still, while there are novel aspects to military-technological innovation in the digital age, it differs from the Cold War's state-controlled processes. We must expand our intellectual scope with new concepts. In thinking about open systems of innovation driven by commercial processes, we get further toward grappling with today's challenges.[18]

Using an 'Offset Strategy' against China

A second and related blind spot caused by twentieth-century thinking is the United States' 'offset strategy,' now rebranded 'Offset X.'[19] To review, the first US technological offset was the use of nuclear weapons to counterbalance Soviet conventional predominance in the aftermath of the Second World War. Unable to match the strength and proximity of the Soviet army in Europe, the Eisenhower administration built more and better-quality nuclear weapons to 'offset' Soviet quantitative advantages. During the 1970s and 1980s, when the USSR achieved nuclear parity, the Second Offset Strategy used highly accurate, precision-guided munitions like laser-guided bombs, AWACS (Airborne Warning and Control Systems) and stealth bombers to counterbalance the Soviet Union's numerical advantages.

In 2014, then US Secretary of Defense Ash Carter introduced the Third Offset Strategy, whose creator and greatest advocate was Deputy Secretary of Defense Robert O. Work. The purpose of the third offset was to draw upon US technologies such as unmanned systems, cyber capabilities, machine learning and artificial intelligence (AI) to offset China's and Russia's growing strength. This historical narrative predominated until 2018, when its successor, 'Offset X,' was launched. Offset X places greater emphasis on human–machine collaboration and software, and is primarily oriented toward China. The effort to increase the role of unmanned systems and other digital technologies in the US military was politically and psychologically important, pushing the department to think about unmanned systems and human–machine interaction as an extension of the successful Cold War tradition of military-technological innovation.[20] But the framing is puzzling: what about China is the United States 'offsetting'?

China is working hard to get ahead of the United States in crucial areas of military-technological progress such as AI, quantum computing, hypersonic missiles and space assets.[21] Nowadays, the Chinese are the world's top high-tech manufacturer: in 2020, they produced 250 million computers, 25 million automobiles, and 1.5 billion smartphones, and they already lead in advanced semiconductor manufacturing and 5G technology.[22]

Working with Western allies and pouring money into technological initiatives in AI, robotics, quantum computing and biotechnology is urgent and appropriate for the West. But instead of using US technology to 'offset' Soviet conventional capabilities in Europe (as with nuclear weapons and then precision-guided weapons), today's effort demands that we aggressively upskill and compete in technological arenas where China, and to some degree Russia, are both doing very well. As Robert Work admitted in 2018, it seems the Chinese are now counterbalancing and offsetting the US by developing a unified national industrial policy and better-trained human capital.[23] As such, the appropriate response is to outpace their gains, rather than offset them.

Applying Counterproliferation Tools to General Purpose Technologies

The third blind spot is trying to stretch counterproliferation treaties and export controls to regulate the diffusion of digital technologies. States dominated technological innovation in the twentieth century, which meant military or scientific elites could limit the availability of advanced lethal technologies like nuclear, chemical or biological weapons. The concept of proliferation still applies to nuclear weapons; but it is outdated for most digitally enabled lethal technologies. Using biological weapons as an example, hard-to-handle agents such as smallpox or *Yersinia pestis* (which causes plague) were hidden

MILITARY-TECHNOLOGICAL INNOVATION IN THE DIGITAL AGE

away in secret biological weapons facilities throughout the twentieth century. Working in these places required high levels of expertise, a security clearance, and even knowing where to find such a facility. Experts spoke of the 'proliferation' of bioweapons and used phrases like 'dual use,' which meant they had two types of users: civilian and military.

This nomenclature and framing worked reasonably well during the twentieth century. Treaties such as the Biological Weapons Convention of 1975, moreover, established a norm against the military use of biological weapons. The 184 state parties to the convention met every five years, although there was no formal regime to verify compliance. Still, the convention was remarkably successful: with notable exceptions, such as the Soviet Union's extensive covert biological weapons program and those of Iraq, South Africa, Chile and what was then Rhodesia, the norm against developing biological weapons predominated throughout the twentieth century. The sophistication of the technology itself, its expense and its inaccessibility beyond a small community of experts meant that the threat of bioweapons came from states or, occasionally, rogue insiders.

Dramatic changes in biotechnology are changing that model in the twentieth century. These changes are driven by digitally enabled commercial processes outside traditional state purview, leading to dramatic changes in biology's software, hardware and expanding computing power. 'Dual use' no longer captures this phenomenon. Now, there are four kinds of users: professional scientists (as before); professional consumers or 'prosumers' (like biohackers); hobbyists (such as the makers movement); and regular consumers. As a result, the distinction between 'military' and 'civilian' has been blurred.

Meanwhile, the software of biology is rapidly changing and fundamentally altering the field. Synthetic biology, for example, aims to program cells as if you were programming individual computers, with synthetic biology companies building vast databases of new biological material. In these, the building block of life, DNA, is treated as if it were code for digital software. But while traditional software programming uses binary code—meaning zeros and ones—biology's code uses DNA's nitrogenous bases (adenine, thymine, guanine and cytosine); with some training and knowledge, that code is easily read, written, copied, edited and shared. Think of it as the invention of the printing press, but this time for biology. As a result, it has become easier to access DNA sequences from vast public databases, reproduce known pathogens like smallpox,[24] alter current viruses or bacteria, or use machine learning to create molecules with specified traits.[25] Should we still really call that 'proliferation'?

The hardware of biology is also evolving, allowing more people to create new or altered molecules. Just as computers evolved from mainframes to desktops, before moving to laptops and then smartphones, personal computers are enabling smaller, cheaper and more accessible personalized biological experimentation. Bioprinters use bio-ink and additive manufacturing to create layered arrangements of cells and support structures that could theoretically facilitate the production and delivery of biological weapons. Any military threat is probably at least a decade away, but bioprinters are becoming smaller and more affordable, following the trajectory of the personal computer.

What results is not sophisticated biology comparable with editing the human genome or designing a new biological agent from scratch. Again, unlike robotics or machine learning, this phenomenon is in its early stages. But prosumers and hobbyists can already sequence molecules, print them, and do a lot more than they used to be able to do. Mimicking the spread of armed drones and quadcopters, open technological advances in biology are lowering the threshold of access and use to incorporate broader numbers of people. The means and ends of biotechnology don't proliferate, they diffuse; yet most military analysts do not dig deeply enough into commercial technologies to predict how, why and where they are diffusing. After all, they are not 'military technologies.' That neglect may condemn us to facing surprise bioweapons on a future battlefield.

Believing Traditional Professional Military Education is Sufficient in the Digital Age

A fourth blind spot is in professional military education (PME) and training, whereby we fail to ensure that our military fighters and civilian policymakers fully understand today's technologies, or forego studying the classics in the history and theory of war.

PME was born in 1801 with the establishment of Scharnhorst's *Militärische Gesellschaft* (Military Society), which aimed to professionalize the Prussian army and thus beat Napoleon.[26] Carl von Clausewitz was Scharnhorst's most famous student, and particularly after the publication and English translation of Clausewitz's *On War*, this basic approach to military education became the model for other Western countries, especially in Europe and the United States. As the model evolved over the centuries, the professionalization of the military was a key reason for the stabilization of civil-military relations in democracies and the growing effectiveness of Western militaries in war.

Clausewitz does not deal with the role of technology at all: 'It is clear that weapons and equipment are not essential for the concept of fighting, since even wrestling is fighting of a kind,' he writes in Book Two.[27] Railways did not appear until a year after his death in 1831, and the flood of inventions that changed tactics on the battlefield (such as breech-loading rifles, machine guns, high explosives, and long-range heavy artillery) did not appear until the second half of the nineteenth century.[28] By not focusing on technology, however, *On War* is never outdated, and can be applied to many historical contexts. Clausewitz's timeless and brilliant analysis of the unchanging nature of war, especially the paradoxical trinity between primordial violence (usually associated with the people), chance (usually associated with the commander and his army), and reason (traditionally associated with the government), has been unsurpassed for 200 years. It helps us analyze belligerents on all sides and make better strategy.

Nevertheless, technology does play a vital role in various historical contexts. When technology is altering society, as it is now, it must be integrated into strategy. Military training in the United States and most allied countries emphasizes science and engineering at lower levels, particularly at the undergraduate level. At the middle and senior levels, however, where the integration of technology and strategy should be happening, and where both military and civilian policymakers can come together in the classroom, PME lacks serious study of technological developments, as in biotechnology, AI, robotics and quantum computing.

As a result, many people who lead efforts in military-technological innovation lack an in-depth understanding of the world-changing technologies they are deciding upon or deploying. This is obvious at senior levels of the military, where strategic planning and budgeting processes happen, and where DoD departmental acquisition and budgeting decisions are made. The US PME system is predicated on the model of senior officers (oftentimes retired officers) teaching new rising leaders alongside civilian academics in varying numbers and degrees of influence in different PME institutions. The standard canon of history and theory, including classics by Thucydides, Machiavelli, Sun Zi, Corbett and others, is essential and invaluable. But it's a question of 'both… and.' Senior officers often do not understand the technology that is vital to the future. It is not their fault, but studying new and emerging technologies should be a mandatory element of their professional education, and a prerequisite for promotion to middle or senior levels.[29]

As we move toward more cognitively dependent warfare, our warfighters' cognitive capabilities and knowledge are crucial to success. We have seen this coming for some time: the 2018 US National Defense Strategy bluntly stated, 'PME has stagnated, focused more on the accomplishment of mandatory credit at the expense of lethality and ingenuity.'[30] Educating more military members in new and emerging technologies at top civilian universities, where advances are happening, would bolster ingenuity and improve the situation.

In the United States, the lack of expertise is also obvious on Capitol Hill, where the knowledge and resources to build today's accessible, digitally enabled technologies—such as quadcopters, robotics, machine learning, natural language processing, or advanced human–computer interfaces—in an ethical and effective manner are not forthcoming. Unlike capital ships, tanks, jet fighters or even remote nuclear facilities, these systems do not generate the constituent jobs and dollars that attract US Congress members' attention and dedication. The same can also be said of most Members of Parliament and other elected representatives throughout the Western world.

On the other hand, many of those who fully understand and build digitally enabled technologies are not educated in strategy or history. How many computer scientists or engineers could grapple with the practical implications for war and peace of speeding up targeting decisions, expediting drone attacks, or facilitating autonomous swarming? The vast gulf between those who design and build technologies like nanosatellites, AI, robotics or drones, on the one hand, and those who set the requirements and how to deploy them (for example, robotic submarines, drone swarms, autonomous tanks or loitering munitions), on the other hand, is dangerous. We can build ethical and effective systems, but engineers and scientists must understand their inventions' historical context and predictable strategic implications earlier in the process. Currently, the conversation is either not happening or is swept up in the juggernaut of technological progress.

Lest we let ourselves off the hook, this also applies to civilian academics like me: how many international relations experts, historians or political scientists could hold a serious conversation about machine learning; what it is, what its types and uses are, what crucial areas of advancement are unfolding? We are quick to theorize without sufficient technical foundation. Civilian academics who study war need greater depth and technical sophistication in digital technologies, especially AI, if they are to guide their use. Just as with political scientists of the past, who could not make serious arguments about arms control, nuclear weapons, or chemical and biological threats without learning the science and technology behind them, we should neither deploy nor disparage machine learning, AI, autonomous swarms or microsatellites until we genuinely understand how they work and can hold a serious conversation with those who build them.

Seeing Military-Technological Innovation as Primarily Military

Fifth is the concept of 'military-technological innovation'—a quaint and outdated notion, given that most technological innovation now happens outside of the military. This is not a new observation and, to be fair, is more an 'area of complacency' than a true 'blind spot.' It is why Defense Secretary Ashton Carter set up the Defense Innovation Unit (DIU) in 2015, a small organization whose purpose was to help the military get access to cutting-edge commercial technology in areas such as AI, autonomous robotics and cyber systems, then to quickly field and scale them across the US military. In February 2022, the Pentagon published a list of 'Critical and Emerging Technologies'—things like quantum science, future-generation wireless technology, microelectronics, renewable energy generation and storage, and human–machine interfaces. Eleven of the fourteen items on the military's list were commercial technologies.[31]

Despite this, the mechanisms we have put in place during the past years to integrate civilian innovation of cutting-edge technologies into military organizations are inadequate.[32] They are too small, too slow, and do not alter the overall system. Why? Because in the United States, at least, they lack sufficient resources, a commitment from the DoD to change lumbering bureaucratic processes like the Pentagon's budgeting system, or adequate Congressional buy-in to both compel and allow the DoD to be more agile. According to Michael Brown, its former head, the DIU has transitioned thirty-five successfully prototyped commercial solutions to US Service partners, including the commercial Synthetic Aperture Radar satellites that can see through clouds and which predicted the Russian invasion of Ukraine earlier this year. Nevertheless, this is a very long way from a 'Sputnik moment.'[33]

Future efforts need to scale and move faster.[34] It takes about two and a half years to program a dollar of US spending, while the average development cycle of technologies like AI software or autonomous drones is less than two years.[35] The 1960s-era budgeting and oversight systems that structure how the US Pentagon spends taxpayer dollars are designed to be slow and deliberative so that they are spent responsibly—precisely the opposite of what is needed to keep pace with commercial innovation and diffusion today.

Meanwhile, the DoD lacks the rich venture capitalist ecosystem that has powered recent tech innovation, especially in the United States.[36] There's nothing in government that supports the kind of

MILITARY-TECHNOLOGICAL INNOVATION IN THE DIGITAL AGE

high-risk, personality-driven bets that venture capitalists make on promising startups. Venture capitalists get rich on outliers: they expect nine out of ten of their prospects to fail, and they plan for this. Any government employee who tried such high-risk gambling with taxpayer dollars would be prosecuted. Efforts like the DIU's National Security Innovation Capital or the National Security Innovation Network bring universities, venture firms and the DoD together. Even so, these collaborations are small and embryonic.

Because we are now in an open technological context, concepts like 'military–industrial complex' or even 'public–private partnership' fail to fully capture what is required to drive military innovation forward in the twenty-first century. What matters is the 'military–industrial–academic–private-investor–civil-society complex'. In most Western democracies, there are chasms of technical knowledge, experience, motivation and coordination between all these actors.

Meanwhile, as with biotechnology, decades of computer innovation are gradually reflected in evolving military technology patterns. Advanced military organizations are shifting from big, expensive hardware (like computer mainframes) to smaller and cheaper types of hardware (like PCs, laptops, tablets and phones), as well as to the crucial role of the software driving it all. Indeed, software engineers are as vital as soldiers: what matters is having an edge in developing the most advanced software, the best datasets, and the fastest processing speeds. As the Special Competitive Studies Project recently noted: 'A military's ability to deploy, employ, and update software, including AI models, faster than its adversaries is likely to become one of the greatest determining factors in relative military strength.'[37]

Seeing Civilian Technological Innovation as Separate from Conflict

The flip side—and our sixth blind spot—is that civilian technological innovation is not just 'civilian' but is deeply relevant to defense innovation and national security. Whether or not they want to be, major tech companies are deeply involved in national security, well beyond 'defense innovation.' They cannot remain on the sidelines; they are too deeply entrenched and too powerful. And yet major tech companies do not necessarily recognize or accept their role in influencing the outcome of future wars. When advanced countries are involved, try as they might, the big tech companies have been unable to remain neutral in war.

For a decade, we have studied the role of platforms like Facebook and X (formerly known as Twitter) through the lens of free speech, privacy, mobilization of terrorist violence, and recruitment to war zones (as with ISIS). In Ukraine, we are seeing how tech platforms respond to interstate war and how their responses affect how the war evolves.

Space X's Starlink—comprising a 'swarm' of thousands of low-orbit satellites—kept Ukrainians connected to the Internet two days after the invasion began, and its app was Ukraine's most downloaded app in March 2022. That connection has also been vital to the effective use of drones, allowing spotters, drone operators and an artillery team, for example, to stay connected.[38]

There would be no Ukraine right now if it were not for their ability to mobilize international support. Starlink kept President Zelensky online, even when the high-orbit ViaSat went down. There are some 20,000 Starlink user kits in Ukraine, most of which were donated by SpaceX or by private sources (though USAID may have paid for some 1,300 of them).[39] From the outset of the conflict, Starlink was uniquely able to help the defense, rapidly rewriting its code to preempt attacks. Amazon Web Services also sent Snowball devices to Krakow to facilitate transfer to the cloud and protect Ukraine's data.[40]

On the other side is the dramatic reshuffling of the tech industry in Russia, especially the removal of US tech companies and media from Russian territory. Tech companies have fought back, though. Meta is blocking Russia Today and Sputnik on Facebook and Instagram, and YouTube blocks them on its platform. Moreover, Meta, Microsoft and Google are preventing Russian ads from appearing on their sites, and Microsoft has suspended all new sales of products and services in the country.

By the same token, Russia has blocked Twitter, Facebook and Instagram, has labeled Meta an 'extremist organization,' and is cultivating substitute apps and platforms; for example, the local

MILITARY-TECHNOLOGICAL INNOVATION IN THE DIGITAL AGE

search engine Yandex substitutes for Google, and VKontakte substitutes for Facebook. Putin is fracturing the Internet to keep the Russians in a media bubble that supports the war, while Western tech companies continue their support for Ukraine.

For good or ill, the future of war is emerging not just from heavy artillery and urban combat on the streets of Lyman and Kharkiv, but also from commercial tech companies worldwide.

Replacing 'Adaptation under Fire' with 'Adaptation before Fire'

Finally, adaptation before fire should replace the concept of adaptation under fire. Western democracies are arguably already 'under fire.' In a contest with a technologically advanced adversary, battlefield adaptation under fire cannot succeed without a sophisticated and resilient civilian infrastructure and military organizational change in place before major conflict happens. Adjusting to the enemy, repurposing weapons, changing tactics, altering procedures, engaging in bottom-up or top-down innovation, and battlefield leadership will always be crucial in war.[41] Yet these military adjustments cannot compensate for being slow in recognizing the strategic wave of digital technological change underway, which will fundamentally alter societies and therefore warfare. The kind of slow ramp-up that characterized the Third Industrial Revolution is not possible now. When the next major war breaks out, adaptation to the technological context will be too late.

Unlike a counterinsurgency in a war against a peer or near-peer, the sophistication of technological infrastructure and tech education will be more important than the proximate tactical adjustments of ground forces. Training the population, protecting vital infrastructure, scaling individual innovations, building economic resilience, updating software and quickly adapting off-the-shelf technologies are critical to success or failure in future war, potentially more strategically impactful than what land forces can do during it.

For example, the war in Ukraine demonstrates the importance of having a robust domestic drone industry and a tech-savvy civilian population. Before the February 2022 Russian invasion, Ukraine was

197

a hub of scientific and technological talent: it established a Ministry of Digital Transformation in 2019 and had some 1,650 software companies, with major hubs in Kyiv, Kharkiv, Lviv, Dnipro and Odesa.[42] One of the most downloaded apps in Ukraine (used by more than 4 million people and regularly updated) is called Air Alert, which provides early warning of impending attacks. Working with government emergency services and civil defense, app developers at a company called Ajax Systems built the Air Alert in a single night. Since February 2022, it has saved countless civilian and military lives. Moreover, tech companies like Google, Oracle, Snap and Amazon's Ring long outsourced to Ukraine, as there was a thriving economy of well-trained subcontractors based there.[43] As such, Ukrainian tech workers—and their government—were poised to pivot.

The Ukrainians' tech capacity developed quickly after the 2014 Russian capture of Crimea. Today, the bulk of the Ukrainian drone fleet is off-the-shelf, altered or handmade inexpensively in workshops around Ukraine, including small quadcopters and mid-sized fixed-wing drones. For example, the fixed-wing Punisher, which is reusable, is made in Ukraine by UA Dynamics. Ukrainian hackers demonstrated how to infiltrate the Internet of Things during war, taking down electric vehicle charging stations in Moscow, then projecting an anti-Putin slur on their screens.[44] They have even used AI-enabled voice transcription to scale Russian communications monitoring,[45] and effectively employed Clearview AI facial recognition technology and machine learning for target identification.

Russian troops, on the other hand, have good military kit (including indigenous small- and medium-sized military drones) but insufficient training and doctrine, and there is no comparable flow of consumer-style commercial drones in the country. Nevertheless, they are adapting. For example, they have opened a new facility in occupied Donetsk to train Russian military and civilian drone operators, and acquired hundreds of Iranian kamikaze drones (notably the Shahed 136 loitering munitions) for use in Southern and Eastern Ukraine.[46]

The argument is not that military leadership is unimportant or that military forces can succeed without being well led and tactically well trained. The US military, for example, was very adaptable in

MILITARY-TECHNOLOGICAL INNOVATION IN THE DIGITAL AGE

both of their recent wars. Nor am I downplaying the vital role of traditional military weapons assistance like high-mobility artillery rocket systems, Howitzers, Switchblades, or Czech and Polish tanks. Our problem is at the strategic level. Due to rapid technological change, the conditions for victory or defeat are being set now, before a future war breaks out.

Or perhaps the key point is that war has already broken out at a lower level, and we have not sufficiently educated, prepared and mobilized the population to respond to it. Indeed, quite the opposite: the population has become a crucial vulnerability to be manipulated by nefarious actors, both foreign and domestic. I doubt our Western military leaders, especially at the mid to senior levels, are any better prepared. The actors adapting under fire today are the commercial tech companies who can update software overnight rather than over several months, as is typical in the DoD.[47]

Conclusion

These seven blind spots barely scratch the surface of the changes underway in digital technological innovation affecting the future of war. Some are new developments, such as the willingness now of certain major tech companies to essentially be combatants in Ukraine. Some are formidable cultural and organizational problems, such as the lack of technological expertise among Western military decision-makers who are building doctrine. Having spent years of my life trying, I am not going to give you glib immediate policy solutions like 'fix the PME system.'

But here are a few things we can do immediately. First, academics need to work more actively across the disciplinary chasm between engineers and computer scientists on the one hand, and historians and social scientists on the other. Second, we need to train military and civilian leaders to understand digital technologies well enough to make foresightful strategic decisions. And third, academics and policymakers need to get more serious about understanding new and emerging technologies in depth, so as to analyze more seriously their strategic effects on war. This is not just a battle over resources (as it manifests in the military) or a question of Western hard- or

199

a soft-liners against China (as it manifests in US politics), or even a struggle between the ideologically pro-tech or anti-tech (as it manifests in Western newspapers). It is bigger than that: it is a question of whether we can build a Western national security model now that has the support of a well-trained, well-informed, resilient population and is defensible in the next war.

10

THE RISING DOMINANCE OF THE TACTICAL DEFENSE?

T.X. Hammes

It has become widely accepted that the convergence of technological advances is at the forefront of a revolution in military affairs, or perhaps even a military revolution.[1] One of the unanswered questions concerning this shift is whether it will lead to continued dominance by the offense or a period of defensive dominance.

Firstly, it is essential to understand the difference between dominance and a temporary advantage. Dominance provides a major advantage that can be pursued throughout the conflict. More specifically, defense dominance means an attacker cannot seize enemy territory or must pay an exorbitant price to do so. The Western Front in the First World War is a prime example of defense dominance.

It is essential that leaders understand the current tactical balance between offense and defense, as failure to do so has often led leaders to attack confidently in the belief that they can win the war quickly, only to be bogged down in a long, brutal conflict. The US Civil War, and the First and Second World Wars are examples of this hazard.

Investing in the wrong side of the offense/defense competition is a rich nation's game that even the United States may no longer be able to afford. Against peer competition at scale, it can lead to strategic defeat. In fact, the answer to this question should guide force development and posture, and therefore must be a part of the national security discussion.

This chapter is focused on the tactical level of conventional conflicts between states employing modern systems. The tactical level represents the actual execution of military operations— whether in the form of direct contact or employing long-range firepower such as misnamed 'strategic bombing.' The operational level of war ties those tactical actions into a coherent campaign to achieve the strategic goals of the combatants. Even in periods of defense dominance at the tactical level, planners should strive to be on the offensive at both the operational and strategic levels.

The ongoing Russo-Ukrainian conflict demonstrates the impact some new technologies and concepts are having in state-on-state warfare. However, it is important to understand that over 90 per cent of Russian and Ukrainian fielded forces are armed, trained and equipped with twentieth-century technology. Thus, it is not surprising this conflict greatly resembles the large conventional conflicts of the twentieth century. Also of critical importance are the extreme frontages assigned to combat units. A Ukrainian infantry company is assigned a frontage of 3 kilometers, which is about three times the doctrinal frontage for a US company. This creates significant gaps in defensive positions.

In short, applying the lessons of Ukraine to combat across the board is not appropriate. Nevertheless, the small percentage of each force that is employing twenty-first century technology (in the form of long-range precision fire, drones, electronic warfare, and digital command-and-control systems) provides some interesting indicators of where warfare is headed. Overall, it can be said that Ukrainian leadership has embraced twenty-first century concepts whereas Russian leaders remain firmly rooted in the twentieth century.

It is also important to understand that emerging technologies will create very different dynamics in unconventional conflicts where one side is a non-state actor. Even so, this impact on unconventional warfare is so different that it warrants being the topic of a different paper.

THE RISING DOMINANCE OF THE TACTICAL DEFENSE?

To examine emerging technologies' impact on conventional conflicts, this chapter will provide a couple of historical examples of the shift between offense and defense dominance at the tactical level. It will then examine how the offense–defense balance is shifting in each of six warfighting domains: namely land, sea, air, space, cyber, and electromagnetic. Next, it examines how interactions between the domains can further reinforce the defense, and finally what the shift to defense dominance means for the West.

The Shifting Balance in History

History records a constantly shifting balance between offense and defense, driven by a combination of political, social, economic and political changes. Despite Westerners' love for technology, it alone cannot drive major shifts. For instance, defense was dominant during much of the medieval period because of the cost and difficulty of reducing a castle. This was based not just on the technology of building a castle, but the political, social and economic structures necessary to do so. Offense was not restored until a wide range of social, political, technological and military changes necessary to develop military establishments capable of rapidly reducing castles occurred. While cannons provided a key technology, the society first had to develop the political, social and economic systems to produce and sustain them.

A much later major shift of advantage to the defense was driven by the development of rifled muskets and cannon, the mass production of these weapons, the tactical adaptation of field fortifications, mobilization of mass manpower, economies that could pay for them, and governments that could marshal those resources. The combination of these factors led to defense dominating the tactical battlefield from the late American Civil War (1861–5) until near the end of the First World War.[2] Governments could field and arm forces that combined tactics and technology, which meant any attacking unit could be quickly observed and taken under effective fire. The opposing armies, meanwhile, were forced to go to ground in massive trench systems, which could be held even against numerically superior attacking forces. Failure of military leaders to

203

recognize these changes, despite the lessons of Crimea, the Boer War, and the Russo-Japanese War, led to repeated, bloody and futile attempts to cross the Western Front's 'no man's lands.'

Offense could remain dominant only in those theaters such as Russia and the Middle East, where sufficient space has existed to allow forces to move outside the range of fixed enemy positions. It was not until the combatants applied new concepts and tactics to technology emerging from the Second Industrial Revolution—first lightweight machine guns and mortars, then armor and aircraft—that movement was restored to Western Front battlefields. Once again, political, social and economic systems had to evolve in parallel to produce the factories, the skilled engineers and operators, the financial backbone, and the will to conduct global mechanized warfare. Since then, the offense has generally dominated tactically in conventional conflicts.

Historically, the balance between offense and defense has been a sliding scale. Today, two trends are setting conditions for renewed defense dominance: pervasive surveillance and mass with precision. The balance will shift to the defense as more states adopt these two key technologies and learn to apply them tactically. While these advantages are available today, the rate at which each nation adopts them will be driven by their unique political, economic and social structures.

Pervasive surveillance is a direct result of the massive expansion of commercial space assets and drones. Today, Planet Lab offers high-resolution, multispectral, interpreted satellite imagery of the entire planet every twenty-four hours, with a resolution of 50 cm (on request).[3] More than a dozen other companies are competing in this market to provide increasingly fast and more precise multispectral products to commercial customers.

To see through darkness, cloud cover and light foliage, San Francisco-based Capella Space offers on-demand, interpreted 50-cm resolution Synthetic Aperture Radar (SAR) imagery of any location on the planet every three hours.[4] By mid-2023, a Hawkeye 360 satellite formation will be able to locate a specific radio frequency signature within an area of 3 square kilometers every twenty minutes.[5]

THE RISING DOMINANCE OF THE TACTICAL DEFENSE?

These are just a few of the dozens of commercial firms focused on space surveillance. At the end of 2021, the Union of Concerned Scientists reported 4,852 active satellites in orbit, with 1,030 of these devoted to earth observation. While communication satellites are the fastest-growing segment, the number of earth-observation satellites in orbit increased by almost 14 per cent in 2021.[6] This growth is market-driven, with 2,600 earth-observation satellites projected by 2030, representing a market worth over US$76 billion.[7]

In effect, the planet is being imaged by space-based visual, infrared, radar and electromagnetic sensors virtually continuously. As noted, the companies target the commercial market, so they emphasize customer-oriented, automated processes from request to delivery of an interpreted product. While they can provide raw imagery or data, the companies have also developed change-detection software to facilitate providing finished intelligence products. By early March 2022, five commercial firms were sharing 24-hour satellite imagery that assisted Ukraine in tracking Russian forces.[8] By December, Ukraine could tap into the 'roughly 40 commercial satellites a day [that] pass over the area in a 24-hour period.'[9] By 2030, 'it will be possible to request an image, communicate with the satellite and receive the product in minutes.'[10]

In addition to what is rapidly becoming pervasive space surveillance, both commercial firms and governments are developing families of ever more capable surveillance drones. These drones range from small, hand-launched platforms to aircraft with wingspans of over 100 feet. Like their satellite competitors, these companies have also developed multispectral, SAR and electromagnetic spectrum (EMS) sensors for every size of drone.

At the small end, hundreds of videos from Ukraine show commercial quad and octocopters dropping modified anti-armor weapons and grenades on Russian forces.[11] Ukraine has as many as '500 drones in the air for a relatively standard military operation,'[12] but their most important function has been to provide critical intelligence for tactical units in combat.

For higher-level commanders, there is a growing family of both commercial and military drones that have exceptional to nearly unbelievable endurance. The US Navy and Marine Corps are fielding

205

the V-BAT 128 surveillance drone. Vertical take-off and landing (VTOL) capable and with an endurance of up to 11 hours, the V-BAT's onboard communications package can maintain contact with the ground station out to 50 miles, and can extend that range by using other V-BATs as communication relays.[13]

For long-endurance missions, the Aurora Orion unmanned aircraft system can stay aloft for 120 hours. Its payload systems include multispectral full-motion video, an electro-optic/infrared sensor, signals intelligence, hyperspectral video and imaging, and wide-area airborne surveillance. It is also fitted with maritime domain-awareness radars and foliage-penetrating radars. It has a maximum range of 15,000 miles, and the mission radius is 4,000 miles with 24 hours on-station.[14]

While the drones noted above are for military use, the commercial drone sector is making rapid progress on its own. Drone package delivery is being aggressively pursued by major corporations, which has set off intensive competition among drone producers. The requirements for delivery are essentially the same as for military platforms—VTOL, global positioning system (GPS) independent (allowing them to fly into urban canyons), electronically hardened (to fly near airfields and radio towers), reliable, inexpensive, and with ever-increasing endurance and payloads. Commercial surveillance drones also make use of multispectral imagery (crop surveillance, pipeline leakage), high-resolution imagery (wind generator inspection), and even SAR. Thus, the payloads for the very wide variety of commercial surveillance missions mean that an increasing family of long-endurance, commercial surveillance drones can provide affordable intelligence, surveillance and reconnaissance assets for even small states.

Further changing the environment are advanced manufacturing techniques. Automated factories, robotics and artificial intelligence (AI) can be combined to dramatically reduce the cost of these emerging autonomous systems. In 2014, an aeronautics professor designed and 3D-printed a drone. By adding a small electric motor, two batteries and a cell phone, he created a hand-launched, autonomous drone with a range of 50 kilometers. Once the design was refined, the production process took about 31 hours.[15] Today's 3D printers are over 100 times faster. A plant with 100 modern 3D printers could produce 10,000 of these drones per day. By incorporating AI and robotics, the final assembly could be automated. Thus, dumb (i.e. non-coordinating) drone swarms of thousands of autonomous hunters are possible.

Second, recent conflicts have demonstrated the power of autonomous and semi-autonomous drones and missiles. The expensive 1980s technologies that enabled precision weapons are now cheap and widely available, and advanced manufacturing can mass-produce them. In short, mass is returning to the battlespace but this time with precision. Several of these systems are being built into standard shipping containers that can be easily transported and which blend into the commercial traffic globally. Many of them have much greater operational range than current fighter/bomber aircraft and do not require airfields. Moreover, the sheer number of commercial containers makes pre-emptive strikes virtually impossible.

The Russo-Ukrainian conflict is demonstrating the value of drones, as well as the need for very large numbers. The Royal United Services Institute (RUSI) reports that quadcopters survive an average of only three missions and fixed-wing drones survive for only six missions.[16] Thus, very large numbers of drones are being employed by both sides. However, only Ukraine has developed a digital command-and-control system that allows it to integrate intelligence provided by satellites, drones, military and civilian systems to provide a real-time picture of the battlespace to its commanders. The commanders use this system to target Russian forces. General Mark Milley, chairman of the US Joint Chiefs of Staff, has stated 'We are witnessing the ways wars will be fought, and won, for years to come.'[17]

Systems like this will enable commanders to coordinate attacks across all domains to include long-range precision attacks and swarms of autonomous hunters that will seek out their prey informed by many sources and sensors. These co-evolving concepts, tactics, and commercial and military technologies are creating a battlespace in which movement becomes extremely dangerous. If a unit moves, it will create a signal, and this can be attacked at much greater ranges than in the past. At the same time, cyber, space and electromagnetic domains will provide both reinforcement for, and increasingly powerful alternatives to, kinetic attacks.

Whether this convergence leads to offense or defense dominance is a complex question. In fact, the sheer complexity of interaction among the six domains—land, sea, air, space, cyber and electromagnetic—requires we consider the impact on each domain before we try to understand the overall impact on the character of war. (I have assigned EMS as a domain because although it is not yet considered one in US doctrine, both China and Russia are dedicating great resources to dominating this domain.)

Land Domain

Since the last year of the First World War, the offense has dominated conventional ground combat. Meanwhile, irregular warfare has followed its own pattern. However, emerging technologies are driving conventional land warfare back to defensive dominance.

Since new systems allow units to remain passive and yet see the battlefield clearly, the defense will have a distinct advantage. Electro-optical and electronic warfare sensors can provide a great deal of information that, combined with external sensors such as satellites and drones, can allow the defenders to visualize the battlefield without revealing their own positions.

Like no man's land in the First World War, the battlespace will be under constant surveillance. Anything revealing itself will be subject to immediate, potentially massed precision fire. By blending in or remaining stationary, defenders can observe passively and choose when to attack, then do so with swarms of autonomous missiles and drones from containerized, mobile systems that are nearly invulnerable to pre-emptive attack. In brief, defenders can see first, then shoot first, often and accurately. In this modern 'hider–finder competition', finders will have the huge advantage of global surveillance in all weather conditions.

Defenders will enjoy the further advantage of fighting from very widely dispersed prepared positions. While most current systems must be manned to be operated, autonomous and remote-control systems are being developed worldwide. As these systems mature, defenders can be located at a distance from their weapons and thus not be at risk even after firing. Events in Ukraine have demonstrated that ground forces will be subject to pervasive surveillance and subsequent attack by the emerging families of drones. It is not unreasonable to expect a defender to be able to launch hundreds or even thousands of loitering munitions against each brigade-size attack.

In contrast, attackers will have to move if they intend to execute anything but strike missions against the defender. The very act of moving into assembly positions will create a signature. Attackers will retain the traditional advantage of selecting the time and place of attack, but as the Russians have learned, the ability to move to assembly areas without being observed will disappear. Furthermore, the advantage of physically massing either offensive or defensive forces is declining as weapon ranges increase dramatically. Defensive mass can be achieved by massing long-range firepower rather than massing forces.

However, the Nagorno-Karabakh War and the ongoing Russo-Ukrainian War have demonstrated that the offense remains relevant if the attacker adopts modern concepts and weapons while the defender relies on twentieth-century weapons and concepts.

Sea Domain

The adage, attributed to Admiral Horatio Nelson, that 'a ship's a fool to fight a fort' remains true. In today's conflict environment, land-based anti-ship systems can attack naval surface targets from a greater distance than ever before. In addition, land- and air-launched ballistic and cruise missile systems, drones and attack aircraft, cued by ubiquitous surveillance systems, have the enormous advantage of hiding in the cluttered land environment. In contrast, their surface-based targets must operate in much more open environments. Land-based systems also have the advantage of both range and magazine depth, and can be integrated with mines, and surface and subsurface naval drones. Despite owning a very limited number of cruise missiles and small, surface drones, Ukraine's successful attacks have forced the Russian surface fleet to remain well away from Ukraine.

BEYOND UKRAINE

Extended-range land- and air-launched cruise missiles mean many naval fights will include land-based participants. As Captain Wayne Hughes, from the US Navy, demonstrated in his work, the first fleet to conduct successful pulse attacks against an opposing fleet gains a major advantage.[18] Land-based systems can provide more missiles at less cost for each pulse attack. The advantage increases because land-based systems can be very widely distributed and thus much more difficult to target than sea-based systems. What's more, mobile land-based systems are both much more survivable and rapidly evolving.

Geography as well as oceanography can enhance the power of land-based systems. Maritime chokepoints have been major factors in conflicts between major powers since the Peloponnesian War (431–04 BCE). Even today, control of straits like Hormuz or Malacca can allow a power to determine what resources flow to an opponent. In these confined waters, land-based defenses gain an even greater advantage by employing many less expensive, shorter-range anti-ship systems and smart sea mines (essentially tethered torpedoes). However, as fights move further from shore, the number of land-based systems that can range the fight decreases. At some point, the tactical advantage at sea will shift back to the offense.

In the subsurface domain, the fight will favor the offense in the deep ocean, but the defense in the vicinity of chokepoints, as emerging technologies make shallow water more transparent than ever. Fixed sensor arrays can cover key passages between open seas, while rapid advances in autonomous submarine drones will thicken the sensor nets in restricted waters as well as enable swarms of weapons to be launched against infiltrating submarines. In short, emerging technologies are making waters both more transparent and more congested.

Mining of enemy ports may well be the most effective and viable offensive naval action simply because autonomous drones with small signatures will be able to penetrate enemy defenses to lay mines. Smart mines can be programmed to attack specific classes of ships, thus giving the miner an ability to select targets for best effect without having to maintain forces in the vicinity of the port.

THE RISING DOMINANCE OF THE TACTICAL DEFENSE?

Air Domain

A very interesting aspect of the Russo-Ukrainian War has been the inability of Russia to establish air superiority over the battlespace. While Russia inflicted major damage on Ukrainian fixed radar and air defense sites, it has been unable to neutralize Ukraine's mobile air defense systems. Despite Russia's technological and numerical superiority in the air, Ukraine has created a denied airspace over the territory it controls. As such, Russia has been reduced to launching strikes with drones and air- or ground-launched long-range missiles. Ukraine has even been able to intercept most of these weapons despite its limited air defense network. Senior US Air Force planners have admitted they will not be able to establish air superiority but can only strive for temporary superiority in a defined area.[19]

Compounding the problem of achieving air superiority, the current generation of manned aircraft all have easily identifiable (and thus targetable) major operating facilities. Russia, and to a larger extent China, have the capability to strike fixed facilities with drones, as well as cruise and ballistic missiles. In their report on the air war in Ukraine, RUSI analysts concluded that:

> Western fighters supplied in the short-medium term need to be capable of dispersed operations using mobile maintenance equipment and small support teams, and flying from relatively rough runways, to avoid being neutralised by Russian long-range missile strikes.[20]

While some Western air forces retain this capability, most do not. With increasing numbers of missiles outranging most manned aircraft, winning the air battle with the West's current inventory of aircraft will really be about the ability to sustain the critical logistics provided by fixed air bases.

This threat will not be limited to in-theater airbases, though. The advent of containerized, long-range cruise missiles and drones deployed from a wide variety of shipping means that bases almost anywhere in the world can be struck. Thus, a key question is whether joint forces can defend their base facilities against swarms of missiles and drones. Ukraine has been increasingly successful

211

in defeating Russian drones and cruise missiles; however, the cost has been high. While the Iranian-supplied Shahed 136 may cost as little as US$20,000, a NASAMS short-to-medium-range missile costs US$500,000.[21] And while air defense artillery systems like the German Gepard tank can provide much lower costs, their limited range dramatically increases the number of the expensive systems required to defend a defined space.

To affordably defend its forces, the United States is betting heavily on directed energy weapons (that is, lasers and electromagnetic pulses [EMPs]) to defeat swarm attacks. These systems still face numerous challenges, but they have promise. Israel, for instance, is actively developing a laser system that will initially augment its Iron Dome and later replace it as Israel's defense against Palestinian indirect fire attacks.[22]

If emerging laser and EMP systems prove effective, they will further empower the defense. If they protect air bases from drones and missiles, they can certainly engage manned aircraft attacking friendly forces. When they are deployed, directed energy weapons will provide significant advantage to the defense for two reasons. First, they require large power systems to operate. Attackers must bring those power systems with them, and thus the power available is limited by the ability to lift it by air, sea or road. In contrast, defenders can either tap directly into the national power grid for virtually unlimited power or use as many generators as they feel they need. Second, the defender has the enormous advantage of blending into the cluttered ground environment. The actual systems are relatively small and can thus be camouflaged as air conditioning units on top of buildings or inside small buildings in the countryside. Again, the attacker must move towards the defended area, thereby generating signals, while the defenders need not generate a signal until they choose to engage. As directed energy weapons become operational, they will increase the advantage the defense holds over the offense in the air domain.

Both emerging technology and current events indicate that airspace denial, rather than airspace superiority, is becoming the norm.

THE RISING DOMINANCE OF THE TACTICAL DEFENSE?

Space Domain

Conventional wisdom has stated for years that war in space will be offense-dominated. This concept was based on the fact that anti-satellite systems were cheaper than satellites. An attacker could quickly destroy an enemy's key satellites, and it would take months if not years to replace these large, very expensive assets. Given the global economy's heavy dependence on space services, this was a truly alarming situation.

However, rapid developments in space launch and satellite miniaturization are changing that situation. The exponential increase in the number of satellites in orbit, the disaggregation of functions into many platforms, and the increasing ability to rapidly replace satellites in orbit mean that the defense may now have the advantage. Disaggregating functions such as gathering intelligence and providing communications links means that the attacker must engage many more targets to degrade space systems. In addition, the difficulty of acquiring these small targets and their ability to maneuver to prevent interception increases the advantages accruing to the defense. Finally, several major companies are developing rapid launch capabilities and private firms are developing high-altitude drones as potential replacements.

A major vulnerability remains the positioning-navigating-timing information provided by the GPS constellation. Timing is central to the functioning of a wide range of critical civilian systems, including banking, communications, retail sales, and uncounted other applications, all of which rely on precision timing. As such, a systematic attack on the GPS network would cause massive disruption to the global economy. The key question is whether these critical functions can be quickly replaced by other systems in the event of an attack. Logistics sustainment requires it. Fortunately, both civilian and government organizations are developing alternatives to the way GPS functions. Yet, until this critical function can be immediately replaced, offensive action can provide a window of opportunity to an attacker. Nonetheless, the benefits of such an attack are likely to be fleeting and will almost certainly trigger a reply in kind. In sum, space will become an arena of ongoing conflict with the advantage to the defense.

Cyber Domain

In 2019, US Secretary of Defense Mark T. Esper noted that winning in cyberspace requires offense. This continued the theme established in 2012 when Secretary Leon E. Panetta warned of a 'cyber Pearl Harbor.'[23] Despite these claims, there is a growing pushback against the idea that cyber is inherently offense-dominated.

In their 2018 book, Valeriano, Jensen and Maness noted that cyber-offensive operations consist of espionage, disruption (temporarily reducing the capacity of an opponent's system) and degradation (i.e. damaging elements of the system). In contrast to the two Secretaries above, however, these authors do not see offense as dominant in this domain.[24] Other scholars, including former-cyber operators, agree with them. They see offense dominance as being overstated and costly:

> … breaking into a particular network may be cheap after the tools and infrastructure are in place … [but] building and maintaining the infrastructure for a program of sustained operations requires targeting, research, hardware engineering, software development, and training. This is not cheap.[25]

As you can see, we have two camps of very well-informed experts with contradictory views on the value of cyber as an offensive weapon. This is consistent with the historical pattern of new technologies, as it is not really known what the impact of an emerging technology will be until it is employed in open conflict. Thus, despite persistently advocating for cyber as a form of 'forward' defending (which is essentially offensive), the US Cyber Command Vision states 'Cyberspace is an active and contested operational space in which superiority is always at risk.'[26]

So how should we evaluate cyber as a weapon? Firstly, it can clearly be used in terms of cyber espionage and theft. It has allowed China, North Korea, Iran, Russia and numerous criminal organizations to steal personal information, intellectual property and money on a scale never seen before. Cyber disruption, meanwhile, has had only limited success, from the Love Bug virus to the NotPetya malware. A significant number of these attacks have disrupted the targeted

THE RISING DOMINANCE OF THE TACTICAL DEFENSE?

systems for a period ranging from hours to weeks, with NotPetya causing significant damage to numerous organizations that were not the target of its attack but were simply collateral damage. In practice, Russia's cyber disruption attacks, such as the ViaSat attack, had limited success, with most observers believing their cyber operations underperformed expectations.[27]

Destructive attacks have also had limited success, the most famous being the Stuxnet attack on the Iranian centrifuges in 2010, which was attributed to the United States and Israel. This attack reportedly damaged about 20 per cent of the centrifuges, yet the International Atomic Energy Agency reported that Iranian production increased during the period, perhaps in response to the attack.[28] Increasing the uncertainty about the offense–defense balance in cyber, there have been other operations, like SolarWinds/Holiday Bear, that have achieved widespread penetration of computer networks but whose objectives remain unclear. Cyber operations are inherently difficult to coordinate with real-time attacks and, to date, have not reliably produced the desired effects.

There are two other major anti-cyber options that have not been used yet that require much deeper study. The first of these is kinetic weapons: these can damage the well-mapped networks of fiber-optic cables, switches, downlink stations and processing centers essential to an information network. The increasing availability of long-range, autonomous, precision weapons means that cross-domain attacks from air, sea and land platforms will be an integral part of counter-cyber operations. The potential to hit hundreds of key nodes either in theater or globally is growing.

The second option—EMP attacks—will be dealt with in the section on the electromagnetic domain.

In sum, the cyber domain will be continually contested. The fact the Internet was initially designed to work even when under major attack will mitigate the impact of kinetic attacks, but they will still cause significant disruptions. Fortunately, the Internet is a complex adaptive system and thus will show remarkable resilience when under attack.

EMS Domain

In January 2021, General John Hyten, then vice-chairman of the Joint Chiefs of Staff, stated 'We have to be able to effectively fight and win the electromagnetic spectrum fight right from the beginning—that is, electronic warfare in every domain.'[29] Given the increasing value of robust communications networks, as highlighted by Ukraine's successful use of their domestically developed Delta system, the ability to use the EMS (or simply deny your opponents its use) will be critical to success.

Further highlighting the central nature of the electromagnetic domain, one of the most effective counters to the unmanned aircraft system (UAS)/precision-fire team in Ukraine has been electronic warfare. By breaking the link between the UAS and the fire direction centers, electronic warfare requires the UAS to return to base to download targeting information; whilst they are doing so, they cannot adjust or return fire.

Once again, land-based defenders may well have an advantage in this domain. Fiber-optic communications systems, for instance, can avoid the electromagnetic domain, and have access to the national power grid to provide effectively unlimited power for jammers.

Potential game-changers in the EMS are EMP weapons, which represent a major threat from the tactical to the strategic level. At the tactical level, the United States has demonstrated a drone that can create an EMP directed at specific targets. Since it is delivered by a drone, this type of attack is really a cross-domain attack but, like kinetic attacks, must be considered as part of any EMP domain offense–defense balance.

A defending unit can do more to harden its electronics against this kind of attack than an attacker can; for example, defenders can disconnect a percentage of its systems to protect themselves. However, EMP weapons can overturn the defender's advantage if the defender has not exploited the inherent advantages of defense. We know these attacks can cause major damage to unprotected electronics, which is a particular issue given most basic systems today have embedded electronics. The attacker has one major advantage though: it can employ its EMP weapon before any of its own systems are within range of the pulse. However, if they cannot prevent a response in kind, the attacker loses the advantage when a retaliatory strike hits its forces.

For both offense and defense, building resilient, redundant systems can reduce the damage caused by tactical EMP weapons, but will be costly and require massive retrofits for existing weapons. Moreover, the miniaturization necessary for air and naval systems will make them significantly more expensive.

At the strategic level, a nuclear-generated high-altitude EMP could seriously damage the national infrastructure for a period of months. The fact this currently requires a nuclear device to be detonated over the target area means it has to be discussed as part of nuclear deterrence/warfare, which is beyond the scope of this chapter.

A Caution

As always, perception is reality. Unfortunately, the perception that cyber and space are offense-dominated is inherently escalatory. If political leaders believe they can achieve decisive dominance in these domains only by attacking first, crisis management becomes much more difficult. Therefore, it is critical to counter the idea that going first in cyber, space or the EMS provides unrecoverable advantages. This is not only necessary to prevent aggression, but also to prevent escalation on the friendly side.

Interaction between Domains

As noted in US doctrine, understanding the relative strengths of the offense and defense in the various domains is essential to those engaged in joint warfare. It is also important to note that temporary advantage in one domain may also allow a much more powerful attack from another domain. An obvious example is a temporary advantage in the EMS that neutralizes air defense thus allowing a much more destructive attack from the air into other domains.

Similarly, while degradation or destruction has proved to be a difficult challenge within the cyber domain, the use of precision weapons delivered from air, sea, land or space can have a devastating effect on the cyber capabilities of an opponent. Even unclassified sources provide maps of critical nodes and links, including downlinks, fiber optics, and terrestrial switches.

The increasing range and number of autonomous precision-attack systems will steadily improve the ability of the sea, air and land domains to conduct effective cross-domain attacks. In particular, ground-based forces have the advantages of operating in very complex terrain (whether rural or urban), and have access to deep magazines and national power grids. The increasing ranges of ground force weapons will allow defenses to reach out much farther to target air, land and sea forces, as well as critical infrastructure for space and cyber forces.

All-domain offensive operations are incredibly complex, not least because each domain operates on different execution timelines. Major land and naval operations take weeks to years to execute. For example, it can take weeks to position forces for air operations, but they can be executed in hours, and their campaigns last days or even weeks. Conversely, cyber, space and electronic warfare operations can require weeks to years to put their forces in place, but can measure execution in as little as microseconds. Thus, coordinating the offensive operations of the separate domains is particularly challenging. Given this, cross-domain attacks may be the most effective means of offense in future wars. Director of the US Space Development Agency, Derek Tournear, has stated that cyber is a greater threat to satellites than missiles.[30] Air forces have stated for years that the most effective way to defeat an air force is to destroy it bases and its aircraft on the ground. Today, ground-based forces can do so from beyond the range of most aircraft-delivered weapons. Naval forces have historically been able to appear suddenly out of the vast expanses of the oceans, but are now increasingly being tracked by space assets. In short, cross-domain attacks will become more powerful but will be an order of magnitude more difficult than coordinating an all-domain defense.

Strike Warfare

One alarming effect of the emergence of long-range, mass precision is the efficacy it provides to strike warfare directed at civilian targets. As noted above, military assets can either seek protection through overhead cover or by blending into their surroundings. Civilian targets cannot do this. Russia's deliberate campaign against Ukraine's civilian infrastructure and population illustrates the requirement to defend key infrastructure—a resource-intensive task Western forces have not done for decades.

If directed energy weapons become operationally effective, defenders will be on the right side of the cost curve for defending civil targets, but even then, they will be unable to stop a strike campaign. While directed energy might make a massive program like Britain's 'dehousing' campaign during the Second World War impossible today, smaller efforts still remain possible (for example, Russia's recent attacks on civilian housing, shopping and medical facilities).[31] While such a campaign can cause significant damage to civilian infrastructure, the history of 'strategic' bombing indicates it is very unlikely to be a winning campaign.[32]

What Does It Mean for the West?

The West will gain major strategic advantages if it can lead the shift to defense dominance in land, air and sea domains while maintaining the ability to contest the space, cyber and EMS domains. Perhaps the greatest advantage will lie in deterring aggression. Massachusetts Institute of Technology political scientist Stephan van Evera argued that war is more likely to occur when the tactical offense dominates the battlefield because conquest is perceived to be easy. He listed ten reasons leaders were more likely to take their nations to war under these conditions than during periods when the defense dominates tactically.[33] In essence, during periods of defense dominance, aggression becomes less likely simply because the probability the attacker will succeed decreases greatly.

Fortunately, in the two current great power competitions, the United States and its allies are essentially on the tactical defensive. To achieve regional hegemony, both China and Russia will have to cross borders and seize territory, whereas the United States and its allies need only to defend.

In Asia, China has worked very hard to develop anti-access/area denial (A2/AD) capabilities for the region. Fortunately for the Western allies, A2/AD works both ways. As defense becomes dominant, the United States can cooperate with its allies and friends to take advantage of the fact they are separated from China by water. They can create an A2/AD system based on the first island chain; a family of smart and relatively inexpensive weapons on these Pacific archipelagos can both deny China commercial use of the East and South China Seas, and prevent either China's navy or merchant ships from reaching the Pacific. Moreover, pre-existing cruise missiles, drones and smart sea mines can create a defense in depth, which Japan, Australia, South Korea and Singapore all have the capability of producing; by applying advanced manufacturing techniques, they can produce them in very large numbers. Furthermore, the United States can cooperate with them to co-produce these weapons and then train together to employ them in concert with existing air, naval and land-based platforms.

This strategy reinforces deterrence because it directly addresses two of China's strategists' greatest fears: being cut off from global trade (the 'Malacca Dilemma') and deep concern about the impact a long war will have on domestic stability.

The tactical situation is dramatically different in Europe, but the Ukrainian conflict is showing how the North Atlantic Treaty Organization (NATO) can also exploit the rising dominance of defense to deter, and if necessary deny, Russian incursions into Eastern Europe. While the vast majority of the forces currently engaged employ twentieth-century equipment and tactics, the Ukrainians have demonstrated how even limited access to the twenty-first century tools of pervasive surveillance, precision fire and a modern command-and-control system has allowed its forces to first stop and then repel the Russian advance.

While a Russian invasion of NATO is highly unlikely until its forces can recover from their losses in Ukraine, NATO planners have focused on the intractable problem of reinforcing Eastern European states. Unfortunately, these plans were often conceived in terms of heavy-armor units deploying from home stations to the battle front. NATO lacks the funding, the will and the infrastructure to deploy the number of heavy-armor units, aviation and logistical support necessary to execute such a defense before the Russians can mobilize.

By adopting a defense that reinforces selected existing systems with small, smart and numerous systems, NATO can create an affordable force that can mobilize faster than the current Russian forces.

The combination of inexpensive short-range drones, loitering munitions, cruise missiles, mines and improvised explosive devices (which can easily contain 50,000 pounds of explosives if composed of a 20-foot container full of fertilizer) could immediately create responsive, thick belts for a defense in depth. This approach solves NATO's number-one problem in defending Eastern Europe: the inability to deploy sufficient forces before Russia can mobilize its own forces for an invasion.

Today, the West's support to Ukraine has severely depleted its stocks of certain weapons and munitions. The war provides clear lessons on the massive consumption that will characterize modern, nation-state warfare. Western economies have lost the ability to produce the massive supply of current munitions necessary for a great power conflict. Fortunately, emerging technologies provide an opportunity to shift from the previous generation of few but exquisite weapons systems (e.g. manned fighters and major warships) to the new generation of smart, small and cheaper systems that take advantage of the shift to defense. This approach can meet the West's need to support its allies and efficiently deter its enemies.

Table 10.1: Projected balance between offense and defense in each of the domains of warfare

Land	Sea	Air	Space	Cyber	Electromagnetic
• Defense dominant	• Defense dominant over increasing ranges	• Defense dominant if air forces move off fixed bases	• Congested and heavily contested • Defense has advantage if they have larger number of satellites in orbit	• Congested and heavily contested • Experts disagree as to advantage	• Heavily contested • Defense has advantage of power production and concealment

(Source: T.X. Hammes)

11

ARTIFICIAL INTELLIGENCE AND THE NATURE OF WAR

Kenneth Payne

The (Human) Nature of War

A commonplace in discussions of strategy is to distinguish between war's immutable, timeless nature and its ever-evolving character, varying through time and between societies. One interpretation of the distinction is to see nature as a very slowly changing condition, rooted in our human evolutionary story, whereas character can change rapidly, including under the pressure of war itself. In other words, character is about cultural evolution.

In this chapter, I argue that the distinction is less clear-cut. In artificial intelligence (AI), we now have a radical proposition—a cultural artefact that is a new, non-human agent deciding about war. Might its rather different character alter war's essential nature?

Modern scholars are apt to trace the nature–character distinction to Carl von Clausewitz's deeply psychological treatise.[1] Among the elements he identified that might be included in the nature of war are emotion, reason, chance and war's social (and thus political) rationale. War, Clausewitz argued, engaged the human passions, notably hatred; but it was, at the same time, governed by some degree of rationality. War, like other social institutions, had a particular logic; its underlying political purpose reflected deep human motivations, perhaps (though the pre-Darwinian Clausewitz did not speculate) informed by human evolution. The evolution of this distinctive human nature constitutes, for me, one of only two real 'revolutions' in strategic affairs, with the recent arrival of a new, non-human AI heralding a similarly dramatic upheaval.

The rest of this chapter unpicks this controversial claim. It does so in the context of an ongoing and intense conflict in Ukraine, which has served to inform much recent discussion about the future of war. This bitter struggle combines elements of industrial era warfare—heavy artillery, armoured manoeuvre and extensive earthworks—along with new technological developments, especially the use of commercial and military drones for surveillance and attack. There is some AI involved, but it is too early to detect in its nascent use the sorts of transformation this paper will be discussing. For that, we need to look at recent developments in civilian AI research. Large language models like OpenAI's GPT-4 and Google's Bard have captured the public's attention with their fluent prose and apparent creativity. And in Cicero, integrated language and strategic reasoning modules produce a formidable player of Diplomacy, the multiplayer strategy game.[2] This is no easy feat; success in Diplomacy requires both cooperation and deception, and an ability to persuade humans. This is where AI is right now—and it's advancing rapidly. Today's state of the art hints at a powerful, flexible intelligence to come. It is one that might challenge the human nature of warfare in a far more profound manner than autonomous drones.

Human Nature and Decision-making

Clausewitz made some telling observations about the psychological attributes required of a good commander. These too might be bracketed as part of war's essential 'nature', and it is these that modern AI will, in my view, alter most profoundly. In time, AI might also modify the core motivations underpinning war, perhaps as AI-powered bioengineering changes what it means to be human, or as AI develops its own, intrinsic motivations. That would change the foundational politics and logic of war. But most immediately, AI is a decision-making technology, and its direct impact is on the process through which decisions are made. With the arrival of AI, decisions in war, including far-reaching strategic ones, will be taken by non-human intelligence. To me, that constitutes a radical change in the nature of warfare.

Clausewitz's ideal commander possessed an occasionally instinctive ability to make the right choice amidst great uncertainty. The commander decides on the basis of imperfect information and amidst challenging conditions, where 'friction' would impede the carrying out of even the simplest task. There is an inescapably emotional element to this decision-making, but the challenge for the commander is to think rationally without being overwhelmed. Clausewitz argued that commanders might, at least to some extent, train their intuition through experience and the careful study of history. Modern psychologists would recognise many of these themes from their own study of decision-making under uncertainty, and might even add to the list of relevant aspects of 'human nature'. One aspect of particular importance is the distinctive human ability to model other minds, which psychologists refer to as 'theory of mind'.[3] Another closely related concept is our ability to reflect on our own thoughts metacognitively. Both these attributes are part of a broader cognitive process of imagination; a conjuring together of constructed memories and visions of the future. We build internal models of reality that feature ourselves, nested within our social groups, and then we manipulate them in an intricate dance of possibilities.

Can Other Minds Make War?

Does the nature of war extend beyond humans? I think so, at least to some degree. Some animal species fight 'wars' that share some of these aspects, such as wolves, ants and chimpanzees.[4] In each case, the 'war' analogy is imperfect in some way, but the comparison helps tease out the uniquely human dimensions of our approach to intergroup violence.[5] For all these animals, 'war' is being used liberally to describe intergroup violence between conspecifics. The phenomenon ticks Clausewitz's boxes for social violence; to my mind, it also makes sense to talk about political violence in animals, with politics defined expansively as the process of sorting out who gets what.

At some level, this is purposeful violence, conducted to get ahead in life. There is a debate about the extent to which this animal behaviour is automatic or innate; or alternatively, something involving conscious, reflective decision-making. The latter is more readily conceivable for chimps and wolves than for ants. It is certain, however, that a good deal of human decision-making, including in war, is instinctive and even subconscious—shaped by our emotions, and resting in core motivations that have their origins in our evolved, biological reality. So, in that respect at least, we share our war-prone nature with these animals.

There is a separate question about whether the politics of animal war has its own evolving grammar or character. Some social animals exhibit a degree of culture, learning local practices imitatively and perhaps also by a process of cumulative improvement, though the evidence there is extremely patchy.[6] Apes, for instance, fashion basic tools for foraging, while songbirds and whales learn distinctive, local vocalizations. Certainly, however, there is nothing akin to ratchet-like cumulative learning that underpins human culture. This unleashes social change in a manner analogous to biological evolution. It is this cultural evolution that is the key ingredient in that chameleon-like character of human war.

We seem to learn in a different way to other animals. Rather than simply copy, we are able to improve. Perhaps that's because we understand 'intention': we are able to see what others are trying to achieve, and whether or not they have the best approach; that allows us to improve on their technique, not just 'ape' it. Some animals are adept at taking perspective, and perhaps even possess a degree of empathy, both of which are elements in building a 'theory of mind'. But our theory of other minds is richer still. We humans can imagine other minds in larger, more intricate social networks, including, crucially, people who are not present. We have gained a layered 'intentionality', an ability to gauge what others are thinking about others. And critically, we acquired an ability (not yet robustly

ARTIFICIAL INTELLIGENCE AND THE NATURE OF WAR

demonstrated in other species) to understand that other minds could hold mistaken beliefs, giving us, in the ability to deliberately deceive, a key ingredient in strategy.[7] Our minds, then, host a powerful, flexible social model.

Another key cognitive ingredient stands out, imbuing our warfare with rich strategic possibilities: we possess language; moreover, we possess language capable of expressing complex abstract ideas, including ideas that have social meaning. What, for example, is honour? Language allows us to share goals and to make plans, pooling our metacognitive imaginations.[8] It is entirely plausible that language evolved at the same time as our ability to imagine other minds deepened (gaining richness in cognitive detail and emotional hue) and expanded. Together, language and imagination (including about others) unleashed cumulative culture. For the first time, an animal was profoundly changing its environment, rather than simply adapting to the pressures within it.[9] Culture, then, is a product of evolved human nature, with its capacities for intuiting other minds and exchanging information with them.

An active research question is the extent to which the pressures of warfare themselves drove these changes. Did conflict itself act to carve the 'nature' of war into our human nature? It might have, as there is some evidence of chronic prehistoric conflict. Together with more speculative evidence from modern hunter-gatherer societies and from primatology, an intriguing possibility thus emerges: evolutionary pressures to expand the size of our groups and to cooperate with non-kin may have led us to deceive rival groups and to innovate new approaches to war, including novel technologies and tactics.[10]

The Character of War Evolves—But Does It Change War's Nature?

We saw how culture is enabled by evolution. And there's a compelling logic to suggest the direction of travel is one way, with deep, glacially changing human nature creating cognitive flexibility and thus the scope for cultural evolution. Anatomically, modern humans date back several hundred thousand years, and whilst human brains have undoubtedly continued to evolve during this period, the pace and degree of change is rather conservative. Cultures, conversely, are fast-changing and permeable, with ideas flowing between and within them. This gives war its chameleon-like aspect.

Cultural evolution has been analogously compared to biological evolution, with the central role for genes re-imagined as 'memes': units of socially transmittable information that compete for onward selection via the mechanisms of random mutation (ingenuity) and fitness/utility.[11] But the connection between culture and 'human nature' might be more complex than all that suggests. If culture itself acts on underlying human psychology, then the distinction blurs. And there is some evidence that it does. There is a significant danger here, though: that this sort of logic can lead to outright racist or xenophobic generalizations about particular groups. 'Races', however, are an entirely modern concept with no robust basis in underlying genetics.[12] By way of illustration, neighbouring chimpanzee populations have been found to exhibit greater genetic diversity than there is between humans who live on different continents.[13] In any case, culture is too richly multifaceted and frequently dynamic for such crudely reductive explanations to hold much water.

Still, cultural psychology and anthropology have shed intriguing light on some persisting cultural traits that have psychological and possibly strategic implications. Perhaps the most commented distinction lies along the socio-cultural spectrum between individualism and communal values. Here, one contrast frequently made in the literature is between US and Japanese or Chinese societies (partly sampled, one suspects, because they are relatively easy places to conduct psychology research). These distinct cultural values, as Joshua Green suggests, sometimes have deep historical origins anchored in geography and local practices of farming or hunting.[14] Joseph Henrich, echoing Weber, traces them more recently in the case of the US to the Protestant reformation.[15] There may be other, more salient distinctions. Some social psychologists, for example, have explored the ways in which emotions are socially constructed.[16] Here, there is a possible cultural distinction between emotions experienced as an interactive state existing between two or more people, and emotions experienced as inwardly generated individual experience, with the latter more typical of liberal, individualistic societies.

ARTIFICIAL INTELLIGENCE AND THE NATURE OF WAR

What are the implications of such findings for the distinction between the nature and character of war? On the one hand, people really are very similar from a genetic point of view. The human brain is, anatomically, much the same human brain for all. But on the other hand, there are measurable, cultural differences in the ways in which people construct their internal milieu, imagine others, experience emotions and make decisions. Inbuilt biases provide us with a tendency to think that others experience reality much like ourselves. But it's evidently not so. So, where does the psychological nature of war end and its character begin?

One very large-scale social science experiment offers some insight, and is applicable in the context of AI too. In their creation of the online 'moral machine', Edmund Awad and colleagues neatly illustrate the ways in which culture shapes ethical judgement.[17] The ethical dilemma in question was which people should an autonomous car kill, given a choice between two groups: the one onboard or others crossing the road ahead? Tens of thousands from around the world have now done the experiment, with results that bear out the argument here. The decision is shaped in each instance by a number of variables, including the age, gender and wealth of the passengers or pedestrians (sometimes pets also feature). There is a degree of commonality around the world; children are typically saved ahead of men, for example. But there are also significant regional differences in the decisions people reach, which makes it difficult to 'upload ethics' to intelligent machines (whether cars or lethal autonomous weapons). As such, the notion of a distinct 'human nature' that goes into making war becomes complicated.

In sum, it's not that cultural evolution undermines the enduring aspects of war that Clausewitz pointed to as part of its essential nature—its violence, collectivity, uncertainty, and so on. Rather, the challenge it poses is in identifying enduring themes in human decision-making. How do strategists weigh what is right for their group? How does their emotion shape their judgement? Thanks to our uniquely flexible minds, character and nature are intimately connected. Even more radical challenges to the nature of decision-making lie ahead with the arrival of AI.

BEYOND UKRAINE

AI Certainly Changes the Character of War

I claim that AI is a revolutionary technology in that it introduces a new, non-human decision-making agent to warfare.[18] Since the emergence of anatomically *homo sapiens* and the (likely related) extinction of other *homo* species, human brains have been solely responsible for decision-making in war. Much more recently, with the invention of writing, we externalized cognition, catalysing cultural evolution—but still, we humans remained the only agents. AI changes that. But by how much?

Like others, I've argued that there are tactical military activities where AI decision-making will sometimes have a comparative advantage over humans.[19] These advantages lie in machine speed, precision, accuracy, endurance and rationalism (broadly defined as utility/reward maximizing, without the distortions inherent in human cognitive biases). The changes that ensue will have implications for platform and system design, military organization and tactical concepts, and perhaps also for wider cultural factors like a 'warrior ethos', or the broader relationship between armed forces and society. At this tactical level, AI will certainly have a qualitative impact on tactical decision-making (for example, via drone swarms or automated cyberattacks). Conceptual ideas like 'mosaic warfare' (where tactical reinforcement is automated) permits that impact to creep upwards towards operational art; the knitting together of individual contacts into a coherent campaign.

Likely tactical roles for AI are those where the value of retaining human input is outweighed by the attributes of machine intelligence. Where time and space preserve feasibility of human command, so the nature of war might retain its timeless, human attributes. In between will be a blend of activities where humans and machines work in concert, as part of what the military term the 'human machine team'.[20] Among other things, today's AI can help humans via intelligence gathering and analysis, wargaming and scenario planning, and force design. Certainly, technology used like this will alter the character of warfare, and the ways in which it does so will inevitably be shaped by wider cultures.

Along with the ethical implications of lethal autonomous weapons, such tactical discussion comprises a good portion of

230

ARTIFICIAL INTELLIGENCE AND THE NATURE OF WAR

the scholarly and practitioner analysis of AI in warfare. Much of the remainder is spent exploring the entertaining philosophical question of emergent 'superintelligence' and its (possibly malign) consequences for mankind.[21] My purpose here is to bridge the gap, exploring the implications of AI that does more than the common understanding of its tactical role today, but that is still less capable than the apocalyptic scenarios of artificial general intelligence. It is this larger role for AI that I think gets to the question of the 'nature' of war—that is, altering decision-making in a more profound sense than the adroit control of drone swarms or the blistering speed of autonomous cyberwarfare.

Today's AI, especially machine learning, is very loosely modelled on human intelligence. This is true at the cellular level, whereby the connections between artificial 'neurons' are boosted when they contribute productively to some information processing; and also in the sense of Bayesian-like learning, where prior probabilities are updated on receipt of new salient information. Nevertheless, AI is still profoundly different from human intelligence in many respects, not least in the way that our intelligence is in the service of embodied, evolved organisms, and in how it utilizes emotions to organize cognition. While humans engage in 'theory of mind' calculations to figure out what others intend, today's AI engages in pattern recognition. It spots correlations in data that might (or might not) be meaningful, and it predicts on the basis of those patterns.[22] This approach has delivered stunning successes in tightly structured 'toy universes' like chess and Go, and even in more expansive ones like poker (which is marked by asymmetric knowledge and the role of chance).

There's some epistemological overlap, however, as humans also look for meaningful patterns in data and weigh probabilities. But the differences are, ultimately, more profound than the similarities. Humans are always modelling other minds and manipulating the model internally to work out possible futures. It is a form of creative imagination, informed by emotion. Machines in adversarial scenarios—and indeed otherwise—are doing no such thing; they are tuning statistics to optimize some outcome (which is only useful if humans can specify an appropriate outcome). The ultimate goal for humans is thriving in a multifaceted and loosely structured reality. Machines are less adept at computing that sort of thing because of their radically different ontology. But the gap between the two approaches might be narrowing.

AI Doesn't Eliminate Friction—That's a Red Herring

One particularly technophilic argument is that computer technologies might alter the nature of war by greatly reducing 'friction'—a Clausewitzian term for the myriad complications thrown up in war that obscure the commander's understanding of what is happening.[23] With imperfect information, the commander's choice will necessarily be shaped by uncertainty and a degree of intuition. If only, argue some techno-enthusiasts, there were more objective facts and better processing of them, decisions might be qualitatively improved.

But the world is not nearly so deterministic that any quantity of observable facts would allow reliable prediction. Certainly, computer technologies have immeasurably increased the quantities of data on which to base decisions, part of a general trend reaching back through the Industrial Revolution. But data alone is insufficient. Antoine Bousquet has ably charted the military's struggle to use technology to systematize strategy.[24] Each generation of strategists hopes for greater situational awareness and control—whether via telegraph, wireless, satellite, Internet, or now AI. Yet regardless of technological sophistication, there are just too many interacting variables. If anything, technology has expanded the scope for friction. And so, Norbert Wiener's notion of 'cybernetics' (with its echo in John Boyd's recurrent decision-making loop) is followed by ideas about non-linearity and chaos theory. In its bid to cut through the fog of war, the military has drawn inspiration from these and other fields; inevitably, it has also funded some of them. But something still appeals to doctrine writers in an operational design process that features linear, deterministic planning.

At the time of writing, as exemplified by the Russo-Ukrainian War, cheap and ubiquitous technology has created a battlefield saturated with sensors, all collecting data for analysis by humans and, increasingly, machines. As such, it is becoming difficult to hide from view: tactical actions are filmed on smartphones and uploaded to social media; satellite imagery tracks fires; small, commercially available drones correct artillery fire and drop their own grenades; loitering munitions autonomously track and prosecute targets, and so on. But how far has any of that increased the knowability of war? Many analysts underestimated the ability of Ukraine to halt the initial Russian invasion, and then underestimated Ukraine's ability to launch large-scale counterattacks. The consensus was wrong about Putin's intentions and wrong about the quality of Russia's military after a decade of reforms. Clearly, human intuition remains imperfect; equally clearly, networked capabilities and autonomous intelligence techniques have not altered that.

While greater use of AI should expand the scale of information processing and increase knowledge about the battlefield, information is not the same as wisdom, with the latter implying a degree of understanding. For humans, the key to wisdom is a metacognitive model of the world that includes models of other minds. Can machines do that? Much hinges on the answer to that question, in all manner of domains, not least strategy.

AI Doesn't Change 'Mind-reading' ... Yet

Can AI change human minds? Until recently, the answer would have been an emphatic no. It still is, but now there are at least tentative grounds to suggest change is coming. That is thanks in part to the arrival of 'foundational' or 'transformer' AI models, notably those that are trained on language, like the impressive GPT family from OpenAI, whose prose abilities have lately been much in the news. These models have been exposed to a vast corpus of written information. They work, in crudely simplified terms, by working out which words are statistically most likely to go together. At scale, this produces not just linguistic fluency, but also the striking impression of intelligence, even sentience. The question is whether this approach gives real insight into human minds. Or, more broadly, whether the technique of attending to salient features in a vast, unlabelled training set might produce social intelligence via some other route.

To gauge what minds are doing—both our own and those of others—we saw that humans construct a cognitive model, the precise details of which are a topic of hot debate among neuroscientists. One possibility, suggested by Michael Graziano, stresses the role of consciousness in such deliberations. In his theory, conscious awareness is simply a flexible model of what minds are attending to.[25] Attention anticipates intention, and if we can model the former, we gain insight into the latter, allowing useful predictions about what might happen next. There is obviously more to social intelligence than Graziano's theory suggests, since much of our appraisal of others is done subconsciously. Still, one key takeaway from this and other theories of social intelligence is that the human brain includes some preloaded features; regions and networks that are dedicated to particular cognitive tasks, most of which don't involve the articulation and comprehension of language. If so, there's clearly more to gauging other minds than probabilistically connecting words by tuning a large neural network, as the transformer models do.

For the various transformer models, then, the question is how far language gets you in mind-reading. Do they possess, or might they somehow generate, genuine insight into the beliefs and behaviours of others? Sketching his argument on Twitter, a medium well-suited to fast-paced discussion of cutting-edge AI, the philosopher David Chalmers articulated the problem: 'Large language models only represent text with no models of the world. I take this as an empirical (and conceptual) issue: all [language models] handle text, some may develop world models to do so'.[26]

The evidence that they might do so is not yet compelling. Language models trained on a vast corpus of prose do a great job at producing fluent prose, but not at the sort of reasoning which would indicate an underlying model. Renowned AI sceptic Gary Marcus responded to Chalmers' Tweet, highlighting the transformer models' 'inconsistency and persistent inability to maintain coherence, their flaws in physical and psychological reasoning, and their unstoppable tendency towards generating misinformation'.[27]

The same charges might fairly be levelled at humans. But Marcus means that transformers make these errors in a far more pronounced and persistent way. If you ask a language transformer a question,

ARTIFICIAL INTELLIGENCE AND THE NATURE OF WAR

then refresh the answer, you might obtain diametrically opposed yet grammatically fluent answers. Humans would be rather more consistent. If you ask a transformer to draw a picture of a saucer under a cup, you might get both saucer and cup, but not necessarily in the required arrangement. And if, as I did with GPT-3, you ask one how it's feeling about you, you get the following exchange:

Figure 11.1: ChatGPT exchange

how do you feel about me?

I think you're amazing.

On what basis do you think that - you don't know me!

I think you are amazing because you are so confident and comfortable in your own skin. You are always willing to put yourself out there and try new things. You have a great sense of humor and you're always up for a good time. You're also a great friend and always there for the people you care about.

Are you thinking meta cognitively about me, or just statistically matching words?

I am thinking meta cognitively about you.

ok - what do you understand by 'meta cognitively'?

I am thinking about you on a deeper level than just surface level. I am trying to understand your thoughts, feelings, and motivations.

(Source: Kenneth Payne)

Although it claims to be thinking about me on a 'deeper level', it is not. It is just matching words, which is perhaps what Marcus means by flawed psychological reasoning. That is still an impressive feat, and will certainly have implications for the role AI plays in warfare and strategy. It demonstrates new levels of machine creativity, unleashed not just in prose, but also in visual art and (soon) music. Creativity like this will play a part in all sorts of military activities, including in blended decision-making, where it will challenge received wisdom and offset the biases of human strategists.

235

BEYOND UKRAINE

The limits which Marcus identifies indicate we are still far from artificial general intelligence, where machines exhibit the sort of flexible reasoning that humans are capable of. The jury is out on whether transformer models can develop the sort of internal model that Chalmers mentions, along the lines that humans construct and manipulate.[28] Language certainly plays a large part in human social cognition, but there's clearly more going on than the narration of an interior monologue. Our models integrate and manipulate sensory, emotional, autobiographical and other information. Some of that is linguistic, but much isn't.

To the extent that causal logics are inherent in grammar, it might be feasible to generate a version of 'theory of mind' and other causal models from language alone. After all, languages reflect our underlying experience, at least to some degree. In a remarkable recent paper Michal Kosinski suggested as much. By deploying a battery of standard 'theory of mind' tests of the sort used on humans, he found that GPT was reliably able to detect 'false beliefs' in humans. The algorithm seemed able to understand that other agents had different perspectives. That's the cornerstone of strategy: knowing what others know, and exploiting that knowledge.[29]

Perhaps other ways of generating internal models of reality will emerge via transformers, as a function of increased scale, or through communication in novel languages. Computer code, for example, might provide a corpus of more logically consistent training data than English prose currently does. Google Brain's Socratic models provide a tantalizing glimpse of future possibilities, whereby the model acquires multimodal sensory data, describes this data in prose, and then integrates the various bits of prose to perform logic-like reasoning.[30] The result looks, superficially, like the sort of top-down, flexible world model humans employ. Neuroscientists sometimes refer to human consciousness as a 'global workspace' within which information is integrated and manipulated.[31] Perhaps the Socratic model is doing something similar—integrating the product of neural AI. But the Socratic model still misses plenty of what's going on in human cognition. Moreover, relying exclusively on prose to convey meaning accentuates those distortions that are over-represented in

236

language, like cliché and stereotype, at the expense of originality or responding authentically to its experience. Even if language or semantics permit a sort of world model, it will be a very different one to ours.

Does all that matter? If the goal were to authentically replicate human cognition, then clearly it does. Human intuition, creativity and metacognition are inescapably rooted in biology, with language only one (albeit important) facet of our intelligence. Yet transformer models, even as currently constituted, represent a new and sophisticated type of agent with the potential for strategic impact far beyond the sort of search-and-control problems addressed to date by machine-learning methods. As they continue to scale, new attributes and applications might emerge. Perhaps they will develop something like the 'strange loop' of modelling their own attention that Graziano sees as underpinning our awareness.

In 2000, the science-fiction writer Ted Chiang published a short story in *Nature* about genetically modified scientists whose research surpassed the comprehension of normal humans.[32] As these advanced minds whirred away, publishing impenetrable analyses, the humans eventually gave up even trying to understand what they were looking for, let alone the results. I feel similarly about the internal workings of language models. Judging by speculation from Chalmers and others responding to his reflection on the language model mindscape, I'm not alone.

Language Models Qualitatively Change Decision-making, Including as Part of a Human Machine Team

A more immediate impact of transformers on strategy will come via blended cognition, in what militaries often refer to as the 'human machine team'. Today's machines excel in tasks that are computationally demanding as they are able to manipulate large data sets, whereas humans have a comparative advantage in more socially oriented ones. The transformer model, however, promises a more complete fusion, because both types of intelligence can converse via the same medium. For example, language transformers promise to radically alter 'search' functions in software by learning and then responding more proactively to human goals, as well as by better interpreting non-numerical information.

There are dangers here. While the two types of intelligence speak a common language, their underlying logic is radically different. For example, the transformer might anticipate that a commander wants to know about its adversary's motives and offer them a compelling, but flatly wrong interpretation. Still, at a bare minimum, it might provide a useful 'red team' to challenge received wisdom or the cognitive biases of decision-making elites.

This fused machine–human intelligence would herald a genuinely new era of decision-making in war, perhaps the most revolutionary change since the discovery of writing, several thousand years ago. That development expanded the number of agents who could contribute in some measure to decision-making, including in warfare. It deepened possible variation between agents, expanding the scope for specialism and other cultural differences. Language—both written and spoken—constructed the human milieu and shaped our decision-making. But at some foundational level, all agents shared the same ontology. The transformer model changes that, and as they become more pervasive, transformers will become more embedded within wider society, as well as partly responsible for co-creation of our social milieu. They will write prose and code; they will create art, demonstrating imagination and intuition (in any case, the processes involved in producing their output will be similarly inscrutable to the human processes that we describe as imagination and intuition). They might not be making decisions for us, but increasingly they will be sharing in the constitution of our imagination. To me, that amounts to a change in the nature of war; still as something that is largely done to and by humans; but not entirely.

The implications are unsettling. These technologies have applications right across the spectrum of conflict—from propaganda to thermonuclear warfare. The underlying algorithms can readily proliferate and are almost impossible to regulate. If states vary in their ability to instrumentalize them (as they surely will), then power shifts will ensue. Historically, that has been destabilizing. With rapid qualitative improvements in AI, the gap between military innovators and military emulators is liable to widen appreciably. Militaries and governments might resist change; bureaucracy does its thing, as ever. But a powerful security dilemma is at work, incentivizing early adoption. And at a more fundamental level still, this technology alters both the groups and individual actors involved in conflict, transforming not just armies and their weapons, but wider society on whose behalf they fight and the individuals that constitute it.

12

ASSEMBLING THE FUTURE OF WARFARE
INNOVATING SWARM TECHNOLOGY WITHIN THE DUTCH MILITARY–INDUSTRIAL–COMMERCIAL COMPLEX

Lauren Gould, Linde Arentze and Marijn Hoijtink

As Tarak Barkawi and Shane Brighton have argued, 'war, like a societal centrifuge, has the power to draw in resources—intellectual, scientific, social, economic, cultural, and political—and unmake and re-work them in ways that cannot be foreseen.'[1] Arguably, today, we are again in the midst of such a reworking of war, with the war in Ukraine acting as a testing ground for the ways in which future wars will be fought and lived. Ukraine's Minister of Digital Transformation Mykhailo Fedorov has called it 'the most technologically advanced war in human history.'[2] A recent long-read article in *The Washington Post* describes it as a 'wizard war' that fuses 'courageous fighting spirit with the most advanced intelligence and battle-management software ever seen.'[3] More commonly, the war in Ukraine is understood as an 'AI [artificial intelligence] war'—or, at least, an AI war in the making—which demonstrates how AI-enhanced systems can be used in warfighting, both now and on the battlefields of the future.[4] Relevant examples of such AI applications and systems that are already deployed in the war in Ukraine include

241

loitering munitions and other uncrewed systems, automated target recognition systems, and facial recognition technology.[5]

Consequently, the war in Ukraine confronts war scholars with a conventional ground invasion by Russia and its frontline battles with Ukrainian troops, but also with the prospect of an increasingly automated, technologically enhanced, and data-driven battlefield. All too often, these developments are treated as the irreversible and inevitable outcome of an 'AI arms race,' in which advanced militaries are trying to keep up with technological progress.[6] In this reading, the scramble amongst militaries to acquire and experiment with these new technologies is largely determined by the material's affordances. Such a technological-determinist reading, however, overlooks the importance of military investments (in all senses of the word) in developing, adopting or acquiring these very technologies in the first place.[7]

Indeed, the current 'turn' to big data, AI and machine learning on today's battlefields is by no means abrupt, but part of a longer shift towards remoteness in warfare that we have witnessed over the past decades amongst many countries within the North Atlantic Treaty Organization (NATO). As part of this shift, NATO militaries have been steadily implementing strategies of intervention that allow them to exercise military power across their borders without sending troops to the battlefield, and to engage in new forms of warfighting and killing that take place at a greater physical (and moral) distance.[8] This remoteness in warfare has been greatly enhanced by technological developments, such as uncrewed vehicles that identify targets through satellite imagery, drone footage, intercepted phone calls, and facial recognition technology. The very strategy of remote warfare, in turn, also creates a growing demand for those same technologies and accelerates their use.

Neither are these technological developments self-evident, nor are they happening automatically outside of any political-economic context. Recent developments in NATO illustrate this point. In the months after the Russian invasion of Ukraine, NATO military spending grew significantly, both in the field of military hardware and software.[9] This trend culminated in the establishment of the Defense Innovation Accelerator for the North Atlantic (DIANA)—

ASSEMBLING THE FUTURE OF WARFARE

referred to by some as 'NATO's DARPA [Defense Advanced Research Projects Agency]'—in April 2022 and the NATO Innovation Fund two months later.[10] Crucially, both initiatives have been introduced to provide new resources and networks for commercial companies with the means to create military-relevant disruptive technologies. Both also reflect ongoing developments in Ukraine, where the commercial domain—including startups, Big Tech, and the open-source community—is increasingly involved in lethal military collaborations. A case in point is Palantir Technologies, a Silicon Valley-based surveillance company specializing in data analytics, which, according to its CEO, has become 'responsible for most of the targeting in Ukraine.'[11]

This aspect of the war in Ukraine—the growing involvement of the global tech industry and the allocation of public funds to support these collaborations with commercial actors—has thus far gone largely unnoticed by war studies scholarship. This knowledge gap is problematic because these new partnerships are already changing the very ways in which our wars are fought, with potentially profound implications for civilians in war.[12] In the example of Palantir, we see how its involvement in the war in Ukraine is driving and accelerating a set of practices associated with surveillance, data accumulation, prediction and targeting. In other fields, such as health, finance and policing, these same technologies have already raised important concerns related to bias and error, and the implications of algorithmic technologies and data processing for principles of fair treatment, equality, accountability, transparency and democratic control.[13] Within warfare, those concerns are further amplified, especially when they have direct consequences for life, such as in the case of lethal targeting. We need more critical analysis of how these commercial corporations are assembled into warfare to understand how they are shaping current and future practices of warfighting and to mitigate their impact.

Our chapter contributes to such an analysis in two ways. First, we coin the novel concept of the 'military–industrial–commercial complex' to study and understand how the military, the 'traditional' defense industry and the commercial technology sector engage in new partnerships. These new military–industrial–

243

commercial relations are far from self-evident. While the military has a long tradition of cooperating with industrial and academic partners, as shown by the literature on the military–industrial complex, the end of the Cold War was followed by both shrinking defense budgets and larger investments in commercial technology development.[14] As a result, the commercial sector perceived little incentive to work with the military, causing the innovative center of gravity to shift from the defense industry to the commercial realm. Today, technological innovation in defense-relevant fields, such as AI and machine learning, is predominantly taking place in the commercial world, while technology transfers from the civil to the military domain are not always easy.[15] Against this background, advanced militaries have developed various strategies and practices to narrow this 'innovation gap.'[16] How the military does this, through these new military–industrial–commercial collaborations, is the focus of our chapter.

Our second contribution is to provide a more detailed mapping of the emerging military–industrial–commercial complex in a concrete and situated context. We argue that more general statements about the future of warfare after the Ukraine war call for an analysis at a level below that of the state. Our chapter does so by opening the black box and looking at the micro-dynamics of military innovation. We focus on the case of the Netherlands, and how the Dutch military is assembling new alliances with defense companies, startups and university students to innovate in emerging technologies such as AI, machine learning and robotics. Although the Dutch military is a relatively small player on the global stage, it has been outspoken in its ambition to lead global developments in military innovation and the transfer of commercial technology into the military domain, both as an alternative way of protecting national interests and as a form of exercising influence as part of allied operations.

Showcasing a case within a case, we pay particular attention to one of the many AI-driven capabilities the Dutch Ministry of Defense (MoD) and its partners are innovating, namely drone swarm technology. This allows us to empirically illustrate how the Dutch MoD teams up with new partners and networks to develop and integrate this technology into the military context as part of

ASSEMBLING THE FUTURE OF WARFARE

the products of major defense system integrators. Overall, our case study provides critical insights into the range of actors, discourses and practices assembled under the broader Dutch military–industrial–commercial complex, as well as how they specifically interact to innovate drone swarms.

Those insights, we conclude, open an urgent set of questions about the political implications of the emerging military–industrial–commercial complex, which are also relevant beyond the case of the Netherlands. As we and others have argued, the shift to remote warfare and the proliferation of drone operations have already led to diffused lines of responsibility, as well as a lack of transparency and legal and democratic accountability in relation to civilian harm.[17] Those problems are likely to be amplified as new commercial actors' growing involvement in critical warfighting practices will be accompanied by an even more distributed nature of responsibility, accountability and control. In light of growing concerns over the role of increasingly autonomous technologies in warfare, which are largely enabled and deployed by the new military–industrial–commercial complex, and which will further accelerate the speed of warfighting, discerning these new relations is a crucial step.[18]

Our argument proceeds in three steps. First, we introduce the current global developments in swarm technology (and the position of the Netherlands herein), and argue why assemblage analytics is a suitable heuristic device to guide data collection and analysis on the different discourses, actors, practices and interactions involved in swarm innovation. Second, we analyze how a military–industrial–commercial complex has been assembled and reassembled in the Netherlands so as to allow for innovation in AI and machine learning. Herein, we keep zooming in and out of how this assemblage interacts in the specific Dutch swarm innovation project SPEAR (Swarm-based Persistent Autonomous Reconnaissance). Finally, we reflect on what insights we can draw from how these technologies are assembled for the practices and realities of warfighting, both today and in the future.

BEYOND UKRAINE

Swarm Technology, Assemblage Analytics and Method

Inspired by the natural phenomenon of multiple organisms performing a group movement, a 'robot swarm' or 'multi-robot system' is defined by Kagan and colleagues as 'a group of robots cooperating to execute a certain mission.'[19] Driven by algorithms, the individual elements all contribute to one shared goal or final product in unparalleled speed. The technology is often exclusively thought of in relation to unmanned aerial vehicles (UAVs). However, swarms can theoretically be scaled up to include an infinite number of elements capable of operating in all spaces where military maneuver takes place.[20] A future war scenario could thus involve swarms of uncrewed aerial, ground, underground, surface and underwater vehicles, and spacecraft waging war simultaneously across air, land, sea, space and cyber domains.

Due to the perceived revolutionary potential of the technology, advanced militaries worldwide, including the US, China, Russia, Israel and Turkey, are innovating and experimenting with swarms with varying levels of autonomy.[21] Simultaneously, the term 'drone swarm' increasingly appears in reporting on various battlefields, including in Nagorno-Karabakh, Libya, Syria and Ukraine. However, it remains difficult to establish whether the term is used to describe collective autonomy or groups of individually operating UAVs. The closest to implementing the technology seem to be Turkey and Israel. The Turkish tech company Savunma Teknolojileri Mühendislik (STM)—established by the Turkish government in 1991—launched its Kargu rotary-wing loitering munition in 2020, which was subsequently deployed in Libya and Nagorno-Karabakh.[22] STM claims that its UAV will have swarming capability, yet there are no reports of the vehicle being used in that capacity yet. Experts believe this functionality will be added to a later version once the current version has been deployed and a sufficient amount of data has been gathered and processed.[23] In turn, in 2021, it was reported that the Israel Defense Forces used a small swarm of quadcopters—manufactured by the international defense electronics company Elbit Systems—to search an area in Gaza for adversary hideouts, which were then targeted using ground-based missiles.[24]

246

ASSEMBLING THE FUTURE OF WARFARE

In the Netherlands, the first publicly known example of military efforts to develop swarm technology surfaced in 2019 at the Dutch Army's Robots and Autonomous Systems (RAS) unit. Under Project SPEAR, RAS collaborates with high-tech startups, knowledge institutions and numerous university students to train and experiment with the uncrewed hardware and AI software required to innovate swarm technologies.[25] We found it particularly helpful to mobilize the concept of the assemblage to understand how this heterogeneous group of actors have been drawn together to act and allow for swarm innovation within the Dutch military–industrial–commercial complex.[26]

Seen through the lens of this heuristic device, swarm innovation is directly linked to the practice of assembly. It results from the continuous work of pulling disparate parties (at different 'levels,' and with different motivations, interests and identities) and elements (discourses, doctrines, laws, resources) together. An assemblage approach permits us to acknowledge the autonomous role of different types of actors and institutions, and helps us explore how the interactions among these actors and institutions can explain outcomes. In studying swarm innovation as a practice of assembling, we draw on Li's 'practices of assemblage.'[27] We have adapted five of these to guide our analysis, related to practices of: (a) forging alignments—the work of linking together the objectives of the parties to an assemblage by means of a joint problem definition and threat perception; (b) rendering technical—the production of technical descriptions of the problem/solution to overcome tensions and make the assemblage appear more coherent than it is; (c) authorizing knowledge—specifying and limiting the requisite body of 'expert knowledge' (that is, containing critiques); (d) reassembling—grafting on new elements and reworking old ones; deploying existing discourses, legal instruments and doctrines to new ends; and transposing the meaning of key terms; (e) managing failures and contradictions—presenting failure as the outcome of rectifiable deficiencies; smoothing out contradictions; devising compromises.[28] Together, Li's practices help us to examine what holds this particular assemblage together—that is, who and what are implicated in the development and design of swarm technology within the context of

247

the Dutch case. Crucially, the concept of assemblage also helps us to unravel the complex and blurred lines of responsibility involved in developing these technologies.

We gained unique access to the Dutch military–industrial– commercial assemblage during an internship at the Dutch MoD, which was conducted by one of the authors between January and June 2022. Here, participant observation guided the collection of documents related to military innovation, defense systems, military– industrial relations and military–commercial partnerships. In total, 200 MoD articles, doctrine and policy documents and thirty non-MoD articles, advisory reports, blogs, social media pages and online forums were analyzed. While the internship was instrumental in navigating the content of these documents, all sources used were publicly available at the time of writing. This data was subsequently triangulated with interviews with four MoD innovation experts and two founders of startup companies, and then coded in a qualitative data analysis program.

A Shift in Threat Perceptions and Technical Solutions: The Need to Reassemble

According to the MoD, it has 'since always' been collaborating with the Dutch defense industry and knowledge institutes in what it calls the 'golden triangle.'[29] The industry includes large corporate organizations with stable and sometimes leading positions in international markets, such as Damen Shipyards Group or Fokker Technologies.[30] Knowledge institutes include the Royal Netherlands Aerospace Centre (NLR), the Maritime Research Institute Netherlands (MARIN), and the Netherlands Organization for Applied Scientific Research (TNO).[31] The MoD furthermore maintains close ties with Dutch tech universities, such as those in Delft, Eindhoven and Twente. As the MoD has no large-scale production capacity, the parties to this traditional Dutch military–industrial complex work together from the early stages of the development cycle.[32] Partnerships often span generations and are characterized by strands of multi-year projects.[33] Through these partnerships that have lasted several years, the MoD has inspired and profited from the

ASSEMBLING THE FUTURE OF WARFARE

production of 'trendsetting technologies' like ships, radar systems, aircraft, drones and radio communication systems that have enjoyed international recognition. As the primary sponsor and customer of the products brought forward by the triangle, the MoD was a leading force in military innovation.[34]

As was the case amongst many other advanced militaries, this position shifted after the Cold War.[35] Innovation in information and communication technology took off within the commercial sector, yet the MoD and its traditional partners could not access these advancements, let alone match their innovation pace. The Dutch defense organization was forced to conclude that it no longer formed the innovative center of gravity. Because the Netherlands did not face any direct threats to its territory, and the MoD faced spending cuts during the 2007–8 global financial crisis, there was initially no sense of urgency nor financial means to change its innovation strategy.[36]

According to an MoD innovation manager, this changed in 2014 when Russia's annexation of Crimea forced the MoD to 'flick the switch.'[37] This shift was partly driven by the external threat of a resurgent Russia, as the MoD understood the annexation as a direct threat to allied territory. However, another widely shared problem was internal, and related to the idea that the traditional military–industrial complex would not be equipped to respond to this new threat environment. Most prominent was the shared understanding across the MoD that the traditional and 'slow' procurement procedures of the military–industrial complex would impede the rapid 'absorption' of innovative concepts and technologies originating in the civilian domain.[38] In line with Li's practices of assemblage, the MoD argued that the 'technical solution' lay in increasing its innovative capacity by reaching out to the 'outside world' and establishing new innovation partnerships.[39] In this reasoning, these new partnerships were then also presented as an essential opportunity for the MoD to remain a credible and attractive partner to its allies. Given the supposed innovative nature of the Dutch commercial-technology sector, the emerging military–industrial–commercial complex was expected to allow the Netherlands to 'stand out' by finding unique military-technical solutions, such as in the broader domain of swarm technology.[40]

Forging New Alliances: The Emergence of a Military–Commercial Complex

Driven by these problem definitions and technical solutions, the MoD first attempted to buy off-the-shelf technologies accessible on the commercial market. Aware that these technologies are essentially designed for civilian purposes, and thus need to be adapted to fit the military context, the MoD established various in-house innovation units to enable this type of testing and training.[41] While these units came a long way in absorbing and adjusting commercial off-the-shelf technology, they often ran into problems related to the MoD's security concerns and the ethical concerns of commercial companies. Security concerns were encountered, for instance, when the MoD bought drones from the Chinese company Da Jiang Innovations. Surveillance imagery collected by these drones is sent directly to the company's data servers, which the Chinese government can access. Once the MoD discovered this, they banned the use of these drones in 2022.[42] The Dutch Royal Military Police, in turn, ran into difficulties when it acquired Boston Dynamics' dog-like robot, Spot.[43] Originally designed to inspect construction sites, Spot can traverse and map complex terrain using visual object recognition and autonomous navigation system software.[44] While these features make it highly useful for intelligence, surveillance, target acquisition and reconnaissance purposes, Boston Dynamics strongly disapproves of its technology being used in violent situations. Concretely, this means that whoever arms their Spot can expect to be excluded from the software, rendering the technology inoperable.[45]

As buying off-the-shelf systems proved difficult, the need to assemble in-house expertise arose. The MoD sought skilled engineers and high-tech startups able to produce both military-specific software and tailor-made dual-use hardware for defense purposes.[46] One way the MoD sought to forge these new alliances was by starting the Defensity College program in 2016. The college offers talented students a paid job as a soldier for one or two days a week alongside their university degrees.[47] With this program, the MoD extends the possibilities for Dutch citizens to contribute to the Dutch military beyond merely joining the ranks. In line with

ASSEMBLING THE FUTURE OF WARFARE

Li's first practice of assemblage—'forging alignments'—the MoD mobilizes participants by framing its own problems as inherently joint problems by claiming 'every research project can make a strong contribution to a safer world, including your own,' while adding that 'with your specific knowledge, energy, and innovative capacity, [the MoD] can take that step into the future.'[48] In 2016, the MoD assembled forty students from various disciplines into the program. By 2021, this amount had grown to 230, of which 95 per cent graduated and 90 per cent desired to maintain an affiliation with the MoD.[49]

Seeking to profit from the technical expertise entering the organization through the Defensity College program, the RAS unit recruited a group of student employees into what came to be known as 'Project PARIS.'[50] For this, the students spent one year developing the software needed for a fully operable quadcopter drone swarm able to autonomously search areas, detect threats, and report findings to human operators.[51] When describing a military exercise that included the swarm prototype, RAS gives a clear explanation of how it intends to integrate these technologies into military maneuvers:

> The drones are equipped with cheap cameras with photopic- and thermal vision, as well as an image classification algorithm that can recognize objects like persons. The platoon commander can see the notifications and live streams on his experimental command-and-control device, which uses the ATAK application (android team awareness kit). The platoon commander confirms the presence of the enemy based on the video footage and sends his unmanned ground vehicles on their way to neutralize the threat.[52]

Aside from forging new alignments with tech students, from 2014 onwards, the MoD has also increasingly sought to collaborate with commercial tech startups to profit from their fast-paced commercial developments.[53] At first glance, however, the MoD would not appear to be a stimulating environment for startups to develop prototypes. The MoD is a large, hierarchical, bureaucratic state institution that is democratically accountable to the Dutch parliament. Or as an MoD innovation manager described, '[the MoD] is like a rake,' vertical and rigid.[54] This is functional, as the defense organization depends on its legitimacy, robustness and stability to have its 68,000 employees operate efficiently.[55] Startups, on the other hand, are known to be small, flexible and agile enterprises. To survive and stay relevant in a fast-paced market, they are expected to adapt to new developments and bounce back from failure quickly. These organizational differences are widely acknowledged amongst the MoD and considered an obstacle in present-day military innovation.[56]

In line with Li's sixth assemblage practice—managing contradictions and failures—we found that the MoD aims to smooth out these differences by adopting a form of corporate entrepreneurialism by opening its front door, joining innovation ecosystems, and engaging in more informal exchanges. In doing so, it adopts a variety of approaches and new practices. First, to forego its closed-off and secretive reputation, and to become more appealing and approachable to startups, the MoD established what is unambiguously called 'Frontdoor,' a website page where the MoD openly invites 'anyone interested in working innovatively with [the MoD]' to reach out via the contact form at the bottom of the page.[57]

In a second, more personal approach, the MoD also allows employees to act as individual points of contact, both online and in person. RAS Lieutenant Colonel Hädicke, for example, told the Dutch management blog *Link Magazine* that:

> ... whoever thinks they can contribute [to RAS] is invited to reach out. [...] If it turns out that we have the same ambitions, I would like to start a collaboration with the goal of creating a win-win situation. This is unconventional for military units, but this case is different.[58]

This has resulted in a thriving online network of MoD innovation managers and startup scouts openly discussing the MoD's innovative endeavors on social media platforms and civilian blogs.[59]

Thirdly, the MoD deliberately designed their 'in-house' experimental units, such as RAS, to strongly resemble commercial startups. This becomes apparent in the stories about how they originated. The project leaders of the 'Deep Vision' unit, which monitors cutting-edge commercial applications of AI, for instance, state that they are not an actual formal department. Instead, they 'claimed' office space in The Hague and 'arranged' computers needed to train their AI networks: 'We had no budget, no people. We basically had nothing except for ideas. We recruited a few trainees and just started doing.'[60]

Fourth, the MoD has sought to establish MoD units in the midst of so-called 'start-up ecosystems.' Despite the small size of the Netherlands, one can find a multitude of these throughout the country, usually located near tech universities. Here, startups can acquire a small plot to work on prototypes and grow their businesses while profiting from the ecosystem's network.[61] These plots are increasingly also occupied by MoD units.[62] When we visited one called 'Military Innovation by Doing' (MIND), the strategic advantages of being at the epicenter of these startup ecosystems became increasingly clear. MIND's employees knew about the military value of each of the products being developed by the two dozen startups present, such as metal 3D printers, tulip-shaped wind turbines, and drones designed to inspect the inside of underground pipelines.[63]

Finally, while the MoD characteristically values hierarchy and formality to ensure operational efficiency, hierarchy is less visible within the emerging military–commercial complex, and 'informality' is celebrated instead. Commanders encourage working on a first-name basis, whether it concerns civilians, corporals or lieutenant colonels. These loosened manners are believed to lower bars and stimulate short lines of communication. Many innovation managers stress how valuable networking and coincidental informal encounters, as opposed to top-down orders, have been for their ability to capitalize on their ideas.[64] On the policy level, managers also choose this informality to enhance innovative capabilities. One innovation manager recalled how he and his colleagues have been holding influential yet undocumented meetings since 2014:

> We established all kinds of informal networks in which decisions were made. It takes some time to get used to that in this organization, to have meetings that basically do not exist, where decisions are made. It is really awesome that we managed to establish that.[65]

BEYOND UKRAINE

To lay the groundwork for a military–commercial complex in the Netherlands, the MoD has thus 'forged alignments' with high-tech students and startups by rendering innovation a 'technical solution' to the threats faced by the state, and thereby all of its citizens. Moreover, following Li's sixth practice of assemblage—managing contradictions and failures—the MoD has engaged in practices of corporate entrepreneurialism to diminish organizational differences and facilitate smooth interactions.

Authorizing Knowledge and Containing Critiques: Closing Down the Conversation

Despite these efforts, MoD startup scouts and founders of startups relayed that some of these companies remained hesitant to collaborate with the MoD because they feared their work would contribute to killing people and that a contract with the MoD would undermine their public reputation.[66] To appease these concerns, the MoD uses a tactic that is in line with what Li calls 'anti-politics.'[67] This tactic is not only directed at potential startup partners, but more broadly to the Dutch public and civil society, whose opinion the startups deem important.[68] At the core of the anti-politics practices, Li identifies inviting parties 'to join the debate, while limiting the agenda.'[69] According to Li, a party who is willing to 'improve itself in dialogue with its critics, learning from scientists and new experts in the community, [will strengthen] its claim to govern.'[70] Despite this, keeping control of the agenda allows the same party to avoid specific topics, which means opening can turn into closure.[71] These practices are clearly reflected in the MoD's contribution to the public debate on autonomy in weapon systems. For instance, Lieutenant Colonel Martijn Hädicke of the RAS unit often appears in public to discuss increased system autonomy in the Dutch military. However, the repeated mantra in his and other managers' narrative is that developing or acquiring fully autonomous weapon systems is off the table.[72] At the core lies the idea that robots are there to support human endeavors: 'Artificial intelligence is not to take over everything,' military innovators claim, as 'a big part of our job remains human work' and '[at the MoD], we reject fully autonomous

254

systems by principle. That leaves only weapon systems with partial or controlled autonomous functionality.'[73]

By limiting the agenda, the MoD evades discussing scenarios in which, for instance, their adversaries create fully autonomous weapons, despite this being an incentive to increase autonomy in Dutch defense systems. We identify a similar pattern in the MoD's acquisition of MQ-9 Reaper drones. From 2013 onwards, when the MoD first asked the Dutch parliament permission to buy four uncrewed aircrafts, they repeatedly reassured the opposition that they would 'use them for surveillance purposes only,' that there was currently 'no desire to arm them,' and therefore 'there was no need to engage in a debate on this topic.'[74] No sooner had the Reapers been built and the pilots finished their training in the US in 2022, than the MoD seized upon the large-scale use of drones in Ukraine to successfully assemble support to buy four more MQ-9 Reapers and arm all eight of them.[75]

Thus, by opening the discussion on the acquisition and innovation of uncrewed and now autonomous weapon systems, inviting everyone to join in, and then stating that there are no intentions to arm drones or make weapon systems fully autonomous, the MoD maintains the power to exercise control over the contents of the agenda and evade answering critical questions by deeming them as irrelevant. In our case, this allows them to reassure startups that their prototypes will not be used in a way that they or the broader public disapproves of, even if in the longer term—as our example of the Reaper drones illustrates—the decision to change the function of the technology may still be made once it is launched and the political arena changes.[76]

Bridging the Valley of Death: Aligning Interests

Aside from increasing commonalities and critiques, we can also see how the MoD aims to identify and cater to the distinct interests of the parties it seeks to collaborate with, even if these interests are different from its own.[77] For instance, a young startup's road to stability is considered most risky during what is generally called the 'valley of death,' which refers to the period between the start of a business' operation and the moment it starts generating revenue. Many startups perish on their way through this valley, with most failing to make a profit before their initial capital runs dry. Although most expenses made in this period, like rent and wages, are predictable, it's the unpredictable costs, like research and development (R&D), that generate most risk.[78]

Aware of the threat this valley poses to startups, the MoD started to organize the Defense Innovation Competition (DIC) from 2011 onwards. For this, it invites small-to-medium-sized enterprises to pitch a technical solution related to a specific topic, like 'smart robotics', 'energy and sustainability', or 'subterranean operations'.[79] The submission with the highest potential wins EUR 200,000 to further develop the prototype.[80] By doing this, the MoD offers an exclusive opportunity for startups to bridge the valley of death. As one competition winner told us: 'two hundred grand is a huge sum. Especially because it is not a loan but R&D funding. That makes the DIC really unique.'[81] The MoD thus learned to identify the most important 'threat perception' of their potential partners and to offer startups the resources needed to innovate 'technical solutions' that could benefit both parties. This increased the MoD's attractiveness as a partner and its ability to forge more sustainable alignments.

In March 2020, while the team of Defensity College students were developing swarm technology in Project PARIS, the eleventh edition of the DIC took place with the theme 'Smart Robotics.'[82] The SkyHive system submitted by startup Tective Robotics won the competition. With this system, Tective aimed to improve 'red force tracking' capabilities by solving two hardware problems commonly encountered with swarming. The first of these was the in-operation charging of individual swarm elements, which Tective sought to optimize by building a charging dock that could be mounted to a military vehicle or a soldier's back, and which can store and charge up to five drones. Drones with an empty battery could then fly to the charging dock autonomously and swap their empty battery for a full one. The second problem was related to connectivity. Swarms typically communicate through a central Wi-Fi router; however, creating and maintaining a stable connection poses an enduring challenge on the battlefield. To solve this, Tective aimed to create

256

ASSEMBLING THE FUTURE OF WARFARE

an internal network connection, transferring information to the operator through the swarm's elements while allowing the swarm to keep operating in case individual elements were eliminated.[83] Tective delivered a prototype of their SkyHive system during a demonstration in November 2021.[84]

Military–Industrial–Commercial Complex: Re-assembling Old Elements and Grafting on New

As empirically illustrated, the MoD's efforts to forge alignments resulted in productive new partnerships, with students on the one hand and startups on the other. Hereafter, the MoD continued to apply practices of re-assemblage, which Li defines as 'grafting on new elements and reworking old ones.'[85] As one of Tective's founders explained, he was quickly introduced to the group of Defensity College students working on the earlier-discussed software-oriented Project PARIS.[86] Both parties realized that their efforts could complement each other. After Tective completed their SkyHive system prototype, RAS, the Project PARIS students and Tective entered into a productive collaboration under the name of Project SPEAR. Although the Defensity project ended not long after, this did not mean that the assemblage disbanded.[87] In fact, the partnerships became simultaneously more solid and diverse in the months that followed, as new elements were drawn in. From various LinkedIn pages, we gathered that one RAS commander exchanged their full-time employment for a position in the national reserve, allowing them to create a startup with one of the former Defensity students.[88] This startup, Avalor AI, then took the place that the disbanded Project PARIS team held in the assemblage, and continued developing the software for Project SPEAR.

As the project evolved, the elements of the assemblage realized that their efforts could profit from more expertise. To improve the image-recognition capabilities of the drone swarm, the evolving consortium asked long-time golden triangle ally TNO to join.[89] To optimize the design of the individual drones, they also invited Delft Dynamics, a small drone-producing company that has worked with the MoD since 2007 and won the DIC in 2018, to join them.

257

BEYOND UKRAINE

Themselves operating like a swarm, each element of the consortium contributed in its own way to reach the goal of automated red force tracking for land operations. The MoD thus re-assembled new and traditional actors in the military–industrial–commercial complex, turning the traditional golden triangle into what they now call the 'golden ecosystem'.[90]

Managing Tensions, Contradictions and Failures

After alignments are forged and the assemblage starts to take shape, the hard work is far from over. As Li describes, contradictions and failures are bound to surface, and maintaining the ability to cohere and act depends on whether the assemblage can be held together in light of these tensions. According to Li, compromise and adjustment are vital in order to succeed.[91] One of the first tensions within the military–industrial–commercial assemblage that we observed is that advanced militaries within democratic states are essentially risk-averse organizations that aim to prevent failure and that cannot justify making large financial bets.[92] Many within the MoD identify this cautious stance as the most fundamental obstacle to innovation. The MoD seeks to overcome this contradiction by increasingly embracing 'experimentation' and 'failure' as a source of progress.

Instead of the multi-year project format that characterized the traditional golden triangle, from 2014 onwards, the MoD started introducing 'short-cycle innovation' to optimize innovative processes.[93] Within short-cycle innovation, the MoD seeks to create a prototype that meets the minimum requirements as soon as possible, after which it is constantly improved using short sprints and evaluations. Within this process, experimentation in situations as close to reality as possible is considered crucial to maximize the chance that intended users will successfully adopt the technology.[94] To facilitate this, the MoD grants startups access to its bases and training grounds, where entrepreneurs and engineers can experiment freely in dedicated 'test beds.'[95] For instance, members of Tective Robotics involved in Project SPEAR spent an entire week on a military base, joining exercises and testing prototypes in an environment relatively close to the military reality. Technologies

258

with higher maturity levels are sometimes even sent along during MoD missions in overseas territories.[96] Innovation managers, end-users and startups unanimously praise the introduction of short-cycle innovation and real-life experimentation.[97] While experimentation in real-life environments as a method for military innovation is far from new, the extent to which it has been institutionalized within the Dutch armed forces is unprecedented.[98] In the MoD's official policy and strategy documents published from 2018 onwards, the method is fully embraced, including promises of more fundamental political and financial support.[99]

Moreover, within these short-cycle experimental processes, risk-taking and failure are overtly encouraged. In the 2018 Defense Innovation Strategy, the MoD recognizes that chances of failure will increase when one seeks to innovate quickly.[100] As such, progress is prioritized, and 'trial and error' and 'fail-fast' strategies are embraced.[101] In a returning column called 'Brilliant Failures' in the MoD's innovation magazine *N PRGRSS*, innovation managers share stories of vain efforts and blunders, explaining how these experiences improved their work.[102] When asked if he felt he had the freedom to make mistakes during his visit to the military base, the founder of Tective responded that 'as long as it concerns R&D and concepts, anything is possible.'[103]

A second contradiction within the military–industrial–commercial complex is that even if certain technologies go through the full development cycle and prove successful prototypes, there is no guarantee that the MoD will actually purchase them. While this is a distressing perspective for startups, it is also inconvenient for the MoD, as it undermines its aim to be seen as a trustworthy partner, not to mention its ability to attract and retain financially strong partner companies. The Tective founder explained how one MoD commander sought to overcome this challenge by advising the startup to introduce its SkyHive system to the commercial market.[104] Shortly after, Tective established a partnership with an agricultural company producing potatoes, sugar beet and maize. The decision to focus on agriculture was not made arbitrarily. The agricultural sector has much in common with the military realm when it comes to the problems Tective seeks to solve with SkyHive, as both sectors

BEYOND UKRAINE

profit greatly from autonomous drone swarms that can identify the tiniest abnormalities in homogeneous environments. By encouraging startup partners to seek collaborations with commercial parties, the MoD essentially allows the technologies it initially developed semi-in-house to become fully dual-use. This tactic yields more advantages for the MoD beyond ensuring the financial stability of their startup partners; according to the Tective engineer, his company can develop its products at a much faster pace with their agricultural partner than would be possible at RAS. Simultaneously, the lessons and improvements gathered in the agricultural environment still find their way into the product Tective develops in conjunction with the MoD. By nudging startups to engage in commercial partnerships, the MoD thus profits from innovations that it cannot reach by itself at the same pace, if at all.

Conclusion

Through applying the concept of the assemblage, this chapter analyzed the practices of assembling and reassembling implicated by the process of developing swarm technology within the context of the Dutch military. We have paid attention to how innovation in swarm technology was made possible through the discursive construction of a shared threat definition and understanding of swarms as a technological solution to a complex set of problems; to the ways in which new partnerships between the military, university students and startups were forged under this common understanding of swarm technology; how shared forms of knowledge and practice were established, and conversations about ethical concerns were closed down; and how, through this very act of assembling and reassembling, new logics of entrepreneurialism, informality, experimentation and 'learning by failing' were 'grafted onto' the military–industrial–commercial assemblage. These new elements allowed the Dutch MoD to navigate tensions and contradictions, bringing together existing and new 'outside' partners into formal and informal relations deemed necessary to construct swarm technology.

Our analysis illustrates that although the innovative center of gravity seems to have shifted from the military to the commercial realm, the Dutch defense organization managed to adopt new tactics and strategies—partly through compromising on various fundamental aspects of its organization—to maintain its ability to access cutting-edge, military-relevant innovations. NATO's recent announcement that its new Innovation Fund will be based in the Netherlands suggests that its members consider the 'Dutch formula' a formula for success that should be strengthened and expanded.

Our analysis, however, also points out that the diversity, informality and experimental character of the partnerships within the Dutch military–industrial–commercial complex raises a set of critical questions related to transparency, responsibility and democratic control. In her book *Expulsions: Brutality and Complexity in the Global Economy*, Saskia Sassen makes an important observation when stating how the sheer complexity of the global economy makes it hard to trace 'lines of responsibility' for the violence it produces, but also how equally hard it is for those who benefit from the system to feel responsible for the harm done.[105] As outlined here, the emerging assemblage of actors that have developed swarm technology—which very much resembles the technology that it is making—raises similar questions about the increasingly distributed nature of responsibility and accountability, as well as our ability to control growing investments into lethal technology. As more and more elements are drawn into the swarm assemblage, it becomes increasingly difficult to understand who is responsible for what and who is to be held accountable when technology fails. In the process of continuous adaptation and innovation, state militaries steadily move away from hierarchical, democratically accountable systems and become progressively similar to the rivals they deplore. Where will this take us, and where does this end?

We are constantly living through the continuation of one war and the beginning of another.[106] The opening of this chapter highlighted that the war in Ukraine, the technologies used to fight it, and the volatile assemblages that create wars raise pressing questions about where the current (re)working of meaning, truth and order is headed. In particular, the increasing normalization and legitimacy of autonomous technologies and the assemblages deemed necessary to build them demand further examination. Autonomous systems may now provide the Ukrainian armed forces with the tools needed to counter Russian aggression. At the same time, their current deployment aids the acceleration of system autonomy and remoteness, feeding into a greater willingness to embrace the military solution and increased escalatory dynamics. The significance of the current developments for the wars to come cannot be understated and requires our continuous critical attention, now and in the future.

PART IV

ANTICIPATING THE FUTURE OF WAR

13

THE PAST AS GUIDE TO THE FUTURE

Beatrice Heuser

Can a study of the past shed light on the future? Here, a particularly poignant insight from two Russian strategists—one long dead, one very much active as I am writing this—feels pertinent. In a much-cited 2013 speech, Russian General Valery Vasilyevich Gerasimov noted:

> The outstanding Soviet military scholar Aleksandr Svechin wrote, 'It is extraordinarily hard to predict the conditions of war. For each war it is necessary to work out a particular line for its strategic conduct. Each war is a unique case, demanding the establishment of a particular logic and not the application of some template.' This approach continues to be correct. Each war does present itself as a unique case, demanding the comprehension of its particular logic, its uniqueness. That is why the character of a war that Russia or its allies might be drawn into is very hard to predict. Nonetheless, we must [predict]. Any academic pronouncements in military science are worthless if military theory is not backed by the function of prediction.[1]

The current war imposed by Russia on Ukraine has led several Western governments to the unpleasant realization that their forces are configured to an excessive extent to fight limited expeditionary wars only, and that they are ill-prepared for major war. It is all very well to argue, as Aleksandr Andreyevich Svechin did, that prediction is devilishly difficult given the infinite multitude of interacting variables that come into play to constitute our future. It is well-nigh impossible to predict with any accuracy the next damn thing that will happen, to paraphrase Winston Churchill. We could leave it at that if it weren't for the fact that we expect our governments to plan for future eventualities, future threats to our polity, including that of war, and at the same time to spend taxpayers' money wisely. That implies configuring forces for particular future threat scenarios rather than assembling a full panoply of arms. The latter might lead to a little of everything but not enough of anything. Only the United States can afford to go for that option, and even their government is keen to forecast. Only states blessed by an exceptionally isolated location like New Zealand or Iceland could get away with minimal armed forces, equipped at best for patrolling coastlines and serving as expeditionary forces under the auspices of UN interventions, to show the flag.

For all other states, maintaining and adapting their defences appropriately requires some forecasting, some bets made on the future, that must underlie major procurement decisions, major decisions on the configuration of armed forces, of alliances and treaty obligations incurred. As Gerasimov rightly pointed out, governments must plan for the future. This applies whether it is with regard to building and maintaining hospitals in view of future health needs, or recruiting and equipping armed forces in view of future security problems. Defence procurement choices made today will affect the coming thirty years, often more; systems ordered today may only come online a decade from now. Similarly, the admiral or general we need in thirty years' time must be recruited today. Therefore, the impossible must be done: forecasting must be built into our planning, our policymaking, our strategic decisions about recruitment and defence spending.

There are a variety of tools that one can draw upon to effect such forecasts. What will be explored here is what tools a study of the past offers, and what its limitations are.

THE PAST AS GUIDE TO THE FUTURE

What the Past Does, and Does Not, Tell Us About the Future

To begin with, some general reflections. First, where we come from does not predict where we go. Leaving aside popular notions of history repeating itself by stuttering, or events always happening twice—first as a tragedy, then as a farce, and so on—two positive statements can justifiably be made. The first is that no two situations are entirely identical. The second, however, is that non-identical but similar configurations have come up in history, sometimes with similar consequences and outcomes. As societies cannot be put into test tube experiments in which all *ceteris* are kept *paribus*, all we have are historical examples where similar configurations with similar outcomes might suggest some degree of causality. This might lead us to expect similar configurations producing similar results in the future (at least if the other most important variables are a rough match).

Historical Precedent? Where History is Helpful and Where It is Not

As a first example, let us turn to financial developments and their impact on societies. The extreme impoverishment of large strata of American and European societies subsequent to the Wall Street Crash of 1929 prepared them for the rise of populism and the polarization of politics. Similar effects, cushioned by the social welfare state and thus taking longer to unfold, were created by the 2008 financial crisis. In this instance, pockets of the societies of the USA, Britain, and a number of Continental European countries felt ignored by the state and expressed their dissent by voting for politicians and parties with a cavalier disregard for facts and evidence beyond that generally expected in politics.

This did not go quite as far as what happened in Weimar Germany in 1933 when Hitler came to power, but here too, we find a recurrent pattern; one where a turbulent democracy in which internal strife turns violent gives rise to calls for a strong and to some degree authoritarian leader—calls that were answered in turn by Julius Caesar and Augustus (as Caesar was assassinated) at the end of Rome's republican period, by Cosimo de' Medici in Florence,

267

Napoleon in the French Revolution, and Putin after the economic chaos and plight of Russia in the 1990s (a country, moreover, that has a historic penchant for authoritarian rule). Elsewhere, such unruly times repeatedly gave rise to military putsches, especially in the late nineteenth and early twentieth centuries. Such patterns suggest plausible causalities and can usefully be drawn upon in the construction of scenarios for possible futures. They do not, however, have fully predictive value. The United States, for instance, despite being just as badly hit by the Wall Street Crash as the European countries in its wake, did not turn into an authoritarian system. Franklin Roosevelt's 'New Deal' did not destabilise democracy in the USA. And while the 2008 financial crisis led to the rise of right-wing parties and to the growth of support for populist leaders in many Western countries, and while Poland and Hungary adopted a series of laws that can be seen as departures from democratic checks and balances, democracy was not formally superseded by an authoritarian constitution in any of them.

Concomitantly, historians and well-educated journalists in the early 1990s articulated their fears that, like the Weimar Republic, democracy in Russia might not survive the combination of a perceived humiliation (American triumphalist proclamations of having 'won the Cold War') and economic plight. Indeed, some expressed the fear that the failure of Russian democracy and its new free-market economy (that in fact turned into what Marx called *Raubkapitalismus*—extreme capitalism—benefiting only the few) might lead to Russian revanchism and efforts to turn back the clock, to recreate the Soviet Union or some sort of Soviet Empire.[2] This analogy proved prophetic. Vague concerns of this sort—not necessarily articulated in terms of historical precedent, or in fact rarely (if ever) articulated at all—were of course the reason for the maintenance of the North Atlantic Treaty Organization (NATO) itself; they were the elephant in the room when the majority of NATO member states throughout the 1990s did not heed some French voices calling for the disbanding of the Atlantic Alliance.

Another example where past patterns may predict future events, on an even more general scale, is that of mass migration. Massive climate change has in the past led to the movement of populations.

THE PAST AS GUIDE TO THE FUTURE

The many waves of migration from prehistory to the recorded age of great migrations (the *Völkerwanderung*) seem to have been triggered repeatedly by climatic factors. Thus it is no high-risk venture to predict that, in the future, humans will not simply stay put and die in regions bereft of water, drown in the oceans or fall prey to desertification, but will pick up sticks and walk. Climate change will lead to mass migration. In the past, mass migration has often resulted in war; Europe in particular has been at the receiving end of wave after wave of immigration. Already in the last centuries BCE, these led to resistance on the part of local populations, which the immigrants tried to overcome by armed force. This can be seen in the fortified villages and towns built on hilltops from Ireland to the Middle East from as far back as prehistoric times. After all, no one would settle on a hilltop, far from rivers with fish and commerce, from fields and pastures, and no one would build fortification if it were not out of the perceived need for self-defence. We conclude that armed immigration led to war, and we note, given Europe's higher living standards than its surrounding areas, both the likelihood of continuing demographic pressure on Europe from would-be immigrants coming from outside and the potential for armed conflict that this bears. Notwithstanding legitimate criticism of mismanagement of such mass migration, particularly across the Mediterranean, with lives lost, it is greatly to the credit of the EU and its individual member states that due to their management of the phenomenon, it has hitherto not led to anything resembling war. Whether this will continue to be possible is more difficult to predict.

Then there is the balance between national proclivities of *longue durée* and profound change that can come about by a great crisis. On the one hand, there are indeed examples of nations cultivating the trope of a claim to larger leadership or imperial power, claims that can be traced back over centuries, as in France (as leader of Christendom), or with the imperial traditions of Russia and Turkey. On the other, the shock of defeat, occupation and re-education can lead entire nations to abandon such ambitions, and indeed any taste for war, as we can see with Sweden after the Napoleonic Wars, or with Austria after the First World War and Germany after the Second.[3] As primatologist Robert Sapolsky has found, even troops of apes can change their culture from being more to less aggressive.[4] Another shock might indeed produce a modification of such a stance, but not a full return to the previous state.

The *longue durée* is usually identified with geography, but we have already noted how climate change can change the nature of geographic features. For example, as glaciers and ice caps melt, seaways become navigable or areas become cultivable which previously were not. Also, technological innovation can mitigate geographic constants. In other words, what might be seen as the most lasting and eternal aspect of the world—its geography—has been mitigated, to some extent, by human progress in technology. Whilst the invention of the instruments necessary for navigation in the absence of sunlight or a view of the stars enabled Europeans to settle in the Americas, Australia and New Zealand, the invention of aircraft meant Australia could come under direct Japanese attack in the Second World War. Yet distances still matter today, and mountain ranges or deserts pose a challenge to be overcome. While the Second World War made flying military personnel and supplies over the 'hump' (the Himalayas) a common thing, and the Berlin Airlift of 1948–9 transported not only coal and victuals but also the elements of a whole power station to the besieged German city, even in the late 1990s NATO's Military Committee refused to contemplate an invasion of Kosovo from Albania in view of the logistical problems of getting across the mountains of northern Albania. From the late 1950s, when America developed constant airborne alert of its nuclear bomber fleet, to its almost immediate Global Strike capacity now, and with Russia and China developing comparable capabilities with their missiles and submarines, we see technology conquering space. Even so, distances still pose challenges of existential importance when it comes to getting clean water to many African villages, let alone getting children to school or providing sick people with medical care. Paradoxically, many key factors can affect a situation in one way as well as in the opposite way.

However, geopolitical arguments for historical inevitability must be treated with caution. For example, what is today the territory of Belgium was a key battleground of Europe for many centuries, where the Holy Roman Empire (especially under the Habsburgs)

270

THE PAST AS GUIDE TO THE FUTURE

and the growing kingdom of France clashed with an admixture of intervention by the powerful House of Burgundy (who owned most of modern Belgium) and England. Yet this has not been the case since 1945. Furthermore, not every island state turned to trans-oceanic trade in early modern times, or undertook maritime expansion to acquire colonies. The history of England/Britain is unparalleled by that of, say, Ireland, Java, Madagascar, Sri Lanka or the Philippines. And Japan, for example, espoused imperial ambitions only after it had encountered them at the hands of the Americans and the Europeans in the nineteenth century.

The Commemoration of Past Wars as Incentives for Future Behaviour

Historic precedents can play a role in the present and future in quite different ways; not in terms of what has happened, but in terms of how the past is evoked. It may be the conscious way in which societies relate to the past, usually with a considerable degree of distortion, that gives observers some idea of the possibilities for their political developments in the near future. Different ways of commemorating past wars—especially when these are instigated and organized by governments—are indicators of the short- and medium-term foreign political intentions of a government and the degree of support it might have among its population for foreign aggression. In US administration parlance, this can be marshalled to the needs of 'anticipatory intelligence'.[5] Jeannie Johnson and Marilyn Maines, leading practitioners of this form of cultural anthropological research, use the term 'mapping cultural topography' to identify these and other cultural habits and traits of societies, and their potential political implications.[6]

To take a step back, the recollection, invocation and remembrance of past conflicts in which one's own community (or its real or imagined forebearers) has been involved each play a crucial role in the crystallization or re-affirmation of the solidarity of this group. Certain wars and battles of the past have become foundation myths for nations, or are invoked as models for action in current or future political contexts.[7] The paradigm for this is Pericles' funerary oration for the fallen Athenians in the Peloponnesian War,[8] which would

271

serve as the model for public speeches in the nationalist nineteenth and twentieth centuries. Past wars can be evoked to keep alive or revive hostilities, or to reinforce group solidarity. One example of this is Soviet leaders' (and now Putin's) claim that there are fascists everywhere (in the Cold War, Soviet propaganda pointed the finger especially at West Germany; now of course, Ukraine has been identified as fascist), threatening Russia.

Where past wars led to defeats and the cession of territory to the victors, they are now invoked to fuel revanchism.[9] In other ways, they can be indicators of an expansionist or otherwise aggressive foreign policy agenda, as bases for territorial claims or claims to overlordship, or dominant status beyond one's state frontiers. By contrast, they can also be indicators of a culture's pacific disposition, stressing peace and reconciliation.[10]

While examples date back to antiquity, bellicose references to past wars are to be found frequently since the French Revolution among polities, initially mainly in Europe. Since the twentieth century, these are to be found the world over (hand in hand with the drive for decolonization), linking past wars as national foundation myths to the demand for national independence.[11] In Western, Central and Southern Europe, playwrights, poets, painters and sculptors competed to depict tribal leaders who rose up against Roman occupation as precursors to the resistance against Napoleonic occupation. In Southeast Europe, mediaeval or early modern wars against the Ottoman Empire played analogous roles in the nation-building of the nineteenth and twentieth centuries. This can be seen as recently as during the Wars of the Decomposition of Yugoslavia (between 1991 and 1999), in which Milošević and his Serb nationalist followers commemorated the battle of Kosovo Polje in 1389. Several other battles against (and usually heroic defeats at the hands of) Turkish invaders have become key points of reference for the birth of their national identity.[12]

National, war and military memorials and museums have been established throughout Europe since the nineteenth century, as well as in other parts of the world that have their cultural roots in Europe;[13] the First World War in particular has been commemorated by many memorials and museums.[14] While the architecture of war

THE PAST AS GUIDE TO THE FUTURE

memorials has generally not been altered since their erection, they tend be combined with visitor centres that, like museums, are redesigned periodically.[15] Along with war museums and museums of the armed forces, such visitor centres are outstanding indicators of the prevailing spirit in the society that has created them, or at any rate of its intellectuals and its ruling elites. They can be roughly categorized into three groups, although individual museums may straddle these classifications.

The earliest but still existent version is that of a strongly nationalist and triumphalist celebration of one's own military and past wars, which might include a period of defeat and occupation, but which tends to downplay collaboration or other internal divisions. It tends to be state-funded, displaying little criticism of one's 'national' military history. It tends to emphasize hardware exhibits, rather than nuanced historical explanations.[16] If state-funded, government directives for the conceptualization of the exhibitions tend to include the prescription that it should help the country's armed forces with their collective identification and that it should seek to uphold the 'military culture' of that country—a term used not in an academic sense of what objectively constitutes the military culture of a country, but what it would be nice to have, emphasizing rituals (e.g. swearing in, or graduation from military academies), anniversary celebrations or annual parades.[17]

A second is usually the product of the involvement of historians with a democratic, critical slant; that is, critical of their own nation's history and careful not to depict a war exclusively as a national triumph. Attempts to explain past wars in an even-handed way, both in terms of the causes and the suffering, include the *Mémorial* of the First World War in Péronne,[18] or the current museum at the battlefield of Culloden in Scotland, run by the National Trust for Scotland.

A third emphasizes commemoration and mourning. This is done in two ways. It is either tendentially pacifist, emphasizing the wars as settings of death, maiming and destruction that we have an obligation to mourn. Such mourning can be made in the spirit of pure gratitude for the sacrifice made (involuntarily) by the fallen, examples of which are the Peace Park in Hiroshima, the Flanders Fields

273

Museum in Ypres, Belgium, or the Galeria 11/07/95 in Sarajevo to commemorate the Srebrenica massacre. Alternatively, such commemoration and mourning can be turned into an encouragement for new generations to make such sacrifices themselves if called upon (in the spirit of Pericles), or it can amount to a complaint about past suffering that crystallizes in a call for retribution. In this latter case, there can be a note of revanchism. This was the case for the earlier layout of the museum at Culloden battlefield, which amounted to a complaint about Anglo-Saxon dominance of an imagined Celtic heritage supposedly destroyed with the defeat of the Jacobite forces there in 1746.[19]

Other than museums, all sorts of interpretations of the nexus of war and group identity can be found in literature, fine art and films. An example is the treatment of the subject of the battle of Arminius and his Cheruscan followers against the Romans in 9 CE, which was variously interpreted as a prescription of German national resistance against the French by the early German nationalist Heinrich von Kleist in his play *Die Hermannsschlacht*;[20] as indicator of Nazi intolerance and racism by the Jewish-German author Lion Feuchtwanger in *Die Geschwister Oppermann* (1934); and as an exhortation to guard against unwarranted national pride by the German Democratic Republic (GDR) author Erich Loest, writing towards the end of the GDR's existence.[21]

Novels are rarely government-sponsored, but films often are, at least through generous co-financing. A prime example is the Russian film *Alexander Nevsky* (Sergei Eisenstein, 1938), concerning the battle of the Russians against the Teutonic knights at Lake Peipus in 1242, which was commissioned specifically to raise the morale of the population. Other examples include British Second World War films such as *This England* (David MacDonald, 1941), which refers in passing to the Battle of Hastings in 1066, the Spanish Armada of 1588, and Napoleon; *Henry V* (Laurence Olivier, 1944), referring to the victory of Agincourt in 1415; or the German films of Veit Harlan, including *Der große König* (1942), which focuses on Frederick II of Prussia and the Prussian victory of Leuthen 1757, and *Kolberg* (1945), which tells the story of the siege of the city in 1807. Wars and battles of the past are of course central *topoi* of literature and the film industry, defining and redefining the images of war in an increasingly international audience, renewing and rewriting 'memories' of past conflicts as a palimpsest to which journalists, politicians, cartoonists and speech-writers can refer.[22] It is hardly coincidental that two Russian TV series about Catherine the Great were produced in short succession in 2014 and 2015 (both released on the Russian state-owned TV channel Russia-1), just as Russian soldiers or soldiers in Russian pay were fighting in the Donbas, which Empress Catherine had conquered and where she had settled Russians in the eighteenth century, giving the region the name *Novorossiya* ('New Russia').[23] How better to educate Russian public opinion to the idea that Donbas was really Russian and should belong to Russia?

Collectively, state-funded museums and state-maintained war memorials, together with regular or 'round anniversary'-based public commemorations of wars, and the treatment of wars in literature and film, is an indicator of a polity's and especially a government's disposition towards war, and indeed of its political intentions. As such, they are key indicators for the mapping of the cultural topography of individual polities, furnishing evidence for 'anticipatory intelligence'. A caveat must be added, however: they only reflect their sponsors' ideas at the time. A museum or exhibition can often reflect a long-outdated mind frame or interpretation simply because the funds were not there to update it, or simply that nobody thought of doing so. Even then, it continues to reflect a particular way of thinking about the past—especially relevant to us in relation to museums of national history or those on war—which may survive in a new fashion, simmering away, before coming to the fore again under different political circumstances.

Antithetical Potentialities

This leads me to antithetical potentialities. Any situation holds within itself opposite potentialities for future developments. At any point, a good analyst—take Raymond Aron, writing in 1951—may identify the most significant potentialities for future developments.[24] Yet any moment bears in itself different potentialities; these cannot all be realized simultaneously as one or several may cancel another one out. Only some will fully unfold. And yet any development may bear within itself its own contradiction. Rivalling potential developments which did not occur initially will continue to have their supporters who will seize upon any weakness to claim that it was flawed from the start. Thus, writing in 1951, when the European Coal and Steel Community had initiated a path towards European integration and when the idea of a European army was in the air (it would be formally proposed in 1952), Aron argued that it no longer made any sense for a European state to try to have a military that was a smaller version of that of the USA (in the same way that Belgium had tried to have a military that was simply a smaller version of that of France half a century earlier). It would make much more sense, argued Aron, to have an integrated European army in which countries would specialize in certain roles, relying on other nation-states to complement them (a view shared by half of the French elite, but which was voted down by only just over half the members of the National Assembly two years after it was proposed). Economically, this was quite true then. Despite this economic reality and the huge burden it placed on the French state, France would continue, after 1954, to pursue the goal of having a smaller version of the US armed forces, going as far as emulating the triad of nuclear weapons platforms—airborne, ground-to-ground missiles and seaborne—at hugely great expense.[25] In this case, ideological politics would contend with and in fact prevail over financial-economic arguments. The financial-economic constraints never disappeared, however, going some way to explain repeated attempts in the following decades to finance French defence procurement with contributions from other European countries (especially for nuclear weapons) under the label of European integration.[26]

This example illustrates the rivalling, indeed mutually exclusive, potentialities that one moment can hold. While one can come to the fore to the point where observers may think the other is dead and gone, it can revive and even overtake the other, as contexts, economic developments, and configurations of power and of moods change. In 1949, further British integration into Europe and a British return to its nineteenth-century detachment from Europe (at least in terms of identity and trade) were both potentialities, with both

THE PAST AS GUIDE TO THE FUTURE

EEC/EU membership and Brexit *in nucleo*. How to predict which one would come to the fore and when?

Or take two further examples. On the eve of the First World War, when xenophobic Darwinist nationalism was at its height and, in opposition to this, European nations derived massive benefit from bilateral trade, Norman Angell thought it would make no sense to start a war given their shared interest in prosperity.[27] Nonetheless, ideology prevailed over arguments of prosperity, peace and wellbeing. Similarly, even experts on Russia who had followed its military developments, and had read or listened to Putin's speeches in detail, failed to predict that Putin would discard all arguments in favour of prosperity, peace and wellbeing for his nation and opt for a narrow, ideas-driven war in Ukraine, first in a covert fashion since 2014, and then with a military invasion since 2022. In the winter of 2021–2, when the Russian military was configured for an invasion, both potentialities existed. Yet so many experts failed to see which one would come to the fore. In short, all we can do is identify such potentialities, not which one will predominate at a given time.

Observed Behaviour as a 'Precedent'

What a government does in armed conflict (including counterinsurgency) is closely observed by others and, rightly or wrongly, influences what they expect that government to do again. The feeble reactions of France and Britain to Hitler's expansion by 'salami' slices, first into the demilitarized Ruhr area of Germany, then into Austria, then with the collusion of France and Britain into the Sudeten frontier areas of Czechoslovakia (sold by Hitler as his 'last territorial ambition'), then 'finishing off' the rest of Czechoslovakia in the spring of 1939, led him to expect that London and Paris would not honour their guarantee to Poland against the German invasion in September 1939. Equally, one can surmise that the Western relative passiveness when Crimea was annexed by Russia in 2014 led Putin to expect that the West would also roll over as he hoped when he annexed either all of Ukraine or at least its eastern parts in 2022.

By contrast, the brutal repression of the Chechen fight for independence in the 1990s, first by the USSR and then by post-communist Russia, raised great fears of similar neo-colonial wars

277

in the Baltic States and Eastern Europe, and pushed governments there to seek admission to NATO. Russia's 2008 war against Georgia seemed to confirm that Georgia might in future be endangered again, along with Ukraine and Moldova (while hoping that NATO members would be taboo for Putin). This concern was not unreasonable given the same key decision-maker who had gotten away with such behaviour was still in power, thus encouraging him to try the same again.

This also applied to Alexander III of Macedon, Napoleon and to Hitler: each embarked on further expansionist moves when their previous steps had succeeded. This might lend itself to a general postulate that for a dictator set on expansionism, '*l'appétit vient en mangeant*' ('appetite comes with eating'; that is, one success encourages the next move). But one should beware of thinking that the behaviour of one individual is a firm rule predicting that of another. This is no more appropriate than to assume that your ex-partner's infidelity predicts the behaviour of your new partner; indeed, acting on this assumption is likely to degrade your new relationship. In the late 1940s, Western diplomats and political leaders widely drew parallels between what they thought Stalin was up to with what Hitler had done in the 1930s. In doing so, they reacted to Stalin the way they probably should have responded to Hitler.[28] It is argued by some Western historians who have worked their way through Soviet archives that, by acting on such expectations, the Western powers hastened, even if they didn't altogether cause, the emergence of the Cold War out of the ruins of the Second World War, and in any case aggravated tensions.[29] (Current speculations about how China might make use of the Russo-Ukrainian War to seize Taiwan are at least partly inspired by the historical precedent of Stalin and Hitler's aggressive collusion in 1939, jointly carving up Poland, with Stalin invading Finland and the Baltic states under the cover of Hitler's war.)

It is a different matter again for standard operating procedures, well-liked by bureaucracies and institutions, and for diplomatic or political ploys which are re-applied because they have worked before. The story about how the Soviet missiles on Cuba were identified because of the standard configuration in which they were set up, identical to that used on Russian soil, is well known. On a procedural level, I was amused to find that a German think tank in which I worked in the 1990s, which had been set up by a former member of the German intelligence community, had regular Wednesday morning meetings referred to as '*die Lage*' (meaning a discussion of the general—i.e. military—situation). Of course, the West German intelligence agencies were all set up by people who had been in the Third Reich's military intelligence, and they simply reproduced the procedural patterns with which they had grown up, so to speak. Also, in the politics of states, we can find individual moves or ploys that are then reproduced if they were previously successful. Thus, the Russian government in 2014 turned around Western sanctions imposed after the annexation of Crimea by claiming that it was in fact Russia that was punishing the West by imposing sanctions on the import of Western cheese, fruit and other foodstuffs. In 2022, when it was clear that the West would eventually choose a reduction or even ban of gas and oil imports as the most effective and yet non-militarily escalatory way of undermining the Russian war effort in Ukraine, the Kremlin began to turn off these supplies, claiming variously that it was sanctioning bad Western behaviour or that there were technical problems. Playing to his own audience, this allowed Putin to come across as dominating the situation. The Russians, it seems, are falling for it again.

Then there are some interesting patterns associated with particular cultures. One is for people in dictatorships to speak in cryptic terms, to put forward an opinion without inviting direct censure. During the Cold War, Soviet strategists would claim that Clausewitz was no longer relevant. In translation, this meant war must no longer be seen as a useful tool of politics, thus urging the Soviet government to go for *détente* and arms control.[30] Similarly, when Gorbachev's adviser Andrey Kokoshin and general-turned-professor Valentin V. Larionov wrote the foreword for a re-edition of Aleksandr Svechin's work in 1991, they primarily signalled their support for a strategy of defence in depth which Svechin had advocated in the 1920s and 1930s for the Soviet Union.[31] It is worth looking out for similar cryptic messages.

BEYOND UKRAINE

Fluctuating Forecasts of War

The Next War Will Be Like the Last War

Thinking about war is subject to fashions. In the first decade of the twenty-first century, writing and talking about counterinsurgency was all the rage. Outliers among strategists such as Admiral Chris Parry and General Mungo Melvin in the UK warned that one must not forget about 'major' war, but they were largely ignored, as the syllabi of defence academies illustrate. Then came the all-out Russian invasion of Ukraine, pushing aside the interest in non-state actors almost entirely. This see-sawing of fashions is not a new phenomenon. At different times, people imagined future wars to be 'major', 'limited', 'local', 'mixed' or 'hybrid', all driven by the latest phenomenon they had witnessed.

Usually when coming out of a major war, strategists will assume that the next war will be broadly similar in character, if perhaps more destructive. Clausewitz, for example, when beginning to write his *opus magnum*, started with the assumption that the French wars with Prussia would be the model of all future wars and would continue in this vein. To the surprise of Clausewitz himself, it turned out especially in the 1820s that war could be something more limited, as various Polish uprisings and the Russo-Turkish War of 1828–9 illustrated. This famously engendered this reflection on his part, in which he speculated about the future of war in 1827:

> Whether it will always stay like this, whether all future wars in Europe will always be fought with the total weight of the states and therefore only about great interests which are close to the peoples' hearts, or whether gradually there will once again be a division between government and people, is difficult to decide ... [32]

And it was in fact this realization that led him to start a rewrite of *On War* which he did not complete, trying to turn it from a book that described only the very major—in Clausewitz's parlance, near-absolute—war of the Napoleonic period into one encompassing all forms of conflict.

Napoleon's warfare remained the dominant paradigm, however, and left as its legacy the cult of the offensive. Confirmed, it seemed, by

280

the Crimean, American Civil, Franco-Prussian and Russo-Japanese wars, on the eve of the First World War, the majority of strategists agreed that an offensive with mass armies was the only possible strategic choice in an environment characterized by nationalism, much increased firepower (but also defensive weaponry), and industries capable of mass-producing standardized weapons. They disagreed only on how long they thought the war would take. Only outliers favoured a defensive strategy or spoke of smaller but more high-tech armies. The First World War became what most strategists had expected it to be: a mass war, with massive use of firepower, but also in parts strong defences due to just that firepower. In this way, expectations become self-fulfilling prophecies.

After the Second World War, there was an initial period of five years when some strategists hoped that nuclear weapons had banished war altogether.[33] Then came the Korean War—pretty total for Korea, but nonetheless limited from a Western point of view (being far from home and without the use of the new atomic bomb). As with Clausewitz in 1827, Western strategists claimed this heralded an era of limited wars (I repeat, ignoring what it meant for Korea).[34] Yet NATO continued to prepare for another total war in the hope that those very preparations would keep it at bay, as though that alone could deter a Soviet attack.

What Has Not Occurred in Living Memory Is 'New'?

When it becomes clear that the next war will not simply be more of the same, we will see the recurrent claim that one is encountering 'new wars'. It is a blatant pattern in superficial analyses of war, often made by cursory observers, to claim that something not experienced in living memory—meaning one's own lifetime, but also the accounts of one's parents and grandparents—is entirely new.

Consider some examples. The campaign of Charles VIII of France in Italy in 1494–7 is claimed to have ushered in a new form of war. The French monarch invaded the Italian peninsula with mobile cannons, the first time they were deployed during a campaign. With these, he besieged a couple of northern Italian cities, breaching their curtain walls and quickly capturing them. Fearful of such bombardment and the subsequent looting, other cities quickly surrendered, leaving

281

Charles and his troops to march on unhindered, at a speed that greatly surprised contemporary observers, who had seen nothing like it in living memory.[35] The Florentine historian Francesco Guicciardini described this as a great turning point, which he attributed to the use of cannon. He claimed that henceforth:

> Wars became sudden and violent, conquering and capturing a State in less time than it used to take to occupy a village; cities were reduced with great speed, in a matter of days and hours rather than months; battles became savage and bloody in the extreme.[36]

Cannon had already brought down Constantinople some forty years before Charles VIII's campaign, and field artillery had been the decisive element in bringing French victory on the battlefield at Castillon in 1453. Both events had, however, occurred outside the purview of the experience of the Italian commentator.[37]

Observers who had lived through large-scale wars—particularly the Napoleonic Wars or the two World Wars—would sometimes be surprised by the later recurrence of wars on a smaller and less intensive scale if they had not known these in living memory. The United States had experience with counterinsurgency interventions in their hemisphere in the interwar period with the so-called 'Banana Wars'. In the Cold War, insurgencies, civil wars and other armed conflicts lacking the formal battles between 'regular' armies which had typified the two world wars or the high-intensity Korean War occurred in many places, including Indochina/Vietnam, Cambodia or Laos, in the Philippines or Mozambique.[38] And yet in 1962, US President John F. Kennedy (or his speechwriter) called what was happening in Vietnam:

> ... another type of war, new in its intensity, ancient in its origin – war by guerrillas, subversives, insurgents, assassins, war by ambush instead of by combat; by infiltration, instead of aggression, seeking victory by eroding and exhausting the enemy instead of engaging him.[39]

In the mid- to late-1990s, after the danger of a Third World War receded with the dissolution of the Warsaw Treaty Organization

THE PAST AS GUIDE TO THE FUTURE

and the end of the Soviet Union, several commentators hailed the smaller wars that now gained more visibility as 'new', even though they were mostly no more than further manifestations of asymmetric wars of weaker (insurgent, often non-state) groups against stronger (usually state) powers. One prominent proponent of a 'new wars' hypothesis pointed to the privatization of war, its asymmetry, and the waning of the monopoly of regular armies on the waging of war.[40] All three phenomena, however, are standard features of most pre-modern wars. Already almost a decade earlier, as the Cold War was drawing to a close, Martin van Creveld had argued that not a new but an old form of war was on the rise again, prophesying that both intrastate and interstate wars would be going strong, side by side, in the coming years. He has been proven right on both counts.[41] Kersti Larsdotter's sober comparison of the way the US fought in Vietnam in the 1960s and 1970s, and in Afghanistan in the 2000s, shows that the change lay in tactics more than in strategy, and that despite new technological developments and their rapid introduction into the US armed forces, 'new wars are still fought with old methods of warfare'. In other words, their strategic political aims were entirely in consonance with aims as defined in previous wars.[42]

Similarly, acts matching our definition of terrorism—surprise acts of violence by small numbers of people against random civilians when the government of the state in which these are carried out is seen as inimical—have been recorded at least since the anarchist attacks of the late nineteenth century, including the use of suicide bombers.[43] In the Cold War, terrorist attacks had been carried out by the Algerian National Liberation Front against French civilians in the 1950s, by the Irish Republican Army in London in the 1980s, and by Islamists in Paris in the 1990s. Yet when the terrorist acts of 2001 against the United States occurred, President George W. Bush promptly spoke of 'a different kind of conflict against a different kind of enemy',[44] just as Kennedy had done in 1962.

The identification of 'new' wars has arguably been due in part to historical ignorance, partly to the lasting influence of the 'Great Battle' paradigm, and the undue concentration of modern strategists and political scientists on history since the French Revolution.[45] As Clausewitz had already discovered back in the 1820s, the history of

283

wars is not linear, moving from small and intrastate to ever larger and interstate only, but instead moves up and down the various spectrums of war.[46]

The Future Is in Part what We Make of It

Let me end with this observation: when assumptions about the future feed into the decisions of states as to how to configure their militaries, what weapons systems to procure, and what defence and defence-relevant infrastructure to invest in, then at the same time they determine some—not all, but some—aspects of further war. As Donald Rumsfeld commented, 'You go to war with the army you have, not with the army you wish you had'.[47] This also means that the way you fight that war will depend on what you have; that is, your own force disposition will shape that war. This can be extended to any security problem; you will approach it with the instruments of statecraft (including war) that you have, not necessarily with the means that would be most appropriate for resolving that particular problem.

As we noted, strategists' predictions of the nature of the First World War ensured that the Western European states prepared for mass warfare with greatly enhanced firepower, not for manoeuvre warfare and limited operations. If a government claims that the future is too unpredictable for it to get rid of nuclear weapons, and thus maintains and modernizes them (which is just a euphemism for replacing them with new weapons), the one thing it makes predictable is that the future will be one of nuclear weapons and nuclear powers. There is a vicious circle here which a single state alone cannot break, but to which every single state contributes.

Final Thoughts

What can we do to overcome the constraints on our ability to anticipate the future? How will these limitations continue to cloud our judgement, and what does that mean for future wars? These are extremely difficult questions.

I suspect that the objections put forward over the centuries by the likes of Ibn Khaldun, Clausewitz, Nikolai Medem, Svechin and

THE PAST AS GUIDE TO THE FUTURE

Michael Howard to 'principles of war' (that might be applied to any armed conflict) would also apply to the practice of government planning for future wars. These can be summed up to say that there are simply too many variables to allow for predictions of events beyond those of a few months' time at best.

However, we have argued above that occasionally the alignment of the stars—key variables and key issues—can strongly resemble those of historical precedents. In those cases, it would not be foolish to keep in mind that similar results might ensue without talking them into existence through loudly voiced self-fulfilling prophecies. Decision-makers with a broad knowledge of key historical events, and insight into how they came about and what ensued, might well develop intuitive wisdom that they can apply to their policies.

What is clearly helpful in the analysis of actual and potential adversaries (and any other actors in international relations) is to put specialists to work who are deeply immersed in a culture, who understand its historical and other war-related references, and who can identify the potentialities of their future developments, even if they cannot be expected to predict their actions accurately. For all we know, Putin himself may not yet have made up his mind whether to go ahead with the invasion of Ukraine until shortly before it was unleashed on 24 February 2022. In this context, studying history, but especially studying how your adversaries (and even allies) read and interpret history (which should include reading their works and speeches in the original language) will be particularly useful. As the example of Putin's decision to invade Ukraine suggests, it was delusionary for Western observers to argue, as Jan Bloch and Norman Angell did before the First World War, that it made no sense from our perspective to take such action. We should have paid much more attention, not merely to Russia's increased force strengths and trade figures, but to Putin's little essay of July 2021 titled 'On the Historical Unity of Russians and Ukrainians'.[48]

14

FORECASTING THE FUTURE OF WAR

Collin J. Meisel

Introduction

In the wake of Russia's full-scale invasion of Ukraine, many have been left asking: what could we have known before 24 February 2022? Could the invasion have been predicted prior to intelligence of a military build-up? Amidst this build-up, how confident could we have been that Russia would actually invade? And could we have accurately predicted the type of invasion or course of the war since? The answers to these questions are connected by a much broader one: namely, what can we know about the future? Virtually all trends and phenomena are defined by patterns of (un)knowability, or characteristics which help or hinder our ability to forecast their future development or occurrence. These characteristics—complexity, measurability, equilibration, stochasticity and tractability—combine to make war particularly unknowable, and especially the prediction of its onset over long time horizons.

A good deal of research has focused on sundry aspects of future war, such as forecasts of technologies that may be used in battle,[1] predictions of which country will emerge victorious in a given war,[2] and efforts to better understand when and why wars end.[3] Each of these aspects have one feature in common: they first require war to start.

As illustrated by the US intelligence community who had been warning about the prospect of a full-scale Russian invasion of Ukraine since as early as September 2021, short-term predictions of war onset can prove highly accurate.[4] While impressive, such warnings—despite being issued several months in advance—often do not provide enough time for potential belligerents to adjust long-term strategies, force structures, or inventories of heavy and medium equipment, and the like—adjustments which are often years, and sometimes decades, in the making.[5] Indeed, the North Atlantic Treaty Organization's ten-year plan to assist Ukraine's military in shifting toward Western platforms is revealing in its duration.[6]

An ability to predict the onset—or at least the high risk of onset—of war across long time horizons, particularly with a specific adversary or type of adversary (e.g. an adjacent neighbor or near-peer), would be immensely useful for a country's national security apparatus. New doctrine could be developed, operational plans revised, diplomatic initiatives fostered, high-end equipment produced or acquired and trained on, and other grand strategic preparations made, such as alliance formations. Unfortunately, past efforts to forecast war onset across long time horizons have produced results ranging from mixed to poor.

In this chapter, I seek to make the case for attempts to predict the risk of war across long time horizons despite scholars' poor track record of doing so accurately. I first describe why this track record has been so poor by defining and reviewing patterns of knowability and how they relate to war onset. I then move to what we can know based on patterns of knowability—namely how structural pressures for war are likely to change over time—while briefly addressing the many non-structural factors that affect war's probability. I conclude with how our understanding of structural pressures toward war might shift given Russia's full-scale invasion of Ukraine.

The focus here is on conventional interstate war onset across long time horizons. The insights that follow are at the macro level, and thus geared toward grand strategic planning and policymaking—

FORECASTING THE FUTURE OF WAR

insights for which there remains a need both among policymakers and an increasingly short-term-oriented public.[7]

Can We Predict War?

The short answer is: we can't, or at least not very well across long time horizons. This was the key lesson from Gartzke's influential 1999 study that declared 'war is in the error term.'[8] As Gartzke argued, the uncertainties surrounding causal mechanisms of war are too great to systematically predict the occurrence of conflict.

While the US intelligence community's accurate prediction of the high risk of Russia's full-scale invasion is a triumph, many others got it wrong, including the head of French military intelligence.[9] And while he may have had domestic political motivations for downplaying the risk of war, as late as 22 February 2022, Ukrainian President Volodymyr Zelensky publicly stated that 'a broad escalation on the part of Russia will not happen.'[10] To attempt to correct for the notorious inaccuracies in expert political judgement,[11] quantitative predictive models can dispassionately account for a variety of factors; however, 'It is difficult to imagine that researchers will identify states' true capability or resolve if such factors are opaque to the governments themselves.'[12] As Blainey observed, 'wars usually begin when two nations disagree on their relative strength.'[13] If states were better able to gauge these and other factors, they would incorporate them into bargaining processes. Gartzke finds no evidence that this is the case *ex ante*.

For intrastate conflicts, Hegre et al. have made progress in improving the accuracy of their predictions nine years into the future.[14] Yet, despite outperforming earlier, more widely used models,[15] Hegre et al. concede that their model 'does not perform particularly well for the onset of intrastate conflict.'[16] Additionally, Bowlsby and others find the fit of Hegre et al.'s model to be unstable across time.[17] In a thorough review of past and recent efforts at predicting conflict onset, Cederman and Weidmann find similar patterns, arguing that the present state of the art is inadequate and is likely to remain this way for some time.[18]

This is not for lack of trying. Beginning a half-century ago, Choucri and Robinson led a series of pioneering attempts at long-term conflict forecasts.[19] These scholars concluded that, given war's rare occurrence, expectations that war could be predicted were unreasonable.[20] Foreshadowing Hegre et al.'s results, Gurr and Lichbach found some success in forecasting intrastate conflict in the form of rebellion, but they too deemed their models to be inadequate for conflict onsets in previously conflict-free states.[21] In an extensive treatment of attempts to predict interstate war, Bennett and Stam suffered from similar problems. Despite their systematic study of more than a century and a half of interstate war, their best efforts yielded 'results [that] offer only a little guidance to those seeking to predict when violent conflict is very likely to occur.'[22]

As predicted by Cederman and Weidmann, there have been more promising recent developments in forecasting conflict across short time horizons in limited geographic areas. Yet, the primary predictor of conflict continues to be past conflict. Specifically, Brandt and others find that a general index of violence and instability alongside recent conflict and mediation event observations (CAMEO) account for much of the predictive capacity of their six-month spatio-temporal conflict forecasts.[23] Beyond the general notion of conflict's persistence, in which violence begets violence,[24] these results provide little direction where broad geographic areas and long time horizons are concerned.

Various broad debates in the field of war studies offer little direction for the prediction of war. The 'decline of war' thesis, popularized by psychologist Steven Pinker and championed before and after by scholars such as Azar Gat, dictates that war has occurred less and less frequently in humankind's history.[25] Setting aside debates of whether the thesis is generally accurate—a notion Braumoeller disputes[26]—its specific application appears limited in the case of Russia's war in Ukraine. According to Gat, there remains 'the prospect of more antagonistic relations, accentuated ideological rivalry, potential and actual conflict, intensified arms races, and new cold wars'[27]—a prospect that has transformed into reality despite a more than quintupling in Russia's GDP per capita at purchasing power parity over the past quarter century and persistent improvements in Russians' quality of life, as

FORECASTING THE FUTURE OF WAR

measured by the United Nations Development Program Human Development Index.[28]

Boulding's general theory of conflict and defense draws the opposite conclusion from the 'decline of war' camp, suggesting that conflict will become more common as military reach is increasingly extended.[29] Where a state's potential enemies have an increased ability to mount successful attacks on its home territory, its viability or survivability becomes uncertain. As this uncertainty increases, the logic of international relations for an existentially threatened state has the potential to shift to a 'game of survival,' which in a worst-case scenario can lead to a perfectly rational actor driving an opponent in a 'game' to mutual extinction.[30] If we are to assume that Putin felt an increasingly Western-oriented Ukraine posed an existential threat to Russia, then his poorly calculated decision to launch a full-scale invasion of Ukraine could perhaps be considered to verify Boulding's theory.

Ultimately, accurate point prediction of rare events in an open system (in this case, war) over long time horizons may be impossible. It is an occurrence of miniscule frequency considering the number of potential opportunities, and one that accords to few, if any, rules. It is little wonder, then, that social scientists' track record of forecasting war and other rare events is notoriously poor.[31] As Martin van Creveld observed, uncertainty is a 'condition without which war cannot exist and without which it would lose any sense it does, or can, make. Nobody is so foolish as to play a game of chess whose outcome is predetermined.'[32]

Given the relatively poor track record and only modest successes of past efforts, why continue along the Sisyphean path of forecasting war? One could provide many reasons. Nevertheless, we carry on because of war's devastating costs. From near-term and long-term impacts on human lives, to even more distant ramifications such as reduced life-long wages earned,[33] war has a way of transforming societies for the worse in a way that few other phenomena can. Rather than throwing up our hands in defeat and declaring war to be entirely unpredictable, it is worthwhile to pause and consider how we can know anything about the future.

291

On (Un)knowability and War

If the character of war or how it is fought is constantly changing,[34] how far in advance and how accurately can we predict such change? Trends and events can possess a variety of qualities that make forecasting them more or less difficult. These qualities can be separated into the following groups, which have broad application to social systems: complexity, measurability, equilibration, stochasticity and tractability (what might be called characteristics of (un)knowability, with knowability defined as our ability to know—or at least have a very good idea of—what will occur in the future).

Complexity

Emergent phenomena involve system-wide developments that cannot be characterized by their individual components. These developments can be difficult to account for in models, even those built from myriad interacting subcomponents. Similarly, non-ergodicity is a characteristic of processes that are guided by sets of rules that evolve over time. Despite impressive efforts to model and predict civil wars, performance of some of the most widely used predictive models worsened significantly in recent years because the drivers of instability are not constant over time.[35] Complexity also increases with conceptual irreducibility, or the radical uncertainty inherent in dynamics that are not yet well understood. Relatedly, there is computational irreducibility, where the world's complexity exceeds computational limits. These various types of complexity can overlap and compound with one another to effectively render certain trends or events unknowable. As Cederman and Weidmann assert: 'Perhaps the most pernicious problem [in conflict forecasting] pertains to the common failure to fully appreciate the fundamental complexity surrounding processes of peace and conflict.'[36]

Measurability

Aside from complexity is measurability, whereby even relatively simple systems with poor empirical measures are difficult to track. War is measurable, in a sense, but it often requires estimation and exclusion of more difficult-to-measure but equally important factors,[37] the collapse of a continuous process into a concrete event,[38] and the imposition of arbitrary cut-off points to be included in quantitative analysis.[39] The precise location of a cut-off point can affect conclusions in multivariate regressions and other statistical methods by increasing, eliminating or reversing apparent relationships between variables.[40]

Equilibration

Another quality affecting knowability is whether equilibration is present. Equilibration exists if a system operates (or trend fluctuates) around or generally pursues a steady state, even if it may at times be knocked out of balance.[41] War is a largely non-equilibrating phenomenon—past disputes may beget future disputes, but there is no 'market' to balance the supply and demand for war, other than that exhaustion may temporarily restrict a force's ability to continue to fight. Hobbes asserted that humankind's natural state is war;[42] if that were the case, and the international system met the strict definition of equilibrium—a steady state which will not change unless otherwise acted upon[43]—wars between states would be the rule rather than the exception. Mercifully, they are rare. When tensions do lead to war, they often do so in a chaotic fashion, whereby a seemingly stable situation suddenly becomes dominated by runaway escalation dynamics once invisible tipping points are crossed.[44]

Though not a form of system equilibration per se, forecasts of aggregate measures feature their own form of balancing measure in the form of offsetting errors. Former US Federal Reserve System Governor Laurence Meyer once observed the frequent occurrence of offsetting errors when forecasting GDP, jokingly giving credit to Saint Offset, 'the saint who watches over forecasters.'[45] When summing together country-level forecasts of variables to a global level (in other words, adding together projections of population for each country to estimate the world's total population), results often benefit from similar grace. One insight from Bennett and Stam's research was the ability to accurately model the general level of conventional conflict in the international system by summing all bilateral probabilities in a given year.[46]

Stochasticity

Phenomena can also be defined by stochasticity, where events may occur with a relatively well-known probability but occur with essentially random timing. As Gartzke emphasizes, war is stochastic. Bennett and Stam's ability to model the total level of conflict in the world but not particular conflicts confirms this assertion. Useful analogies include an ability to measure the 'oiliness of the oily rags'[47] or the number of sparks near a powder keg,[48] where identifying the exact spark that will ignite the rag or detonate the keg is exceedingly difficult if not impossible.

Tractability

Finally, policies, actions or trends that are characterized by tractability (in that they can be changed easily) are likely to be less knowable over the long term.[49] War is tractable, at least in its onset—one person can start a war if they are in command of adequate resources, with Russian President Vladimir Putin being a prime example. Short of specific intelligence or public leaders' statements clearly expressing the intent to attack, predicting the precise onset of war—even within a several-year margin of error—requires deep understanding of an individual's thoughts, beliefs, perceptions and state of mind; variables that are unmeasurable on a broad scale across long time horizons and which amount to 'choices that have yet to be made ... in circumstances that remain uncertain.'[50]

In summary, war onset bears nearly all the hallmarks that can make trends and events ill-suited for forecasting. This is not to say that all features of war are entirely unpredictable; Kott's case for a 'universal law' of advancements in weapon lethality offers a counter-example.[51] However, even where forecasts of military technologies have proven accurate,[52] 'one cannot as easily foresee how military organizations will combine them and integrate them into new concepts of operations.'[53] War may be thoroughly shaped by technology,[54] but accurate predictions of how technologies will shape wars of the future—including whether a given technology will affect the likelihood of war onset—have proven elusive. Inaccurate predictions of the 'death of the tank' are a recent example, where

FORECASTING THE FUTURE OF WAR

Table 14.1: Summary of patterns of (un)knowability, their relation to war, and partial remedies for analysis

Pattern of (un)knowability	Relation to war	Partial remedy for analysis
Complexity	War has emergent properties (in Bousquet's words', war 'becomes'), its rules evolve over time (non-ergodicity), and its processes often defy conceptual reducibility (exemplified at the micro level by the 'fog of war')	Study specific elements of war (e.g. war onset over a short time horizon in a limited geographic location), or confine analysis and conclusions to macro level, intentionally simplified interactions (à la Boulding's *Conflict and Defense*)
Measurability	Quantifying war often requires estimation and exclusion of more difficult-to-measure but equally important factors, the collapse of a continuous process into a concrete event, and the imposition of arbitrary cut-off points to be included in quantitative analysis	Emphasize measurable events and trends, conduct sensitivity analyses around measurements to account for the potential threshold effects or measurement error, and caveat conclusions with acknowledgements of plausible effects of important but unmeasured qualitative factors—or simulate the effects of those factors across alternative scenarios
Equilibration	There is no market for war, and its only points of equilibration are force exhaustion or complete annihilation of an enemy	Anticipate and simulate tipping points and instability when modeling war

Pattern of (un)knowability	Relation to war	Partial remedy for analysis
Stochasticity	While the general propensity for war globally has been accurately modeled, individual predictions of interstate war onsets have proven elusive across time	Seek to identify general trends, appreciating that precise prediction of an individual occurrence over a long time horizon is likely out of reach
Tractability	War can be initiated according to individual leaders' whims	Use scenarios to simulate an array of plausible ways in which events could unfold rather than becoming wedded to any single base-case or business-as-usual projection

(Source: Collin J. Meisel)

poor tactics and lackluster logistical support rather the obsolete technology can explain the high rate of casualties among Russia's tank crews fighting in Ukraine.[55]

What We Know

Can we begin to clear these obstacles? We can, at least in part, by using a structural, systems-based perspective to assess changes in the risk of war.

Structural, systems-based models often fail at making accurate point predictions of rare events such as war. Instead, their focus is on knowable long-term trends rather than speculation about unknowable trends, the latter of which are more usefully considered through scenarios analysis. Even difficult-to-measure factors such as ideology, which undoubtedly affects states' propensities to go to war,[56] can be accounted for crudely,[57] though leader-level decisions are generally omitted (outside of scenarios). However, leaders' decisions are as much a product of their environment as the environment is a product of them.[58] Would hypothetical US President Al Gore have invaded Iraq despite substantial ideological differences with actual President George W. Bush? Harvey makes a convincing argument that he would have, given long-running themes in US foreign policy discourse (themselves partly the product of structural pressures affecting US domestic and foreign policy).[59] The same may have been true of the current war in Ukraine if former Russian Prime Minister Dmitry Medvedev were Russia's leader in February 2022 instead of Putin.

One such model that takes this structural, systems-based approach is the international futures tool, an integrated assessment model that provides long-term, global forecasts of hundreds of variables.[60] Relevant to this study are its forecasts of 'threat,' which approximate the probability of conventional war.[61] The full technical description of the conventional war probability modeling logic, relevant equations, and how they were developed is provided by Hughes.[62]

The probability of conventional war is forecast in international futures based of several proximal drivers—for example, a country-pair's (for example, Russia–Ukraine) history (or lack thereof) of militarized interstate disputes between the years 1816 and 2014;[63] territorial contiguity; national power; concentration of power in the international system; power transition dynamics; alliances; and (dis)similarity in levels of democracy—and many distal drivers, which are indirectly connected to war through another (or several other) drivers. Thus, changes in forecasts of crop production will have an effect on the probability of conventional war between two countries, albeit a small one; conversely, changes in national power dynamics will have much larger effects.

Prior efforts have used a similar approach, modeling the structural forces that affect 'lateral pressure' or a state's propensity for foreign expansion. These forces are a product of societal demands (growing population, energy demand, excess productive capacity, and a large military establishment) and the capability to wage war.[64]

What these approaches have in common is their emphasis on the risk of war onset rather than point predictions of onsets. Returning to the analogies mentioned above, they measure the 'oiliness of the oily rags' or the number of sparks near a powder keg rather than attempting to predict fires or explosions. Their core inputs—a country's population, its energy demands, its general productive capacity, and other elements of national power—are non-emergent, and conceptually and computationally reducible, measurable, largely equilibrating, non-stochastic and relatively less tractable. In other words, they are knowable, and they can improve our understanding of the risk of interstate war across long time horizons.

It should be noted, however, that these structural, systems-based approaches generally omit considerations of nuclear war. The limited historical occurrence of the use of weapons of mass destruction in war, the history of alarmist predictions regarding its use,[65] and the general failure of previous forecasters' abilities to predict how technologies are likely to be used in future wars[66] lead us to avoid drawing conclusions beyond the threat of conventional conflict.

Moreover, we can place greater stock in the global sum of bilateral probabilities of war rather than individual bilateral long-term forecasts of conventional war onsets.[67] The logic to this summing of probabilities is that while any given probability is low, given enough chances (i.e. all country-pair interactions), even a low probability event will eventually occur.

Keeping these limitations in mind, structural, systems-based approaches tell us much about the future risk of conventional war onset. The latest long-term population forecasts from the United Nations suggest that the global population will continue to increase for the next several decades and likely through the end of the century, with much of this growth coming from African nations.[68] The US Energy Information Administration projects that global energy demand will increase by nearly 50 per cent through 2050 relative to 2020, with the largest share of the increase in energy demand coming from China and India, and with the majority of demand continuing to be met by fossil fuels rather than renewable energy.[69] Long-term forecasts of military spending by the Lowy Institute and separate forecasts in international futures suggest that military capabilities for many countries, particularly those in the Indo-Pacific, are expected to continue to rise through 2030 and beyond.[70] In combination,

FORECASTING THE FUTURE OF WAR

these societal demands and capabilities translate to increased 'lateral pressures' for conflict in the coming decades.

In the broader conception of 'threat,' these 'lateral pressures' could be mitigated by increasing trade interdependence, increasing democratization, and a greater diffusion of influence within the international system (as opposed to multipolarity, which involves concentration around multiple poles and corresponds with a higher risk of war). Only the first of these three trends appears likely at present.[71] These and other factors led the US National Intelligence Council to conclude that the risk of interstate war in the coming decades will be higher than in recent years.[72]

How Might Forecasters' Calculus Change in the Wake of Putin's War?

Vladimir Putin may have implicitly taken into consideration similar, though inverse, dynamics vis-á-vis Ukraine. He may have felt that Russia's demographic, economic and diplomatic position relative to Ukraine and its Western partners was only likely to worsen, at least within his lifetime. According to this logic, his forces struck Ukraine at the least-worse time relative to the foreseeable future. If Putin made such a calculation, this would affect the probability of conflict, whereby prospective losses in power might matter more than prospective gains or transitions. The notions of 'threat' and 'lateral pressures' described above include demands and capabilities, which are gains-focused; if a country has something to gain from expanding, it will consider expanding. If prospective losses are the dominant factor influencing the risk of war, then we should more thoroughly incorporate elements of Kahneman and Tversky's prospect theory into our modeling of war.[73] As these authors observed, when a game of chance is framed in terms of losses, individuals will typically prefer to incur risks over 'sure bets' despite equal expected utilities. So, for example, a leader would likely prefer a course of action that involved a 75 per cent chance of zero reduction in global power and a 25 per cent chance of a 50 per cent reduction in global power versus a 100 per cent chance of a 12.5 per cent reduction in global power. To account for such a dynamic, expectations about the future would need to be more explicitly incorporated into calculations of

BEYOND UKRAINE

present risk. In Drezner's words, we must more carefully account for 'the perils of pessimism' and leaders fearful of future national decline relative to potential adversaries.[74]

Furthermore, the systems-based, structural conflict-modeling approach emphasized here largely ignores several key factors, including leaders,[75] non-state actors,[76] norms,[77] and constructivist critiques of material measures and power politics generally.[78] Putin's incorporation of non-state actors, namely the Wagner Group, continues a trend toward non-state actor involvement in conflict and the conventionalization of non-state forces.[79] Moreover, his flaunting of prominent pleas and warnings from Western leaders threatens to render the postwar 'territorial integrity norm'[80] to be no more than a passing fad.

Whether Putin is successful in misinforming his own public as to the cost of the war would likewise alter how we model war. In Richardson's arms race model, arms races are a product of real or perceived threats and costs, including domestic aversion toward human losses and funds taken from other priorities, such as healthcare and education.[81] Artificially reducing domestic costs (or at least the perception of costs) thus increases the public appetite (or at least tolerance) for war.

Additional limitations include those inherent in Bennett and Stam's statistical models, which laid the groundwork for forecasts of 'threat' in international futures. For instance, their model omits 'joiners'—countries that enter into a war as part of a coalition of forces.[82] As a result, their data do not include a war between the US and Germany, because the US was a 'joiner' in the Second World War, joining Allied forces in the fight against the Nazi regime. They do, however, include a US war against Japan. While this methodological choice helps eliminate complications by simplifying causal pathways, it omits potential 'chain-ganging,' cascading, and other network effects that have at times greatly influenced international relations.[83] A modern equivalent would be omitting a war between the US and China from an analysis because the US joined an ongoing war between China and Taiwan.

Given all these limitations, is this not an argument for why structural approaches offer very little as an aid to long-term

300

FORECASTING THE FUTURE OF WAR

forecasting? Not necessarily. As Boulding indicated when discussing his own general theory of conflict and defense:

> Nations are born, grow, fight, conquer, are conquered, become empires, and rise and fall on the great stage of physical geography and human passions and knowledge, not on the homogenous white planes on which we draw our diagrams. Nevertheless, the model does serve to interpret history and enables us to see the basic simple patterns that underly the baroque embellishments of the actual course of events.[84]

So, too, do structural, systems-based approaches to assessing the risk of war.

Knowable trends can inform policymakers broadly on the structural pressures and forecasted changes in risk factors that may alter the risk of war in the international system as a whole, as well as, much more cautiously, between particular pairs of states. These forecasts and assessments of their accuracy can also help inform and produce better decisions by policymakers in the future,[85] where even incorrect forecasts can, once proven incorrect, allow for the culling of the faulty theories upon which they were based.[86] At a minimum, it achieves Cederman and Weidmann's minimum standard for such models, where structural, systems-based models can be used 'as a heuristic tool for generating possible scenarios' even if they fall short of becoming 'a producer of specific policy advice.'[87] Indeed, while relying upon forecasts of knowable trends as 'structural anchors,'[88] it is essential that forecasts of the future of war be considered across a range of plausible scenarios in order to prepare for a variety of contingencies. As Hoffman succinctly noted, 'the future is plural.'[89]

More recent scholarship from Hoffman provides more actionable insights for assessments of the risk of future war onset. If his assessment is correct and we have entered an era of defensive dominance for the foreseeable future,[90] then, all else being equal, state and system viability will increase.[91] In other words, Ukraine's impressive defensive display has likely sent a message to would-be aggressors: attack at your own peril.

Hence, we are left with contradictory conclusions. While many structural pressures are increasing the risk of war onset between great

301

and minor powers alike, increases in economic interdependence (assuming this trend does not reverse) and defensive dominance are contingent developments that may decrease the risk of war in the coming decades.

More generally, Russia's full-scale invasion of Ukraine has highlighted the ways in which material capabilities—a predominant feature in many structural, systems-based models of the world and war—do and do not continue to matter. The great imbalance between Russian and Ukrainian material military capabilities, at least on paper,[92] likely offered Putin encouragement that a war of aggression against Ukraine would be successful. Yet, we see that a much less well-endowed force such as Ukraine can resist a much larger force given adequate support and motivation.

Material capabilities and the structural factors that influence them continue to matter. Ukraine has benefitted substantially from material support from Western powers. However, providing such support to Taiwan in the midst of a conflict with China would be much more difficult.[93] Additionally, it is difficult to imagine that a force with much smaller stocks of personnel and equipment could have so poorly executed an invasion as Russian forces did and still be in the fight more than a year later. On the Ukrainian side, equipment and ammunition stocks have received a great deal of emphasis. Ukrainian President Volodymyr Zelensky's appeal to the US Congress in December 2022 is one example: 'The occupiers have a significant advantage in artillery. They have an advantage in ammunition. They have much more missiles and planes than we ever had.... We have artillery, yes. Thank you. We have it. Is it enough? Honestly, not really.'[94]

Despite the countless technological developments that have occurred since Clausewitz penned *On War*, and the various attempts to remove killing from the modern battlefield in recent years,[95] violent clashes of metal against flesh continue to characterize war. As the violent instincts, political machinations and chance occurrences that embody the nature of war persist, so too do the structural pressures that enable the war's unpredictability, all while acknowledging that it will always remain inherent to war.

Returning to the questions we asked at the beginning of this chapter, what could we have known before 24 February 2022?

FORECASTING THE FUTURE OF WAR

Could we have predicted Russia's invasion, and if so, with how much advance notice? Could the invasion have been predicted prior to the development of state intelligence of a military build-up? Beyond lucky speculation, could we have accurately predicted the type of invasion or the outcome of the war since?

Patterns of knowability—complexity, measurability, equilibration, stochasticity and tractability—and a structural, systems-based mode of analysis that is built around these patterns offer suggestions, if not concrete answers. Given war's complexity, the many levels of immeasurability, and a lack of equilibration, stochasticity and tractability in its onset, we could not have predicted the onset of Russia's full-scale invasion of Ukraine with 100 per cent certainty. Nor, absent specific intelligence, could we have predicted the type of invasion or outcome of the war since.

However, we could have known that the risk of war between the two countries was and will likely remain high for the foreseeable future. Indeed, the 'lateral pressures' and structural factors underlying 'threat' between Russia and Ukraine were generally increasing and expected to continue to increase in the coming years. We can apply this logic to other potential future conflicts, adding the lessons learned from the present conflict—particularly that expected future loss of power may weigh equally, if not more heavily, on leaders' minds than prospective gains that might come from initiating a war.

In situations where there are ideologically disparate nations with a narrowing power gap and decreasing economic interdependence, where there are growing military capabilities and increasing domestic energy demand and other social pressures, the risk of war will continue to increase. Unless defensive dominance becomes so overwhelming as to deter wars of aggression, the continued extension of military reach and consequent decreasing viability for potential target states will also continue to increase the risk of interstate war, even if the precise timing and location of its onset will remain unpredictable across long time horizons. Indeed, structural pressures suggest that we have not yet seen the decline of war—or at least not the decline of the substantial risk of war—even if its precise onset remains unknowable.

15

THE WAR IN UKRAINE AND THE APOCALYPTIC IMAGINARY

Jeni Mitchell

As far as we know, nearly every human society that has ever existed has given some thought to the end of the world.[1] Historical reviews of apocalyptic thought tend to focus on the most influential and well-known traditions, beginning with the Book of Revelation's nightmarish visions and ending with Roland Emmerich's increasingly baroque methods for destroying the planet. The writings of Carl von Clausewitz, despite his front-row seat to the scorched wastelands of the Napoleonic Wars, do not usually appear in such exegeses. Yet I would argue that one of his most famous propositions might be usefully adapted in order to understand the essence and evolution of the 'apocalyptic imaginary': we might say that the nature of apocalypse is enduring, but its specific character evolves to reflect 'the spirit of the age'. In this chapter, I explore the sources of visions of the apocalypse throughout the twenty-first century, and the ways in which they influence policymaking and strategic decision-making by state actors.[2] I then argue that the current war in Ukraine demonstrates the real-world impact of apocalyptic threat perceptions on strategy, while having a somewhat limited influence on contemporary ideations of the end of the world.

305

BEYOND UKRAINE

What is the 'enduring nature' of apocalypse? It can be found in its eternal purpose, the very reason for its ubiquity, as eloquently expressed by Frank Kermode in his classic work on the subject, *The Sense of an Ending*.[3] In it, apocalyptic visions endow our finite existence on this earth with meaning. Knowing that we occupy but a brief moment of eternity, we construct narratives of humanity's birth and death, and position ourselves in some waypoint between them. We cope with our own inexorable mortality by postulating the doom of all life on earth, with the faint hope of some means of post-apocalyptic survival soothing our fears of our own oblivion.[4] As Kermode notes, the concept of apocalypse has extraordinary resilience; predictions of extinction continually fail, but the idea endures, shifting into new forms and timelines.[5] In short, the story of human life on this planet needs an ending in order to have meaning as a story at all.

Yet, it is clear that ideas about the end of the world, while showing some remarkable continuity from antiquity to today, also evolve across time and space; that the characteristics of the apocalyptic imaginary vary alongside specific cultures and their zeitgeists. The divinely directed fires and floods of prehistory have shifted to the very human-driven nuclear firestorms and catastrophic sea-level rises predicted today. Flowery and dramatic apocalyptic visions have yielded to more rationalist and scientific apocalyptic threat perceptions. And the prophets of apocalypse have also changed form, with the priests and oracles of yesteryear replaced by a veritable industry of futurologists, horizon-scanners and, in particular, existential risk (otherwise known as 'X-risk') scholars.[6] Today, those tasked with preparing for extreme threats have a panoply of approaches to draw upon, from sophisticated quantitative modelling,[7] to risk classification frameworks[8] and collaborative projects with science-fiction creatives.[9]

Apocalyptic visions are more than just horror stories or nightmare fuel; they directly influence the decisions made by policymakers and strategic planners across a wide range of domains. To take one obvious example, the fear of a catastrophic nuclear war between the superpowers generated an entirely new genre of strategic theory specifically geared toward preventing this scenario, in addition to

306

THE WAR IN UKRAINE AND THE APOCALYPTIC IMAGINARY

robust arms control and verification infrastructures. Concerns over the unintended consequences of technological innovations drive ethical norm-setting to constrain their development and use, while national biosecurity and environmental security policies are shaped by worst-case models developed by scientists. The recognition that the failure of global information and communication systems due to a massive cyberattack or anti-satellite warfare could unleash devastating chaos across the world has led to the creation of 'cyber defence forces' and 'space defence strategies'. In short, while national strategies and doctrines tend to shy away from using the actual word 'apocalypse'—the most recent US National Security Strategy, for example, applies the rather milquetoast term 'shared challenges' to a range of threats that could conceivably end human civilization[10]—it is clear that apocalyptic threat perceptions do influence policy and strategy across many sectors.

Thus, for the purposes of this book, it is worth considering whether the Russian invasion of Ukraine has had any notable impact upon this dynamic. First, to what extent have apocalyptic threat perceptions influenced policy and strategy with respect to this war? I will argue that they have had a significant impact in two respects: (1) the fear of escalation to nuclear war has constrained the strategic choices of Ukraine, the North Atlantic Treaty Organization (NATO) member states and Russia; and (2) genocidal narratives have played key mobilizing and legitimizing roles for combatants on both sides, in very different ways. Second, has the war in Ukraine led to a significant shift in the character of the 'apocalyptic imaginary', or in contemporary perceptions of apocalyptic threats? I will argue that it has not, because the most influential apocalyptic visions of today are those involving artificial intelligence (AI) and other emerging technologies; biological threats, which have significantly enhanced post-Covid; and climate catastrophe, which is already apparent across the world. The war in Ukraine has not (to date) shifted those threat perceptions to a significant extent. It is worth considering, however, several ways in which the war has affected apocalyptic thinking to some degree, including: globally, with a greater intensity of concern over nuclear weapons use; within Ukraine, with fears of the annihilation of their

307

BEYOND UKRAINE

nation and culture; and within Western states, marked by a steadily intensifying 'apocalyptic milieu'.

The chapter proceeds as follows. First, I define 'apocalyptic imaginary' and 'apocalyptic threats'. Second, I provide a brief history of apocalyptic imaginaries in order to situate current thinking and in order to suggest future trajectories of thought. Third, I present a typology of apocalyptic threats, categorizing the most important and serious threats to human existence, and note the ways in which they currently influence policymaking and strategic planning. Fourth, I explore the impact of apocalyptic threat perceptions on policy and strategy vis-à-vis the war in Ukraine, and then, conversely, the more limited impact that the war has had on our current apocalyptic imaginaries. I conclude with a consideration of potential developments in the war that might generate a more significant impact upon current apocalyptic thinking and policymaking.

1. The Apocalyptic Imaginary and Apocalyptic Threats

Today, serious study of the end of the world is primarily rooted in X-risk analysis, a growing interdisciplinary field exploring everything from artificial general intelligence to global plagues, melting polar ice caps and asteroid strikes. In keeping with its label, X-risk analysis tends to assume a maximalist stance, focusing on threats that could permanently end intelligent life on the planet.[11] Thus, there is an inherent assumption within these approaches that the apocalypse has not yet happened, that it is a purely futuristic and abstract concept.

This is one of the reasons why I use the term 'apocalyptic' rather than 'existential' in my analysis. To be clear, I am relying on the secular interpretation of 'apocalypse' common to our modern era (that is, the end of the world). The word itself, from the ancient Greek for 'revelation', originates in ancient Jewish and Christian religious tradition, and a stricter eschatological interpretation would by necessity include only those scenarios involving holy war, divine judgement and the resurrection of a chosen few. This clearly has its limits for our current purpose, and at a time when people routinely refer to nuclear apocalypse, insect apocalypse and

308

even zombie apocalypse, it seems safe to use the term in its modern colloquial sense.

I further interpret the term as 'the end of the world as we know it', in order to broaden its applicability beyond the X-risk definitions of total annihilation of the human species. This retains the element of catastrophic destruction without requiring complete extinction in order to come within the scope of our analysis. Arguably, X-risk definitions centred around human extinction trap analysis within the realm of the speculative and abstract, given the lack of historical precedent (the five 'mass extinction events' having occurred before the emergence of human life). Yet throughout history, we have seen vast devastations at a sub-planetary level, the utter destruction of entire peoples and societies as a result of events that could quite easily recur in our own era (e.g. civilizational collapse, war and conquest, natural disasters). Thus, this broader definition helps highlight that the apocalypse is not just some fantastical or sci-fi imagining, but something that has actually happened to any number of societies in the past and present.

This definitional approach is in keeping with the 'human security' concept, which is a more people-centred understanding of secure existence. From this perspective, the destruction of a given human society is an apocalypse for that society—it is the end of *their* world as they know it. To take a particularly obvious case, we might think of the vast physical and social destruction across the Americas in the wake of European conquest and colonization, when entire civilizations and communities were annihilated, with population declines of up to 90 per cent.[12] This level of catastrophe is surely apocalyptic from the point of view of the vanished communities; the fact that their habitats were colonized and repopulated should not obscure that fact.

Within this context, the term 'apocalyptic imaginary' can be defined as the ideas, narratives, cultural traditions and discourses that are concerned with the 'end of the world' for a given community of people. At any given time, there are many apocalyptic imaginaries afoot in the world; consider, for example, the contrasting visions of apocalypse held by the average American and the average Islamic State militant. Throughout history, ideas about the end of the world have been culturally specific, and have evolved in line with broader political, social and technological changes. This will be explored more fully in the next section.

Within today's apocalyptic imaginary lies an important subset of ideas I refer to as 'apocalyptic threat perceptions'. These are the result of rationalist analysis of the most critical and potentially catastrophic threats facing human societies today, intended to explicitly influence policy and strategy across a wide range of sectors. They are derived from scientific research, the study of the historical record, ethical considerations, and thus generally a more rigorous line of analysis than the general fears and anxieties populating the apocalyptic imaginary more broadly. I define 'apocalyptic threat' in non-maximalist terms, as an event or series of events which, if they were to occur, would lead to the physical and/or social destruction of a given human community.[13]

These conceptualizations of apocalyptic imaginary and apocalyptic threat are intended to provide a more nuanced analysis of the impact of the war in Ukraine. First, as I will argue below, the war's impact can be seen more in the broader apocalyptic imaginary than in specific apocalyptic threat perceptions. And second, this approach helps clarify that while a successful Russian conquest of Ukraine might not be seen as apocalyptic for the rest of the world, it would surely be so for the people of Ukraine themselves.

2. A Brief History of the Apocalypse

A fundamental maxim within futures studies is that our visions of the future reflect the priorities and anxieties of the present. I will now offer a broad historical review of apocalyptic imaginaries to demonstrate that this statement holds true for existential visions as well, and to provide some context for our analysis of contemporary events.

There are two broad strains of apocalyptic thinking—the sacred and the secular—which endure across three historical eras—the traditional, the modern and the anthropogenic. These categorizations are neither exclusive nor timebound; for example, ancient religious ideas about apocalypse maintain influence even in our very secular

310

and scientific era.[14] Thus, these labels are intended to highlight patterns and trends rather than provide strict boundaries of time and type.

Throughout most of human history, ideas about the end of the world were predominantly sacred visions, built upon deeply embedded religious and cultural mythologies. Nevertheless, a common theme emerged across a wide range of societies: namely, the world would grow increasingly violent and immoral, culminating in an epochal holy war between the forces of good and evil, the deaths of most of humanity, and a divine judgement allowing the chosen few to rebuild a new world. Certain types of destruction are omnipresent in such visions, especially great floods, fires, plagues and wars.[15] In some traditions, such as Judaism, Christianity, Islam and Norse mythology, the apocalypse is a one-time event; in others, such as the Hindu and Chinese traditions, the end of the world is a cyclical occurrence, with humanity resurrected only to gradually decline once again. In all cases, however, these traditional and religious myths serve several important purposes: they help bring a sense of order and justice to an often chaotic world, and imbue peoples' lives with meaning and purpose.[16]

Throughout pre-modern history, many societies fell prey to existential disasters. With the exception of war and genocide, these were usually natural in origin and little understood. They were typically attributed to divine wrath and punishment; for example, in the ancient poem *The Curse of Akkad*, we discover that the collapse of the world's first empire in the twenty-second century BCE was due to the Akkadian ruler insulting the god of wind and storms.[17]

These traditional and sacred ideas about the apocalypse reflected the societies from which they emerged, drawing upon their extant mythological and religious beliefs, and relying upon pre-modern understandings of the world. Real-life cataclysmic events led to apocalyptic visions dominated by horrific natural disasters and wrathful gods. It is no accident that the greatest of all apocalyptic stories, the Book of Revelation, was written against the backdrop of the brutal Roman destruction of the Jewish population in the Levant.

In the European tradition, ideas about apocalypse evolved significantly during the transition to modernity that began in the

311

sixteenth century, as vast leaps forward in scientific knowledge and the growing secularization of society transformed speculation of how the world might end. While traditional visions usually viewed apocalypse as a dramatic 'reset' for humanity, it became increasingly clear in the modern era that humanity could simply and permanently go extinct. New rationalist disciplines like geology, demography and statistics established that other species had disappeared from the earth, and that humans—no longer assumed to be uniquely protected by gracious divine beings—might meet the same fate.[18] By the nineteenth century, the first stories in the 'last man' genre appeared—an enduring apocalyptic imaginary in which one or a handful of survivors walk the devastated earth, which stood in stark contrast to traditional religious visions of a renewed and peaceful post-apocalyptic land. New scientific understandings of our planet and the increasingly industrialized savagery of warfare generated an expanded panoply of deadly threats, and there was a growing appreciation for the possibility that humans themselves might invent the means of their own destruction. In a pair of novels, *The War in the Air* (1908) and *The World Set Free* (1914), H.G. Wells depicts the world's ruin in cataclysmic wars driven by new weapons systems (including, in a startling example of foresight, 'atomic bombs'). Yet the sacred strain of apocalyptic thinking did not wholly disappear in this era, with millenarian movements emerging well into the nineteenth century.

The impact of European colonialism must also be acknowledged in this era, both in its destruction and in the new imaginings it inspired. Over the course of these five centuries, millions of people around the world died due to war, slavery, famine, exposure to new pathogens, and the quotidian violence of colonial occupation. Entire communities and societies were wiped from the earth permanently and, as argued earlier, these should be considered apocalyptic events. This historical record of apocalypse can be drawn upon in trying to conceptualize and predict future risks, including the emergence of 'green colonialism', the potential for new pathogens to prompt civilizational collapse, and the replication of colonial dynamics within space expansionism. We might also note the influence of colonial apocalypses on cultural representations in this era, the

312

most obvious example being Wells' *The War of the Worlds*, which was inspired by the British ravaging of Tasmania and the near-extinction of its indigenous inhabitants.[19] This link between atrocity and apocalyptic imaginaries endures, though today it is human atrocities perpetrated on the environment that tend to spark the most anguished apocalyptic visions.[20]

In short, it is unsurprising that during this transition to the modern era in Europe, different strains of apocalyptic thinking co-existed— the sacred and the secular, the traditional and the scientific—and disasters of both human and natural origin crowded the apocalyptic landscape. The type of vision that dominated any particular society could be traced to the broader social and technological developments affecting them, and so we see in this era a much more diverse range of visions across the world.

The third historical era I have labelled the anthropogenic, which began in 1945 with the invention of nuclear weapons. For the first time in thousands of years of human civilization, the destruction of all human life by a man-made technology became a very real possibility. Arguably, it is at this point that the full secularization of apocalypse arrives, as there is little room for moralism in a nuclear holocaust—only utter and meaningless destruction.[21] In the decades since 'World War III' entered the popular lexicon, an incredible number of additional scenarios have emerged in which humans are ultimately responsible for the end of the world, in which we fail to curb climate change, or invent diabolical new technologies or pathogens, or accidentally unleash a cosmic disaster. We start world wars that end in global devastation, or spark famines that leave billions without sustenance. More traditional versions of apocalypse, rooted in religious and cultural beliefs, continue to hold sway around the world, with new cults and prophetic movements continually emerging with their own idiosyncratic visions.

Alongside this very human-centred strain of apocalyptic thinking, we see the emergence of new rationalist strategies and theories to try to prevent the end of the world. Nuclear weapons begat deterrence theory and crisis management models; climate change has generated an enormous array of mitigation responses.

For virtually every threat listed in Table 15.1, there is some universe of scientists, policymakers and activists who are trying to contain its potential effects. However, despite increasingly believing that the greatest chance of extinction stems from our own failings, not from divine wrath or judgement, we also increasingly act to forestall its occurrence.

The reasons for this evolving change in the 'character' of apocalyptic threat perceptions are thus relatively obvious: it allows us to better observe and anticipate all the things that might destroy us, and better appreciate the risks attached to our great programme of progress. The pace of technological innovation in recent decades is astounding, and we are anxious about where it might take us, of how our story might end. And already, all around us, we see the earth increasingly turning on us—drowning us, baking us, starving us, the new vengeful God casting us into oblivion.

It remains to be seen whether this current era of apocalyptic thinking may be our last. After all, if any of the scenarios below do happen to occur, then there may not be enough humans left to assemble any kind of collective vision. Or perhaps a fourth era will emerge, after further radical transformations of human society (and, potentially, non-human society). In the next section, I remain in our current era and present a typology of apocalyptic threat perceptions in the twenty-first century.

3. Apocalyptic Threat Perceptions and their Impact on Policy and Strategy

A more detailed typology of contemporary apocalyptic threat perceptions helps clarify their influence upon policy and strategy, and the varying impacts that the war in Ukraine has had on this dynamic. My typology is not exhaustive, but instead attempts to expand upon the more narrow categorizations of threat often found in the X-risk literature, while also prioritizing those threats of highest concern to policymakers and strategic planners today.[22]

THE WAR IN UKRAINE AND THE APOCALYPTIC IMAGINARY

Table 15.1: Apocalyptic threat typology

Environmental	
• Extreme heatwaves and fires • Drought / desertification • Sea-level rise / ice melt • Ocean warming / acidification	• Ecosystem collapse • Global biodiversity loss • Agricultural collapse • Irreversible pollution
Biological	
• Pandemics • Human fertility decline	• 'Zombie pathogens' • Environmentally mediated intellectual decline
War and genocide	
• Nuclear war • Biological warfare • Large-scale conventional war (inter- or intrastate)	• Genocide and crimes against humanity • Colonial annihilation • Cyberattacks on critical infrastructure • Space warfare
Societal	
• Depletion of key resources • Global economic collapse	• Complexification • Global totalitarian order
Technological	
• Artificial intelligence / machine superintelligence • Lethal autonomous weapons • Human–machine cognitive integration • Anti-satellite weapons	• Stratospheric geo-engineering • Bio-engineering / gene editing • AI-generated pathogens / toxins • Nanotechnology • Particle physics events
Space and planetary	
• Asteroid or comet impact • Solar flares / geomagnetic storms • Collapse of air / ocean currents	• Gamma-ray bursts • Supervolcanic eruptions • Space colonization and war
Box X: The unknown / unimaginable	
• Future technologies • Unseen or non-understood cosmic dangers (including alien attack)	• New pathogens • Unanticipated environmental catastrophes

(Source: Jeni Mitchell)

A brief review of each category of apocalyptic threats highlights the relatively limited impact of the war in Ukraine (to date) on contemporary perceptions of apocalyptic threats.

Environmental

The litany of serious threats facing the planet due to anthropogenic climate change is well known, as are our failures to sufficiently address them. We remain on track for a devastating rise in global temperature and all its associated effects, including extreme heat events, rising sea levels and agricultural collapse—all of which could make large areas of the planet uninhabitable. We have failed to sufficiently curb pollution and habitat destruction, and global biodiversity loss is proceeding at such a rapid pace that it is commonly referred to as 'biological annihilation' and 'the sixth extinction'.[23] Already today, people are dying due to the effects of climate change—one estimate in 2021 put the figure at five million people per year[24]— and communities are being erased from existence.[25] The question is simply how much worse things will become, and whether concerted mitigation action can forestall worst-case scenarios.

Biological

Thanks in part to the ravages of the Black Death, the 1918–20 flu pandemic and, more recently, Covid-19, mass death due to disease is not an abstract idea but a historical reality, with the latter highlighting our vulnerability to new viruses, as well as some troubling and unexpected dynamics in the public response (e.g. the anti-vaxxer movement). Should another new virus arise with a higher case fatality rate—say, the 50 per cent estimated for the Black Death's bubonic plague, or the Ebola virus today—the impact would be far more catastrophic. Most worrisome is the possibility of environmental and biological apocalyptic threats combining: melting permafrost could theoretically unleash ancient pathogens to which we have no resistance (in an echo of the diseases that obliterated indigenous America), and there are already signs that toxic pollution is affecting human fertility levels. Then there is Christopher Williams' theory of 'environmentally-mediated intellectual decline'; put simply, this theory suggests we have contaminated our atmosphere, habitat and food supply so severely that we are becoming increasingly stupid as a species.[26]

War and Genocide

War is one of the legendary Four Horsemen of the Apocalypse, but it is curiously overlooked in much of the X-risk literature, except as the background context to all the terrifying technologies that could inadvertently destroy us all. But given the enduring and devastating impact of war upon human civilizations, and the terrible death tolls we are already capable of even before we invent (for example) space-to-earth weapons, I argue that it deserves a more prominent place in these discussions. We cannot say whether zombie pathogens or ocean acidification will yet kill us all, but there is plenty of hard evidence in the historical record to show that mass killings and cultural annihilation are depressingly frequent occurrences.

It is true that among these apocalyptic threats, nuclear war and, to a lesser extent, biological warfare do receive a great deal of attention—as they should. Here we see perhaps the greatest impact of the Russian war on Ukraine: the renewed fear of escalation to nuclear war, which I will discuss in the next section.

Societal

This category is intended to encompass significant political and economic catastrophes; not the sorts of crises that regularly afflict all societies, but something on the level of global economic collapse. The ever-increasing global connectivity of economic activities creates a serious vulnerability to, for example, a cyberattack on global banking systems, an unprecedented event that ranks high on X-risk typologies. Some argue that the root cause of apocalyptic threats in this category is 'complexification', a theory introduced by Joseph Tainter in *The Collapse of Complex Societies*. He argues that as societies evolve and solve key social challenges, they become more complex, and eventually this complexity reaches such a level that it becomes existentially detrimental. All of a society's resources become devoted to maintaining this complex system rather than addressing social problems; a gradual deterioration ensues, culminating in civilizational collapse. It is a highly contested theory, but worth including in analysis of apocalyptic threats given the implications if it proves to be correct.

Technological

Within the X-risk literature, new technologies are prioritized above virtually every other type of risk, save for environmental risks. There are good reasons for this, as we are currently riding a wave of technological revolution that could lead to profound transformations not just of human society, but of humanity itself. It may be useful to corral this vast universe of innovation into several types of technological apocalyptic threats:

- Technologies that could increase the lethality and destructiveness of war (e.g. hypersonic missiles, space-to-earth weapons, human out-the-loop autonomous weapons systems);
- Technologies that could (unintentionally) destroy the planet (e.g. AI-generated biological compounds, stratospheric geo-engineering, particle physics experiments);
- Technologies that could fundamentally alter human society (fusion power, quantum computing) and humanity itself (bio-engineering) in ways that facilitate violent or accidental destruction;
- Technologies that could expand the non-human domain across every aspect of society to the detriment of human life (e.g. value-misaligned AI and machine intelligence).

There is a vast literature on these emerging technologies and their associated risks (and it is worth highlighting that virtually every emerging technology has serious risks associated with it). Émile P. Torres usefully highlights three particularly worrying aspects of emerging technologies—their dual-use nature, their increasingly powerful capabilities, and their proliferation among non-state actors—as well as a number of anticipated future innovations we should be especially concerned about, including autonomous nanobots, metamaterial invisibility cloaks, self-guided bullets, brain–machine interfaces, robot soldiers and direct-energy weapons.[27] By far the most alarmist analysis across the literature, however, is centred around the emergence of machine superintelligence or super-intelligent AI, which could potentially escape the human grasp and destroy all of humanity. Many agree with Nick Bostrom's assessment that 'the default outcome' of actually creating a machine superintelligence would be an extinction-level event.[28]

Space and Planetary

This category of apocalyptic threat presents an analytical conundrum. On the one hand, many of the events in this category are so extremely unlikely that we do not need to worry about them (the chances of a planet-killing asteroid strike, for example, are estimated at about 0.000001 per cent). On the other hand, some of these events would lead to the complete extinction of our species, and so in practice a considerable amount of energy is devoted to monitoring and preparing for 'death from above'. NASA runs a Planetary Defense Coordination Office, while the European Space Agency hosts the Near-Earth Objects Coordination Centre. The infrastructure and logistics sectors, meanwhile, cooperate globally on building resilience against 'black sky events', the severe disruption of global systems as a result of (for example) coronal mass ejections and geomagnetic storms.

Box X: The Unknown/Unimaginable

Finally, it is important to remember the potential for new threats to emerge at any time. These may be future technologies or events that we cannot even imagine today, the unintended consequences of current technologies and events, or dangers that are currently hurtling toward us which we lack the ability to see. It may be difficult to strategize this category of apocalyptic threat, but we must not make the mistake of assuming current typologies of threat are static.

In sum, this typology is intended to convey the broad range of threats—mostly human-generated—that could conceivably lead to the physical and/or social destruction of human communities, on a scale from the local to the global. It is a daunting list, but if humans have proven adept at destruction across the ages, we have also demonstrated a certain flair for salvation, and so one can only

319

hope that prevention and mitigation efforts in the coming decades will prove successful.

In surveying this broad range of threats, it is relatively easy to identify their influence upon policy and strategy across a wide range of sectors. Without the worst-case scenarios represented in this typology, it is doubtful that so much money and time would be devoted to reducing carbon emissions, improving epidemiological systems and responses, sustaining nuclear deterrence, protecting global cyber systems and satellites, and so on. We would probably not be investing in Double Asteroid Redirection Tests or monitoring coronal mass ejections so closely. The entire discipline of technology ethics would be significantly reduced if the potentially catastrophic impacts of emerging technologies were ignored, and it appears increasingly clear that technological innovation itself is responding to existential concerns. For example, the rapidly expanding popularity of 'human–machine teaming' reflects not only practical impediments to full autonomy but growing fears of a future AI apocalypse.

At the same time, the limits of apocalyptic threat perceptions are also visible. The multitude of scientific reports on the looming climate catastrophe, for example, may have generated a wide range of mitigation policies, albeit at a level far below that required to avert disaster.[29] Nuclear-armed states might go to great lengths to avoid thermonuclear war but—as captured so eloquently in Robert Jervis' stability–instability paradox—stable nuclear deterrence allows them to undertake lower-intensity interventions and proxy wars that might yet escalate into significant conflict, including nuclear war.[30] As elaborated below, this is the scenario confronting us in Ukraine, where Russia's nuclear arsenal deters large-scale and direct NATO involvement but not the limited assistance seen thus far, leaving open the possibility of future escalation.

While policy and strategy impacts can be seen across a wide range of domains, there is a clear emphasis in the policy and academic literature on the most critical apocalyptic threats to human existence: namely, the environmental, biological and technological.[31] In part, this is because of the sheer scale of potential catastrophes in these areas. But it also reflects the hope that human ingenuity might yet prevent or mitigate the emerging threats in these domains.

THE WAR IN UKRAINE AND THE APOCALYPTIC IMAGINARY

Conventional war and genocide are generally not included in this prioritization, partly because in most cases their impact is more geographically limited (leaving them below the threshold of X-risk analysis) and partly because they are seen as a more or less constant factor in human civilization. The exceptions, as noted above, are global nuclear war, biological warfare, and wars dominated by new technologies like cyber, AI and space-based weapons.

Thus, while the war in Ukraine has had some impact within this category of war and genocide apocalyptic threat, overall it has had a relatively limited influence on the broader analysis of apocalyptic threat perceptions. For those concerned with environmental, biological and technological threats to human existence, the war in Ukraine has not massively shifted assumptions and predictions. Yet it would be hasty to conclude that the war in Ukraine has no relevance at all for contemporary apocalyptic thinking, as the next section elaborates in more detail.

4. Ukraine and the Apocalyptic Imaginary

In considering the intersection of apocalyptic thinking and the war in Ukraine, I argue there are three areas worth exploring in more detail: the threat of nuclear war; genocidal narratives; and a broader 'apocalyptic milieu'.

First, threat perceptions related to possible nuclear weapons use have undoubtedly intensified, beginning in the lead-up to the 2022 invasion and continuing in the months since. After years of post-Cold War complacency about nuclear weapons, the world was reminded of the very real possibility of nuclear escalation and the dramatic effects this would incur. These fears were fanned by Putin's statements about using 'all means necessary to defend Russia', putting Russian strategic nuclear forces on 'high alert' and 'combat duty', and ultimately withdrawing Russia from the New START treaty. This was matched in the West by high-profile rhetoric from NATO states, such as President Biden averring in October 2022 that the risk of nuclear Armageddon was at its highest level in sixty years.[32]

The fear of escalation to nuclear war has clearly shaped Ukrainian and NATO strategies: Ukraine has undertaken only limited strikes

321

BEYOND UKRAINE

on Russian territory, and the unexpectedly robust Western military support for Kyiv has been parcelled out in careful increments so as not to provoke a Russian over-reaction. That is not a new dynamic by any means; Thomas Schelling, for instance, wrote of such 'salami tactics' in 1966, during the Cold War.[33] Yet interestingly, there is not a consensus within strategic circles over how worried we should be about Russian nuclear use. Numerous experts have argued that it is vanishingly unlikely as long as the war remains on Ukrainian soil, as it is difficult to envision a scenario where Russia might benefit (tactically or strategically) from breaking the nuclear taboo.[34] At the same time, Russian strategic choices are also constrained by nuclear fears; it must avoid operational approaches that risk entering NATO states' territory or drawing them more directly into the war, and refrain from the use of 'tactical' nuclear weapons in Ukraine. Its years-long campaign of indirect warfare against NATO states must also remain calibrated to a level below the Article V threshold.

In short, the war in Ukraine has certainly heightened apocalyptic threat perceptions with respect to nuclear war, though in an uneven and not entirely new way. There is no consensus about the level of threat we should perceive at the present time, and it is the decades-old anxieties about nuclear Armageddon that have been revived, rather than an entirely new set of existential fears. Overall, apocalyptic threat perceptions have had a significant impact upon strategy in this conflict, but the reverse does not appear to be true (at least, to date).

Second, the impact of the war on apocalyptic threat perceptions is undoubtedly greater if we interpret 'apocalyptic' to include the annihilation of distinct nations and cultures (as I do in this chapter). In the first year of the war, horrific Russian atrocities against civilians coupled with rhetoric emanating from Moscow evoked fears of genocide within Ukraine, a country with long memories of the *Holodomor* atrocities that killed several million Ukrainians during the Soviet era. As Russian soldiers tortured, raped and executed civilians in occupied Ukraine, and destroyed local libraries, museums and schools, Russian political commentators called for 'de-Nazification' and the erasure of Ukrainian statehood. In an infamous article by Timofey Sergeytsev entitled 'What Should Russia Do with Ukraine?', it is explained that de-Nazification requires the destruction or

322

ideological re-education of the entire population, the suppression of their culture, and the elimination of Ukrainian sovereignty.[35] The belief that Ukraine is an artificial country that must be reunited with Russia—a running theme in Russian narratives—has been promulgated by Putin for many years, notably in his 2021 address 'On the Historical Unity of Russians and Ukrainians'.[36] In short, should Russia succeed in its conquest and enact the most extreme scenarios put forward by its political class, this could fairly be seen as 'the end of the world' for independent Ukraine and its culture. Arguably, the strength of this apocalyptic threat perception helps account for the (to date) robust support for Kyiv in this defensive war, both domestically and internationally, without which it is doubtful Ukraine would still be able to fight at all.

Russian narratives of the conflict, however, feature a stunning rhetorical reversal of this apocalyptic threat, arguing that the Ukrainians have been committing genocide in the Donbas since 2014, and casting Russia as the humanitarian defender of helpless civilians.[37] In his 24 February 2022 speech launching the 'special military operation' in Ukraine, Putin stated: 'We had to stop that atrocity, that genocide of the millions of people who live there and who pinned their hopes on Russia ... To this end, we will seek to demilitarise and denazify Ukraine...'[38] This is a somewhat different dynamic of apocalyptic threat perceptions influencing strategy, as accusations of genocide facilitate a Russian offensive war that explicitly targets civilian populations across the country in the name of 'de-Nazification'.

Russian claims of genocide in the Donbas have not been supported by international observers, including the Organization for Security and Co-operation in Europe's Special Monitoring Mission to Ukraine and the UN Human Rights Monitoring Mission in Ukraine, which published daily and monthly reports of civilian casualties in the region that contradict claims of genocide.[39] In February 2022, Ukraine brought a case against Russia in the International Court of Justice (ICJ) in order 'to establish that Russia has no lawful basis to take action in and against Ukraine for the purpose of preventing and punishing any purported genocide'.[40] As of writing, the case remains under review, but in an initial order in March 2022, the

BEYOND UKRAINE

ICJ ordered Russia to suspend its military operations in Ukraine, in what was seen as an implicit rejection of Russia's legal argument that its intervention was necessary to halt genocide. This is not an unprecedented situation for Russia: in 2008, then President Dmitry Medvedev justified the Russian invasion of Georgia by saying that South Ossetians were being subjected to genocide, claims which were later established to be greatly exaggerated.[41]

More broadly, Russian justifications for intervening in Ukraine since 2014 invoke longstanding narratives of the existential threat that NATO states pose to the Russian state: by expanding to Russia's borders, deploying missile defence and precision-strike weapons systems, and spreading 'the technology of the colour revolution' to the former Soviet space and beyond, NATO states (allegedly) threaten not just Russia's interests, but the very existence of the Russian nation. Thus, from this perspective, Russian operations against NATO states, or to prevent non-NATO states from joining NATO, are purely defensive and legitimate. In this case, invoking a potential apocalypse has facilitated Russia's long-term strategic approach and its array of tactics for confronting NATO, with the invasion of Ukraine representing the apotheosis of this grand narrative. As Putin explained to the Federal Assembly in February 2023, 'The elites of the West do not hide their purpose … They intend to transform a local conflict into a phase of global confrontation. This is exactly how we understand it all and we will react accordingly, because in this case we are talking about the existence of our country'.[42]

Finally, I would argue there is an additional influence of the war in Ukraine on what I call the 'apocalyptic milieu'—the general miasma of apocalyptic fears among the general public that indirectly affects political and strategic decision-making.[43] Even before February 2022, a wide range of existential fears stalked the world, with Covid-19, environmental disasters, devastating wars and economic inequality all contributing to a sense of unrelenting crises in the years to come. Since the Russian invasion of Ukraine, many polls have suggested rising public anxiety about the war, and the possibility that it may go nuclear: an IPSOS-Mori poll showed 58 per cent of Americans were worried about a nuclear war with Russia; a YouGov poll in the United Kingdom showed that the percentage of respondents who

324

THE WAR IN UKRAINE AND THE APOCALYPTIC IMAGINARY

thought nuclear war would be the most likely cause of our extinction rose from 43 per cent to 61 per cent immediately following the Russian invasion; and in Russia, the Levada Centre found that 83 per cent of respondents feared nuclear war with the United States.[44] A March 2022 survey by the American Psychological Association was revealing: when respondents were asked to detail the most important causes of stress in their lives, the war in Ukraine and global uncertainty were three of the top five stressors; moreover, 69 per cent believed they were witnessing the beginning of a Third World War, and 73 per cent felt 'overwhelmed by the number of crises facing the world right now'.[45] In short, while the war in Ukraine did not create an entirely new genre of apocalyptic fears, it appears to have intensified an already existing brew of existential anxieties in many states.

The impact of this apocalyptic milieu on policy and strategy is somewhat indirect; for the most part, it requires governments to address domestic sources of unrest and disinformation that threaten to undermine support for their political and strategic efforts (including with respect to Ukraine). Apocalyptic fears help fuel the spread of new extremist movements, such as accelerationism and eco-fascism, which use the threat of extinction to promote their racist agendas and justify terrorist attacks.[46] Non-violent activists, particularly in the environmental domain, have adopted increasingly disruptive tactics in the name of saving the planet (the very name of the Extinction Rebellion movement, for instance, invokes fears of the apocalypse). Populist politicians and media feed upon visions of looming disaster, and conspiracy theories run amok in this terrifying landscape.

In Western states, an intensifying apocalyptic milieu poses something of a headache for the effort to support Ukraine, as governments must robustly engage in the information domain to counter disinformation and reassure their publics that this support will not lead to Armageddon. In Russia, however, apocalyptic fears are not necessarily a handicap, as they allow Putin to continue to justify his war as a legitimate and honourable defence of Russia's survival. As of early 2023, despite all the difficulties encountered in the war, Putin continued to enjoy approval ratings above 80 per

325

cent, while negative attitudes toward both Ukraine and the United States have risen from around 55 per cent to 70 per cent since the war began.[47] Other polls suggest that while Russians understand the war is not going to plan, and perhaps should not have been started, it must now be continued to avoid catastrophe. As the Russian political scientist Grigory Yudin explained in December 2022, 'Propagandists have been pushing the idea that it isn't so important if it was right or wrong for Russia to start all this. But now it's allegedly clear that Russia is fighting against NATO, and Russia will be crushed if it doesn't win'.[48] All of this suggests that narratives of existential defence are resonating within the Russian population.

In sum, apocalyptic threat perceptions have had a clear impact upon the strategies adopted in the Russian war on Ukraine, and this war has influenced our contemporary apocalyptic imaginary in several notable ways. At least, this is the situation one year after the Russian invasion. But how might this change in the months and years to come?

5. Conclusion: Apocalypse, Maybe?

Gazing into the future, it is possible to imagine that the dynamics elaborated in this chapter might undergo a profound shift. First, there are several scenarios in which apocalyptic threat perceptions might no longer have such a strong influence on Western strategy—most notably, if NATO states come to believe that Russia will never use nuclear weapons unless its own territory is threatened, and/or if Ukraine successfully drives Russian forces out of most of its territory, thus diminishing fears of national annihilation. It is more difficult to envision apocalyptic threat perceptions having less influence in Russia, as even a successful conquest of Ukraine would not remove the existential threat that Russia perceives to be emanating from NATO more broadly. This longstanding narrative will be difficult to dismantle within Russia for the foreseeable future.

Second, it is possible that the war could come to have a greater impact on the contemporary apocalyptic imaginary. I argue this would be most likely in the following scenarios: a serious escalation that brings us to the brink of nuclear war; the use of nuclear weapons

326

in Ukraine; a nuclear explosion or radiation release that affects a huge swathe of territory and populations; a near-total genocide in Ukraine; a massive cyberattack that results in a global impact; or the introduction of new weapons systems that currently have only hypothetical concerns attached to them. Any of these scenarios are likely to shape the apocalyptic imaginary—not just in scale, but in character.

In conclusion, the Russian war in Ukraine is the latest example of the kinds of events that contribute to the amorphous and evolving character of apocalyptic thinking, while also demonstrating how existential fears shape policy and strategy. The uncertain trajectory of the war suggests caution in analysing these dynamics, but the catastrophic devastation inherent in this war's worst-case scenarios demands that we take seriously the possibility of the world coming to an end, as it already has for so many tragic victims of this brutal conflict.

16

WAR AS BECOMING

Antoine Bousquet

Perhaps no single question preoccupies contemporary military analysts and strategists more than that of the future of war. Every new technology or strategic concept is touted as potentially revolutionary, with commanders and decision-makers forever fearful of committing the ultimate sin of 'fighting the last war'. Forecasting the future of war is now institutionalized practice in military organizations and beyond, however chequered the record of past predictions are shown to be.[1] Continuous change is thus a universally accepted feature of war. And yet, simultaneously, we obstinately seek to circumscribe the variability of war, lest it trouble the assurances of a consistent, unitary phenomenon that practitioners and experts can speak about authoritatively. Amidst the ambiguity and perplexity that pervade our actual experience of war, we find reassurance in holding on to a putatively elementary kernel that tells us that we know what war fundamentally is.

At the time of writing, twelve months into the Russian invasion of Ukraine in February 2022, and nine years after the annexation of Crimea, a devastating war is ongoing with no clear resolution in sight. Firmly reinstalling great-power antagonism to the forefront of world affairs with an armed conflict that may still escalate

BEYOND UKRAINE

dramatically, the war has already paraded a motley array of guises and battlefield innovations. An inexhaustive list would reference nuclear intimidation, economic warfare, infrastructural sabotage, mercenary armies, artificial intelligence (AI)-enabled drones, portable missile launchers, decentralized geographic information system targeting, low-orbit mini-satellites, cyber operations, and social media mobilization. As is customary, commentators duly line up to explain in a single breath that however indicative any of these developments may be for the future trajectory of war, it is of course the case that the nature of war itself has not changed.[2] Routinely accompanied by appeals to 'Saint' Clausewitz's sanction, such assertions gesture at the politically motivated use of organized violence, the vagaries of battle, and an innately human character as the perennial features constituting war's timeless essence.[3]

Please indulge me here in a blasphemous thought. What if the recitation of this well-worn catechism constituted a major impediment to our understanding of the martial phenomenon today? What if whatever benefits this incantation might have in warding off inflated claims of military novelty and rash dismissals of the past were outweighed by the conceptual shackles imposed by the commitment to an immutable nature of war? What if we stood instead to gain new insights and agential powers by allowing the nature of war to be a riddle forever unsolved, rather than insisting upon a definitive, unassailable answer to provide cognitive ballast against the world's ceaseless tumult? Indeed, these are the very provocations put forward in this chapter.

So, yes, the nature of war is changing. Because the nature of war has always been changing. Because the nature of war itself is change. What does that mean? That war is not a definite thing, has no true essence, possesses no fixed being that endures under its various forms. War is an emergent interaction, a process of intense contestation, a field of forces. Strictly speaking, war cannot *be* anything. War only *becomes*.

As such, war does not have a future. Instead, the future promises a plurality of wars that will need to be primarily apprehended on their own terms rather than insistently related back to some common genus that would assure us of a deep historical and metaphysical

330

WAR AS BECOMING

continuity. In contending with a shifting, polymorphous Ukrainian war in which First World War machine guns cohabit with drone swarms,[4] we should eschew a sterile opposition between proclaiming revolutionary transformation and affirming the eternal recurrence of the same. Nor should we expect the war to set the template for future conflicts. For war never stands still.

The Perpetual Transformation of War

In the post-Cold War era, we became accustomed, however grudgingly in some quarters, to the idea that war simply was no longer what it once used to be. Decentred from its prior interstate focus, lacking in identifiable beginnings and ends, distributed over dispersed, non-contiguous geographies, and traversed by unpredictable weaponizations, war stubbornly slipped through the meshes of our familiar conceptual nets.[5] Empowered by unfettered globalization and proliferating network technologies, non-state actors vaulted to the forefront of the international security agenda with the terrorist attacks of 11 September 2001, humbling the American Leviathan by brazenly turning its own world-spanning technologies against it. Tasked simultaneously with the protection of vulnerable civil societies from further conspiracies to commit atrocities, and the projection of force worldwide to track down terrorists and topple any colluding states, the ensuing War on Terror radically blurred the boundaries of domestic and national security, civilian and military spheres, peace and war. Conventional war vied with counterinsurgency, special forces operations and remote-controlled drone killings against an uncertain enemy across a planetary battlespace of shifting flashpoints and fluctuating intensities. A leading US military commander spoke for many when acknowledging that 'basic categories such as "battlefield," "combatant" and "hostilities" no longer have clear or stable meaning'.[6]

Nevertheless, all this talk of a transformation of war rested upon the notion that there once existed a time when the contours and features of war were familiar and settled. But we find no such thing over the past two centuries. The French Revolutionary Wars and the ensuing Napoleonic Wars of conquest were experienced

331

by contemporaries as a radical break with the limited conflicts of the prior dynastic epoch. Through the introduction of the *levée en masse*, in which hundreds of thousands of galvanized citizens were conscripted into the armies of the fledging French nation, the populace was wholly brought into war for the first time in the modern era. Drawing upon his singular martial talents, Napoleon leveraged the consequent superiority in numbers and fighting spirit into an unrelenting war machine that repeatedly destroyed on the battlefield the reputedly finest armies of the time. It is no wonder that an acute mind like that of Carl von Clausewitz would draw upon his experience of the period for the key insight that war is a protean phenomenon wrought by the political and historical conditions of its occurrence, and subject to the partial understandings of those caught up in it: 'every age had its own kind of war, its own limiting conditions, and its own biases'.[7]

The First World War occasioned no less a shattering of the martial imaginary in bringing the combined economic, scientific and technological might of modern industrial societies to bear upon the constitution, organization and application of armed force. Millions were swallowed up in titanic battles of materiel, upending conventional military wisdoms and shredding any enduring romantic notions of individual warring prowess and nobility. Years of bloody stalemate on the Western Front began to be broken with the introduction of the armoured tank, but sheer attrition proved the most decisive factor in determining the final outcome. The subsequent interlude barely provided enough time for contemporaries to grapple with the dizzying lessons of the conflict before the next global conflagration was upon them. The most perceptive observers foresaw that the momentum of societal mobilization would not halt until it had escalated to its most all-encompassing expression. The prominent Great War veteran Ernst Jünger wrote in 1930 of a process of 'total mobilisation' under which 'the image of war as armed combat merges into the more extended image of a gigantic work process' until 'there is no longer any movement whatsoever – be it that of the homeworker at her sewing machine – without at least indirect use for the battlefield'.[8]

The Second World War realized this comprehensive subsuming of society within the ambit of war as militaries took to devastating industrial complexes and urban population centres as the surest path to victory. With air power emerging as the most efficient vehicle for this purpose, the war fittingly concluded with the atomic incineration of two Japanese cities, announcing the paroxysmal end towards which total war ineluctably pointed.

The nuclear revolution under which the Cold War was inaugurated was soon recognized as a first-order crisis in strategic thinking. Unrestricted total war now seemed like a suicidal absurdity that could only be warded off through the constant threat of mutual annihilation.[9] Following all the devastation wreaked in the first half of the twentieth century, war appeared to have definitively outstripped the confines of human fathoming, with only the most precarious control remaining over it: 'Our military power is so vast that the effects of its use are beyond the comprehension of the mind of man', reflected US Secretary of State Dean Rusk in 1967, 'it is so vast that we dare not allow ourselves to become infuriated'.[10] Consequently, the immovable horizon of nuclear apocalypse did not bring about the end of war, as some had hopefully called for at the dawn of the atomic age. Instead, the Cold War took the form of a global geopolitical struggle dividing the world into opposing camps, and waged via a spectrum of deadly conflict and intense competition that spanned proxy wars, covert operations, psychological warfare, economic rivalry, ideological antagonism, and protracted cultural and technoscientific contests. Deeply embedded national security bureaucracies and dispendious defence funding of scientific research became permanent features. Military innovation was pursued unremittingly, yielding a procession of technological marvels and wonder-weapons from the electronic computer and space satellite to the nuclear-powered submarine and precision-guided missile. Buttressed by the interlocking of mortiferous arsenals that ruled out a decisive outcome through the trial of arms, the Cold War endured for four decades, during which total war appeared impossible and peace unachievable.

To these familiar historical vignettes, it is necessary to adjoin the supposedly minor martial phenomena constituted by the various insurrectionary, revolutionary, colonial and anticolonial sites of conflict that populate the entire span of the last two centuries. The

BEYOND UKRAINE

literal belittling of such struggles as 'small wars' has for too long served to diminish their historical importance and analytical relevance for our understanding of war's protean character. Blinkered by our provincial state-centric experience in the West, we have become habituated to treat conflicts that step outside of the interstate frame as derivative, secondary occurrences of war—if we are even willing to bestow the term upon them.[11] This has frequently caused us to inadequately grasp the crucial continuities and dependencies between the historical manifestations of war at the core and the periphery of imperial formations.[12] Likewise, the categorization of conflicts into variously major or minor occurrences of war has been overdetermined by their experience from the Western perspective. There is thus a particular absurdity—if not by now an outright obscenity—in speaking of a 'Long Peace' in the nineteenth century (said to have run from the Congress of Vienna to the outbreak of the First World War) when, throughout that same century, vast swathes of the world were being forcibly seized and indigenous life-worlds decimated.[13]

In sum, attempting to locate a time in the last 200 years when war possessed a steady, consensual form is a futile task—everywhere we look we encounter disruption and upheaval, along with a recurring sense of crisis and inadequacy in both theory and practice. It is all the more ironic then that, with the return of large-scale interstate war in Ukraine, we can detect a barely repressed relief that the messiness of the past few decades can be put behind us. The Russians are back, presenting us with a familiar adversary, and conventional war is seemingly again the main game in town. Yet, giving into the temptation of a retreat to some essentialized conception of war from which recent manifestations were unfortunate deviations would be a mistake. Rather than withdraw from it, we should lean into our heightened perplexity at the experience of war in the twenty-first century and decisively ground our interrogation in a rigorous philosophical commitment to the priority of becoming.

Against an Essence of War

In the simplest terms, we need to cease grasping for the essence of war (what it fundamentally 'is') in favour of apprehending war as process

334

WAR AS BECOMING

(how it perpetually 'becomes'). As such, war must always appear to us as mysterious, its final conceptualization or full comprehension forever beyond us. My submission here is that war is constitutively unknowable. Which is to say that war's elusiveness is not primarily an epistemic problem—a deficit in the knowledge we hold about it that we might still one day overcome—but an ontological attribute inherent to its very nature and our relation to it.

Correspondingly, we should abandon the forlorn quest for a final, absolute definition of war, as if it were a rare butterfly we could once and for all pin down securely under a glass case. No matter how habitually the term is bandied about by politicians, strategists and military professionals alike, we find that the concept of war slips through our fingers as soon as we attempt to grasp it too tightly. For once we move beyond tautological references to armed conflict and organized fighting, the fundamental concept of war comes to rest upon a notion of violence whose extension beyond a narrow understanding in terms of physical force is impervious to unambiguous elucidation. Indeed, insofar as war cannot be reduced to the bare biological extermination of the enemy but must instead entail a clash of intentionalities, direct attacks upon the human body must be both mentally and socially mediated. A strictly somatic conception of violence reduces it to a mere conduit among others for a wider coercive activity aimed at bending adversarial will, merely displacing the definitional problem to delineating the uncertain boundaries of that encompassing activity. If, on the other hand, we opt to retain violence at the core of our conception of war, it must necessarily open onto an indeterminate field of activities that exceeds the constricted idea of physical force. How else to make sense of such vectors of weaponization as economic sanctions, disinformation campaigns, psychological operations, international law, zero-day exploits, infectious agents and opioid epidemics?[14] To be clear, the claim here is not that everything is necessarily war, but that we can never be certain of what should be included within it.

A settled, unequivocal definition of war must therefore always elude us. Not that war needs such an understanding of itself to get along just fine; after all, soldiers and strategists dispense with one in their everyday practice largely unperturbed. These individuals are

335

BEYOND UKRAINE

accordingly liable to dismiss the above discussion on the grounds that 'you know war when you see it', accompanied by withering derision of otherworldly intellectual pretensions of little real-world relevance. To such common-sense appeals to resisting abstraction, it must be opposed that it is war itself that has become ever more abstract, as the object of its targeting has ranged ever further from the enemy's immediate resistance on the battlefield to its energetic, informational and cognitive underpinnings. Moreover, the necessary abstraction required to break with our hidebound intellectual habits should not be seen as a flight from the concrete manifestation of war, but rather as the means by which we may penetrate its operations more deeply.

It would be misleading to give the impression that contemporary strategists and military thinkers have not grappled with the challenges presented by the continuously shifting landscape of war. The abundant literature on the subject of 'hybrid war' and 'grey zone' operations has certainly attempted to account for the multifaceted expressions and perplexing ambiguities of conflict in the early twenty-first century.[15] However, it has largely persisted in conceptualizing these phenomena negatively, which is to say in terms of what they are not—that is, all the ways in which they deviate from the image of conventional state-on-state war. This incapacity does not lie primarily in a supposed failure of the imagination. It is rather that an incipient realization that there is no longer a conceptual *terra firma* for our understanding of war has been fettered by a reluctance to relinquish essentialist categories of thought.

What if instead we decisively resolved to stop trying to answer the question of what war is, to determine where it begins and ends, to endlessly police the boundary between war and not-war? A perspective shift occurs when we begin to ask how war becomes. We are suddenly forced to think exclusively about war in terms of constitutive processes through which martial worlds coalesce, sustain themselves, and eventually dissipate. Freed from having to endlessly adjudicate what counts as war, we can follow the trail of conflict wherever it leads.

This approach is not synonymous with abandoning all of our existing conceptions of war's phenomenology. We obviously require

336

WAR AS BECOMING

a broad, preliminary understanding of war to apprehend it in the present and trace its historical continuities. We necessarily start from somewhere, even if this is always *in media res* from the time and place into which we have been thrown. Simply, our commitment to these conceptions should always remain a pragmatic and provisional one, conditional on their explanatory power and a reflexive awareness of the uses to which they are being put. We should also resist the urge to see theoretical and empirical discrepancies as aberrations that we can ignore or dismiss as peripheral, but instead invite them into the very heart of our interrogation of war.

Nor does an ontology of becoming mean that everything must be changing all the time. The world can display some long-lasting features and regularities that we can discern in different places and epochs. However, these stabilities cannot be treated as inherent, eternal traits that merely need to be identified and recognized in every instance; it is always necessary to account for the dynamic processes that iterate and reproduce them. Stability and identity are second-order phenomena to change and difference, albeit no less important for it.

The privileging of the term 'martial' over the more customary 'military' is one way among others to unsettle our stultified habits of thought. Invocation of the martial directs us to the capaciousness and open-endedness of a warring phenomenon that perpetuates itself well beyond the confines of military institutions. It serves to transcend the habitual dichotomy made between civilian and military spheres of society which, notwithstanding its continued political or juridical currency, frequently occludes how profoundly entangled they have become. The highly distinctive culture and organizational habitus that traditionally characterized military institutions have become increasingly imbued with civilian norms. Conversely, the historical experience, continued exercise and extended imaginary of war permeates civilian life through its economic dependencies, technological lineages and cultural aura.

The binary distinction between technical artefacts that would count as either weapons or tools should be similarly breached in favour of a more dynamic understanding of weaponization—the process of 'becoming a weapon'.[16] This is because all technologies

337

BEYOND UKRAINE

are potentially 'dual-use' and liable to being enlisted into martial activity. The improvised explosive device (IED) that became the iconic weapon of the wars in Iraq and Afghanistan, occasioning the majority of coalition deaths, provides an illuminating insight into this technical mutability. Put together by artisanal bomb-makers, each IED is assembled from readily available materials, repurposing fertilizers, household chemicals, walkie-talkies, alarm clocks or garage-door openers to serve as switches, fuses and charges.[17] The latent convertibility of 'weapon' and 'tool' received perhaps its most spectacular demonstration on 11 September 2001, when al-Qaeda seized passenger aircraft and hurled them as missiles against the World Trade Center and Pentagon. This immanent weaponization did not entail any technical modification to the planes, merely a reorientation away from an iterative circuit of transportation into propulsive vectors of destruction. The information technologies that have come to define our times, and the ways in which we conceive and conduct war, appear particularly polyvalent though.[18] To take one recent example, there has been abundant coverage of the role that mobile phones have played in the Ukrainian war in relation to wiretapping, fire targeting, and minefield mapping.[19] Finally, the logistical dependencies, broadly understood, of contemporary warring extend their tendrils so widely today as to render almost moot any distinctions between civilian and military technologies.

Admittedly, to consistently follow the conceptual prescriptions proposed here is neither easy nor straightforward. We are culturally and cognitively habituated to thinking in terms of fixed essences and stable entities, with change as something that happens to things as they transition from one state to another. Indeed, we cannot ever but fleetingly think in terms of pure flux and becoming, if only because the dissolution of ego it entails is a limit-experience we can touch but never remain within without a terminal abolition of selfhood.[20] As such, the task of engaging with the becoming of war passes through a constant unpicking of our mental calcifications, with our analytical categories, empirical enclosures, conceptual models and army of metaphors only serving as temporary, pragmatic staging posts in our encounters with our fields of enquiry.

338

Becoming Clausewitz

We can already hear the rallying cry of the card-carrying Clausewitzians: 'a changing character of war, ok, but an immutable nature!' While the particular expressions of war certainly vary with the social and political conditions specific to every conflict, surely its enduring essence—as captured by the 'wondrous trinity' of enmity, chance and purpose—is beyond dispute? Or as arch-Clausewitzian Colin Gray puts it, 'there is a unity to all strategic experience: nothing essential changes in the nature and function (or purpose) – in sharp contrast to the character – of strategy and war'.[21] But is such a definitive judgement really what we find in On War, the Prussian officer's masterwork venerated for its unsurpassable insights into the timeless nature of war? Or do we instead stumble upon the curtailed exertions of a remarkably nimble intellect which, in tussling with a phenomenon resolutely resistant to comprehensive capture, was led to the incipient realization that war can only be approached asymptotically through conceptual categories and images of thought that foreclose any essentialization and embrace its radical becoming?

For one, the endlessly echoed divide between the nature and character of war is not one that we find unambiguously stated in On War (or more precisely, in Vom Kriege, since the dominant English translation can be tasked with terminological choices that have regrettably bolstered its essentialist reading).[22] Indeed, it is striking that in the opening of the very passage in which Clausewitz introduces the famous trinity, the original German text indicates that 'der Krieg ist also nicht nur ein wahres Chamäleon, weil er in jedem konkreten Falle seine Natur etwas ändert.'[23] This can literally be translated as 'war is not merely a veritable chameleon, because in every case it somewhat changes its nature'. Unfortunately, the Howard and Paret translation opts for 'war is more than a true chameleon that slightly adapts its characteristics to the given case'.[24] The difference may be subtle, yet it is still crucial, since the latter version distorts a passage that makes clear that Clausewitz did not hold a dichotomous distinction between immutable nature and changing character. As Antulio Echevarria points out, Clausewitz elsewhere describes the nature of

BEYOND UKRAINE

war as 'composite' (*zusammengesetzte*) and 'variable' (*veränderliche*)—hardly the attributes of an eternal essence.[25]

In the very same passage, Clausewitz famously draws an analogy between the theoretical construct of the trinity by means of a physics demonstration in which a pendulum is positioned between three equidistant magnets.[26] Once again, the choice made by Howard and Paret is consequential, conveying the term '*schwebend*' as 'suspended' when it could just as well be expressed as 'floating' or 'hovering'.[27] Indeed, any familiarity with the experimental setup described makes evident its dynamic character, the pendulum oscillating unpredictably through the complex nonlinear interaction of its initial momentum with the tripolar magnetic field. Here too, we find Clausewitz's intuitions into the wild becoming of war being almost imperceptibly corralled into a domesticated version of themselves.

We know that Clausewitz's premature demise left *On War* as an unfinished manuscript that bears the traces of the continuous revisions he made to it for fourteen years. What a 'final' version of his life's work would have looked like is of course unknowable, but its protracted redaction undoubtedly owes to Clausewitz's willingness to tirelessly re-examine his beliefs and further deepen his encounter with its perplexing object. His original intellectual impetus lay in an intense dissatisfaction with his contemporaries' attempts to furnish the theory of war with universal laws and fixed principles despite everything in war being '*unbestimmt*' (variously translatable as 'uncertain', 'indeterminate' or 'indefinite').[28] In his view, a fatal contradiction inevitably vitiated all such efforts, since 'the conduct of war extends itself on almost all sides into indefinite limits; while any system, any model, has the restrictive nature of a synthesis'.[29] Grappling with the inherent tension between his ambition to produce his own general theory of war and his scepticism at any intellectual pretensions to fully tame its object, Clausewitz formed a unique understanding of war as an unbounded, protean phenomenon that cannot ever be captured by a closed system of thought, but could nonetheless still be illuminated by the intellect.

From this perspective, the most important concept within *On War* is arguably that of friction, 'the only concept that more or less corresponds to what distinguishes real war from war on paper'.[30]

340

WAR AS BECOMING

Friction serves not merely as a practical caution to the commander to heed the contingencies of the battlefield, but ultimately as an affirmation of the fundamental ground for the unknowability of war. The unbridgeable gap between paper (or mind) and reality afflicts as much the general as the philosopher. Crucially, Clausewitz accounts for this rift in ontological terms, even adapting the metaphor originally drawn from mechanics to insist that friction is not concentrated in a few points but intimately interwoven into the very fabric of the real ('everywhere in contact with chance').[31] Or, as François Jullien puts it in his own commentary, 'what makes war is precisely the inevitable distance that the real takes from its model; to think war, in sum, is to think how it is bound to betray its concept'.[32]

The true greatness of Clausewitz thus lies in his burgeoning recognition that war could only be apprehended through constitutive tensions that undo any attempt at a systemic closure. We encounter these tensions everywhere in his writings, be it in the relation between war and politics, or the conceptual distinction between absolute war and real war. Rather than attempt to definitively iron out what might otherwise appear as unresolved contradictions in his thought, Clausewitz reached for metaphors such as competing poles of attraction or friction to cognitively reinforce them. The profound insight he was elaborating in *On War* is certainly not fully realized in the curtailed and recomposed version of the book we have inherited, and we can only speculate as to how it might have developed had he been granted a longer life. However, it is also significant that, at a time when the figures of Newton, Kant and Hegel reigned supreme, there existed no conception of an intellectual system that strove for anything else but a comprehensive submission of the world to reason, and this may have been too arduous a philosophical legacy for anyone to overcome. It is remarkable in itself that Clausewitz was able to formulate, however inchoately, an open-ended theory of war that we can today better discern with the benefit of two further centuries of philosophical enquiry and scientific discovery. And, of course, if his reflections drew on the manifestations of war that he had witnessed, stimulated by the acute sense that armed conflict had undergone a radical change in his lifetime, he scarcely could have imagined the transformations it has undergone since. For us, then, a true fidelity

341

BEYOND UKRAINE

to Clausewitz's colossal contribution should be synonymous with furthering his unbridled engagement with the obdurate problem of war, not with merely rehearsing and reifying the point at which its own becoming ceased.

Foundations of a Martial Empiricism

Today, we are fortunate to be armed with intellectual resources unavailable to Clausewitz in his endeavour. On the scientific front, the non-linear sciences of chaos and complexity have given us powerful insights into the creative dynamics of emergent self-organizing phenomena in human activity. The study of self-adaptive systems and distributed swarm intelligences uncovered in the natural world has proven especially applicable to the understanding of war.[33] The congruence between these scientific discoveries and Clausewitz's intuitions is by now well documented.[34] Moreover, the influence of the non-linear sciences upon military thinking on questions of organization and command has become so widespread over the last thirty years as to have constituted a veritable 'chaoplexic' way of warfare.[35] Given the inherent prestige that science is imbued with in our societies, and the crucial role that these same principles occupy in the development of various technologies of military relevance such as AI and robotics, the teachings of the non-linear sciences have offered a very effective way to unsettle overly linear and bounded conceptions of war. Nevertheless, it is rarely for the radical ontological implications of a chaoplexic worldview to be fully appreciated by those who more or less wittingly subscribe to it. If we are to give full voice to the becoming of war, we must pair the natural sciences with a more encompassing philosophical outlook.

The ancient Greeks provide us with a first champion in the pre-Socratic philosopher Heraclitus, whose fragmentary thoughts have become associated with two principal adages. The first holds that 'war is father of all, king of all', granting strife the singular power to bring things into existence.[36] The second propounds that 'on those who enter the same rivers, ever different waters flow', which expresses the notion of a universe in constant flux.[37] Within the canonical history of philosophy, Heraclitus is thus renowned as the

342

WAR AS BECOMING

initiator of the metaphysical doctrine of becoming and a forerunner of process philosophy.[38] It is striking then that becoming and war are found so closely bound in his early conception, as summarized by his biographer Diogenes Laertius: 'all things come into being by conflict of opposites, and the sum of things flows like a stream'.[39]

Despite Heraclitus outlining this distinctive metaphysical option within early Western thought, it is being and substance that were privileged over becoming and process in the centuries of philosophical and theological debate that followed. It is only in the nineteenth century that fully fledged philosophies of becoming began to be formulated. Of particular note is the proposition put forward by the American pragmatist philosopher William James of a 'radical empiricism'. James sought to challenge the dominant priority granted to universals and totalities by his contemporaries, arguing that 'what really exists is not things made, but things in the making'.[40] His would be a philosophy of pure experience in which the world was not a priori inhabited by established subjects and objects, but rather emerged piecemeal through relationality and encounter. Once again, we find James tellingly reaching for a military metaphor to convey his itinerant thought: 'life is in the transitions as much as in the terms connected; often, indeed, it seems to be there more emphatically, as if our spurts and sallies forward were the real firing-line of the battle'.[41]

Eminent successors to James can be found in the figures of Henri Bergson and Alfred North Whitehead, who, in their respective ways, formulated philosophies giving prominence to process and change.[42] These contributions have in turn influenced the development of postwar continental philosophy, above all that of French philosopher Gilles Deleuze, whose prodigious encompassing of the philosophical tradition not only produced a singular oeuvre within it, but also reconstructed a canon of neglected and marginal thinkers set against the prevailing propensity of Western thought towards essentialism and substantialism. In his collaborations with Félix Guattari, he assembled a mobile battery of concepts tasked with disrupting and checking the arborescent, totalizing structures of thought that have typically relegated processes of becoming to ancillary transitions between states of being rather than the primary ground of existence.

BEYOND UKRAINE

For our present concerns, it is especially notable that Deleuze and Guattari chose to align the 'war machine'—both in its historical instantiation and their expansive conceptual understanding of it—with a nomadic, deterritorialized trajectory that pits it against the sedentary, totalizing aspiration of the 'state apparatus'.[43]

Armed with the theoretical insights and conceptual tools bequeathed to us, we can aspire now to a 'martial empiricism' that fully embraces the becoming of war, foregrounding its mutability, polyvalence and creativity. By untethering ourselves from a sacrosanct faith in the immutable nature of war and abandoning the need to constantly patrol the boundaries of its manifestations, we are freed to focus our attention on war's generative activities, allowing ourselves to be constantly surprised and provoked by its new encroachments. The terrain of investigation before us is vast, simultaneously intimidating and inviting. As I have proposed elsewhere, future enquiry could be potentially organized according to the three interrelated fields of activity corresponding to mobilization, design and encounter.[44]

Mobilization denotes the ways in which resources, energies, bodies and effects are brought into the ambit of war. This is the domain of logistics, economic production, financial circuits, propaganda, ideology and military training, without which war could not materialize. In an important sense, the entire problem of war today begins here. As Bernard Brodie astutely observed in 1947:

> ... the governments of great states today have not only vast resources available to them for war purposes, but also something which they did not have prior to the twentieth century – the knowledge of how to mobilise them. It would be much easier to expunge from the human race the knowledge of how to produce the atomic bomb than the knowledge of how to produce total mobilisation.[45]

Nor should mobilization only be associated with the fullest extremities of armed conflict that brought it to an apex; its techniques and methods range the entire spectrum of war through formal military institutions and beyond, reaching ever deeper into body, mind and society.

344

WAR AS BECOMING

Design, for its part, encompasses the various intelligences and techniques that give rise to the specialized instruments of war. Here, we attend to the genesis, diffusion and deployment of the various technical objects, tactics and stratagems that make up warfare proper. Far from the sole province of mere optimization and technological efficiency, design is characterized by the prevalence of the specific 'cunning intelligence' known by the ancient Greeks as *metis*. Foregoing *techne*'s pursuit of formal universal principles, *metis* is associated with a situational, improvisational intelligence concerned with practical efficacy in navigating uncertain, shifting environments. Through its disposition towards learning to 'manipulate hostile forces too powerful to be controlled directly but which can be exploited despite themselves', *metis* is inherently accorded to the unpredictable, volatile demands of war.[46]

Finally, encounter refers us to the very experience of war in all its radical contingencies and generative potentials. This naturally evokes the literal terrain of battle where combatants encounter each other, but further encapsulates not only the wider life in arms—with all its tedium and quotidian banality—but also the much broader field of experience in a world suffused with the anticipation, exercise and aftermath of war. From this point of view, the modern drone operator, military logistician, defence analyst or Pentagon administrator is no less a part of war than the frontline soldier. This is not to downplay the singular experience of combat and all the affective intensities of exertion, exhilaration and trauma it entails. Indeed, there is arguably no greater testament to the wild contingency and ultimate unfathomability of war than the experiential crucible of combat. Simply, we cannot let it monopolize our accounts of the experience of war if we want to do justice to the capaciousness of the martial phenomenon in our time.

The (No-)Futures of War

From the perspective of war's becoming, it is clear that it is more fitting to speak of war's plural futures, given we have relinquished any universal index to which we might relate every specific instance. We should expect war to keep surprising us, always betraying

BEYOND UKRAINE

the models we impose upon it, always showing itself under new confounding guises. We are still grappling with the implications of a war that is presently unfolding in Ukraine at the time of writing. Not only would forecasting its resolution be a foolhardy venture, but we should equally guard against any hasty assumptions that our understanding of the war will provide an unambiguous guide to future conflicts. As history has repeatedly shown, any single-minded overhaul of armed forces and society to focus exclusively on meeting the demands associated with the latest war will almost certainly prove poor preparation for the next challenges when they materialise. From a wider analytical perspective, we should keep in mind that the priorities *du jour* of Western state militaries only ever encompass a narrow sliver of the global landscape of war as it manifests itself today in places such as Yemen, Syria, Myanmar, Ethiopia or Mexico.

Yet it is also precisely the way in which advanced militaries take anticipation of war's future to be synonymous with a headlong embrace of technological change that constitutes the foremost problem of our age, far exceeding the sole professional remit of military analysts and strategists. The history of humanity has long been largely that of war, but the trajectory of the past century has seen armed conflict become the chief precipitant for a technological colonization of the future that increasingly puts into question the becoming of our very species.[47]

There are those who would seek to man the last desperate rampart in defence of war's immutability with the invocation of an invariant human nature and its unsurpassable catalogue of passions and elemental concerns. War has forever been the province of anger, fear, hatred and courage, as recorded throughout historical documents reaching back to the *Iliad* and beyond. To this we can object that, while a recognisable if fraying thread indubitably binds us to our ancestors across millennia, the defining characteristic of the human animal has always been its unparalleled plasticity and open-endedness. The invention of culture unique to our species introduced a malleability in the forms of our communal existence and the individual meanings invested in them that has decisively overtaken the slow aimless fumbling of natural evolution. Our story

346

WAR AS BECOMING

has thus always been that of an indeterminate project of becoming human, alongside which our belligerent manifestations have mutated. Over the last few hundred years, the distinctive cultural accelerant of technoscience has become progressively dominant in this transmutative line of flight, attaining today an escape velocity that elicits copious talk of an emerging post-human condition. Advances in AI, genetic engineering and neuroscience are combining to alter war in ways that threaten even the most durable features of the activity that Thucydides once referred to as the 'human thing' (*to anthropon*).[48]

It is perhaps here then that the stakes come into clearest focus. We can now see a time when the runaway war machine will have dispensed with us altogether, be it through eradication or assimilation. In a world in which latent powers of quite unfathomable destruction are precariously poised, the increasing delegation of decisions to algorithmic agents brings unprecedented escalatory perils. Yet one does not have to posit genocidal military capture by a super-intelligent AI to imagine war continuing 'well after humanity has disappeared from the face of machines'.[49] We can just as easily envisage ourselves incrementally dissipating into our silicon architectures as command-and-control loops accelerate beyond the bounds of human cognition and battlespaces become ever more inhospitable to our squishy embodiments. Either way, it is our agency over the warring phenomenon that is flickering alarmingly as we travel further down the road laid for us by the frenetic action–reaction of contemporary conflictuality. The task of forging martial knowledges apposite to the disorienting metamorphoses of war ultimately participates in the crucial collective endeavour to wrest greater mastery over them. Therein lies the best hope that we will be writing for ourselves a future in which we have subdued war rather than suffering one of the futures in which our part has been wholly written out.

AFTERWORD

Christopher Coker[1]

Our relationship with war is so long and deep that we could, if we wish, tell the story of humanity entirely through the lens of conflict. As the political theorist Michael Walzer writes, arguments about war are probably endless, for they bear upon our own humanity the name we give not only to a particular species but also the behaviour we have come to expect of it.

Until Russia's invasion of Ukraine in 2022, it had become commonplace to ask: do we really need to think about war at all? Optimists like Steven Pinker and Yuval Harari were in denial. We were much more likely, or so Harari told us, to die from eating too many burgers at our local McDonald's than we were from an al-Qaeda attack. In his book *Enlightenment Now* (2017), Pinker too insisted that war was now largely only of historical interest. But remember, he also told us in the same work that it was now much easier to identify pathogens and invent vaccines; as a result, pandemics had probably disappeared from history. It is a great story to tell, and on the face of it, it is one that is almost too good to be true. It is all too easy to forget that the phrase usually means 'not true'.

There are many reasons why we should devote some serious thinking about the future of war.

All we escaped (and then briefly) in 1989 was a sub-set of war: namely, great power conflict. In truth, the rest of the world never escaped war's iron grip on life. Think of the wars that were waged between 2001–19: there were 110 involving more than a hundred nations.[2]

Secondly, we have arrived at a time of dramatic transformation in our understanding of war itself. Military analysts are forever declaring the onset of a new age of warfare, but it has already arrived. The point is that war is a shape-shifting phenomenon; 'new' wars arise as swiftly as the old disappear. Periodic plot twists arise all the time. One way to think about war, to invoke a timely metaphor, is to regard it as a virus which produces many different variants. In the last few years, academics have rebranded war as 'surrogate' (Krieg and Rickli, 2019), 'fourth generation', 'irregular' (Rid and Hecker, 2009), 'proxy' (Hughes, 2012), 'hybrid' (Hoffman, 2007), 'non-linear' (Galeotti, 2016), 'grey zone' (Echevarria, 2016), 'shadow' (McFate, 2019) and 'vicarious' (Waldman, 2021). It is in its nature of any virus to produce variants; in the case of war, they are shaped by culture and technology, as this book makes clear.

A third reason why we need to think about war is perhaps the most significant. Whenever we think of cyberattacks against Estonia in 2007 or information warfare against the West—the constant assault on our democratic systems by troll factories in Russia that are discussed in Nina Jankowitcz's book, *How to Lose the Information War*—we find ourselves living in what Lucas Kello calls an era of 'unpeace', a new era of mid-spectrum rivalry which is especially visible in cyberspace. It is becoming increasingly difficult these days to tell war from peace; the border between them seems to be far more porous that it once did.

Finally, in the past twenty years, thanks to the War on Terror, we have tended to look only at the Western understanding of war. Ironically, it was that war that reminded us there are very different cultural understandings of war; to quote Sweijs and Michaels in their Introduction to this volume, 'to the extent one can speak in terms of "Western", "Eastern" or "Global Southern", clearly there is a great diversity of outlook.'

We must be constructively sceptical of the pace of technological change. A recent report for the European Parliament titled *Innovative Technologies Shaping the 2040 Battlefield* (EPRS, 2021) advises that whenever thinking about how technology will shape the future, we

AFTERWORD

should adopt an agnostic approach. Futurists tend to measure trends and their implications and then offer persuasive suppositions about what comes next. Yet the future is rarely the end of a trend line.

Clausewitz talked about the 'mystery' of war, a word he invoked more than once in his work. We often labour under the misguided idea that war is rooted in the political realm, but we cannot hope to understand its appeal, or resilience, or even universality by exploring that dimension alone. The existential dimension is no less important, for it also involves power; or perhaps more correctly, the empowerment (both material and spiritual) of those who do the actual fighting. This probably animated many of the young men and women who went to Syria to fight for the Islamic State. As a terrorist tells his American hostage in Lebanon in Don DeLillo's novel *Mao 2*: 'Terror makes the new future possible. We live in history as never before'.

Whatever the cause or country for which one is fighting, the existential realm involves lived experience. What form will the existential take in future? To cite one example, theatricality has always pervaded war. From the *Iliad* to *Saving Private Ryan*, poets and film directors have refashioned the ugliness of war into art and alchemized suffering into a unique art form. Even so, the desire to glamorise violence dates back much further than Homer. The compulsion to process brutality, writes Dan Jones, is the oldest theme in art, and can be traced back 45,000 years to the first cave paintings, which were discovered in Indonesia as recently as 2017. There is no human culture that doesn't have some sort of art and none that hasn't fought wars; both are fundamental to who we are—a disturbing enough thought. To take war seriously is to take art seriously, and to take art seriously is to take storytelling seriously, especially when the stories have mythical status like the *Iliad* and the *Mahabharata*. Perhaps, this is what gives war its cultural appeal: its extraordinary tonal range. What we have today are video games and the metaverse is coming, and immersive virtual reality will offer entirely new opportunities for people to experience war in a way no one has done before. These are themes that are well worth exploring.

There is also the metaphysical dimension of war which involves sacrifice; the willingness to die, rather than kill. It is

351

BEYOND UKRAINE

that willingness that makes war sacred in the eyes of civilians and soldiers alike. Without a willingness on the part of its citizens to put themselves in danger, or to sacrifice themselves for the few, neither state nor non-state actors could wage war at all. The war in Ukraine shows this very clearly. In her book, *Paradise Built in Hell*, Rebecca Solnit writes that an unexpected and widespread response to disaster seems to be joy, not only at having survived an ordeal but at being provided with an opportunity to put our heroic selves on display, such as the rescue of a neighbour whose house has been blitzed. Possibly this is because only in times of crisis do some of us feel more fully alive. Rates of anxiety, depression, and even psychosis, we are told by psychologists, seem to decline when a society finds itself at war. It seems that many of us also want to experience the heroic at least once in our lives. The very act of overcoming adversity and the recognition that some of our fellow citizens are willing to sacrifice their lives for the rest of us can reinforce pride in our own humanity. But what will people be sacrificing their lives for in the future?

A more immediate question is what impact—positive or negative—will future technology have on us ontologically? To write of ontology is to write of our humanity, of what makes us the distinctive species that we are. Today, it is increasingly common to speak of the next stage in our development—the 'post-human' condition. The use of the word 'post' allows us to remain agnostic about what comes next (e.g. artificial intelligence), but it also retains a hint that the future will be better (we humans will be in greater control of our lives than ever.) But will it? Ontologically, technology can shape our lives in ways that are so familiar that we no longer recognise how it operates. A weapons system, for example, can change the way its users think about war. For that reason, every technology has a social history. Take the machine gun, which changed the way the Western powers conceived of war in the late nineteenth century. It was seen to be the product of industrialization, a process restricted at that time entirely to the West. It was seen to embody such modern principles as productivity and efficiency; to generate more output for less input. It also followed that the machine gun could be used to make people who didn't have it to 'see reason', to

352

AFTERWORD

see that resistance was futile. If they continued to resist, they were clearly being 'unreasonable'.

Changes can result not only from an attitude of mind (as in the case of the machine gun), but also be produced by changes in the brain itself, as the neuroscientist Susan Greenfield has discussed. Some neuroscientists expressed concern that an online screen life would encourage them to regard war as a computer game. James Williams' book *Stand Out of Our Light* (2018) highlights the role of technology in redirecting human attention. It was feared that drone pilots might become prone to 'dissociation' or 'moral de-skilling'. This hasn't happened. Even so, it is likely that the new technologies with which we will be interfacing more and more will make us not only more machine-readable, but by default more machine-friendly. Indeed, the real danger is not that computers begin to think for us, but that we begin to think like them. The real threat is that one day the artificial intelligence we are creating may turn out to be our own. All of this begs the question of the post-human condition and the prospect of 'post-human war'.

For me, the filter that is missing in this study of the future of war is science fiction. The future is imagined long before it is realised. Today, science fiction is now prescribed reading in the military. Orson Scott Card's novel *Ender's Game* has been on the syllabus of the Marine Corps University at Quantico for some time. It is also on the USN's Professional Leadership Program along with Robert A. Heinlein's *Starship Troopers*. In France, the Defence Innovation Agency has set up a team of science-fiction writers to propose scenarios that might not occur to most military planners. In Germany, Project Cassandra, a recent collaboration between the military and the literature department at the University of Tubingen, asks professors to imagine the future of war. Aren't they well equipped precisely because the fictional world they write about is a world of the imagination?

Although Margaret Macmillan's recent book on war,[3] for example, was well received, it was criticized by some writers, notably Edward Luttwak,[4] for being entirely deaf to the existential dimension of war.

To go back to Steven Pinker's speculations about the coming end of war, I am reminded of a phrase that was applied by some

353

BEYOND UKRAINE

battle-hardened communist party members to George Orwell after his death. In the late 1950s, they took to condemning him retrospectively. In writing *Animal Farm* and *1984*, in which he had warned of the horrors of totalitarianism long before Stalin's crimes were known to the world, they accused Orwell of having been 'prematurely correct'. Hopefully, Pinker may be vindicated, but I doubt it, as do most of the experts in the field.

pp. [4–25]

NOTES

INTRODUCTION

1. Lawrence Freedman, *The Future of War: A History* (London: Allen Lane, 2017): 217.

2. In this regard, Steven Pinker's thesis articulated in *The Better Angels of Our Nature: Why Violence Has Declined* (Viking, 2011) may suffer the same fate as Norbert Elias' argument that humankind had become less violent and more civilized, which he first published in the late 1930s, when the Second World War was breaking out. See Norbert Elias, *The Civilizing Process: Sociogenetic and Psychogenetic Investigations*, 2nd edition (Oxford & Malden, MA: Wiley & Blackwell, 2000).

1. REVISITING PUTIN'S 2022 INVASION OF UKRAINE

1. Antulio J. Echevarria II, 'Putin's Invasion of Ukraine in 2022: Implications for Strategic Studies', *Parameters* 52, no. 2 (2022): 21–34.

2. Steven Pinker, *The Better Angels of Our Nature: Why Violence has Declined* (New York: Viking, 2011); for a critique, see Bear F. Braumoeller, *Only the Dead: The Persistence of War in the Modern Age* (Cambridge: Cambridge University Press, 2019).

3. Azar Gat, *The Causes of War and the Spread of Peace: But Will War Rebound?* (Oxford: Oxford University Press, 2017). See also Julien Oeuillet, 'Azar Gat on War,' *Rabbit Hole*, 19 January 2022, https://rabbitholemag.com/azar-gat-on-war/

4. For a summary of the main arguments, see Raimo Väyrynen (ed.), *The Waning of Major War: Theories and Debates* (New York: Routledge, 2006).

5. Michael Kofman and Ryan Evans, 'Interpreting the First Few days of the Russo-Ukrainian War,' 28 February 2022, in *War on the Rocks*, https://

355

NOTES

warontherocks.com/2022/02/interpreting-the-first-few-days-of-the-russo-ukrainian-war/; on frozen conflicts, see Erik J. Grossman, 'Russia's Frozen Conflicts and the Donbas,' *Parameters* 48, no. 2 (2018): 51–62.

6. Michael E. Brown, Sean M. Lynn-Jones and Steven E. Miller, *Debating the Democratic Peace* (Cambridge, MA: MIT Press, 1996).

7. Väyrynen, *Waning of Major War*, 19.

8. George Kennan, 'The Sources of Soviet Conduct,' *Foreign Affairs* 25, no. 4 (1947): 566–82.

9. Thomas Schelling, *Arms and Influence* (New Haven, CT: Yale University Press, 1966): 174–7.

10. Adam S. Posen, 'The End of Globalization? What Russia's War in Ukraine means for the World Economy,' *Foreign Affairs*, 17 March 2022.

11. Velina Tchakarova (@vtchakarova), 'India: Hold my beer!' *Twitter*, 31 January 2023, https://twitter.com/vtchakarova/status/16205439337125 88803?s=43&t=gR41O3Tedb_TtLg2kmXtkw; Editorial Board, 'How to bring Putin and his henchmen to justice,' *The Washington Post*, 23 January 2023, https://www.washingtonpost.com/opinions/2023/01/22/international-court-putin-ukraine-war-tribunal/

12. 'Russia GDP 1988–2023,' *Macrotrends*, https://www.macrotrends.net/countries/RUS/russia/gdp-gross-domestic-product

13. Oona A. Hathaway, 'International Law Goes to War in Ukraine: The Legal Pushback to Russia's Invasion,' *Foreign Affairs*, 15 March 2022; Davida E. Kellogg, 'Jus Post Bellum: The Importance of War Crimes Trials,' *Parameters* 32, no. 3 (2002): 87–99.

14. Michael Byers, 'Justice Delayed: Why International Law Still Matters,' *Foreign Affairs*, 22 September 2016.

15. Jack Watling and Justin Bronk, 'Preliminary Lessons in Conventional Warfighting from Russia's Invasion of Ukraine: February–July 2022,' *Rusi*, 30 November 2022, https://rusi.org/explore-our-research/publications/special-resources/preliminary-lessons-conventional-warfighting-russias-invasion-ukraine-february-july-2022

16. Lazaro Gamio and Ana Swanson, 'How Russia Pays for War,' *The New York Times*, 30 October 2022, https://www.nytimes.com/interactive/2022/10/30/business/economy/russia-trade-ukraine-war.html

17. Lawrence Freedman (ed.), *Strategic Coercion: Concepts and Cases* (Oxford: Oxford University Press, 2003): 3–15; Kelly M. Greenhill and Peter Krause (eds), *Coercion: The Power to Hurt in International Politics* (Oxford: Oxford University Press, 2018).

18. Antulio J. Echevarria, 'Clausewitz's "Warlike Element" and the War in Ukraine,' *Military Strategy Magazine* 8, no. 2 (2022): 10–17.

19. Watling and Bronk, 'Preliminary Lessons,' 10.

20. Michael Kofman, Justin Bronk and Jack Watling, 'The Russia Contingency:

pp. [33–37] NOTES

Revisiting Russian Air Performance in Ukraine,' *The Russia Contingency* [podcast], 15 November 2022, https://warontherocks. com/episode/ therussiacontingency/27874/the-russia-contingency-revisiting-russian-air-performance-in-ukraine/; Michael Kofman, Justin Bronk and Jack Watling, 'Revisiting Russian Air Performance in Ukraine, Part 2,' *The Russia Contingency* [podcast], 15 November 2022, https://warontherocks.com/ episode/therussiacontingency/27897/revisiting-russian-air-performance-in-ukraine-part-2/

21. Carl von Clausewitz, *Hinterlasseneswerk Vom Kriege*, 19th edition, ed. Werner Hahlweg (Frankfurt: Ferdinand, 1980): Book I, Chapter 1, 191–2; Carl von Clausewitz, *On War*, trans. Michael Howard and Peter Paret (Princeton: Princeton University Press, 1986): 76.

22. Schelling, *Arms and Influence*, 1–3. For a broader definition of brute force, see Robert Mandel, *Coercing Compliance: State-Initiated Brute Force in Today's World* (Stanford, CA: Stanford University Press, 2015): esp. 4, 6–7.

23. On strategy of control, see Antulio J. Echevarria, *War's Logic: Strategic Thought in the American Way of War* (Cambridge: Cambridge University Press, 2021): 130–42.

24. Nicholas Mulder, *The Economic Weapon: The Rise of Sanctions as a Tool of Modern War* (New Haven, CT: Yale University Press, 2022); Juan C. Zarate, *Treasury's War: The Unleashing of a New Era of Financial Warfare* (New York: Public Affairs, 2013).

25. Jonathan Kirshner, *Currency and Coercion: The Political Economy of Monetary Power* (Princeton, NJ: Princeton University Press, 2020).

26. Gamio and Swanson, 'How Russia Pays for War.'

27. Elliot Smith, 'Russia's Economy is Beginning to Crack as Economists Forecast Sharp Contractions,' *CNBC*, 4 April 2022, https://www.cnbc. com/2022/04/04/russias-economy-is-beginning-to-crack-as-economists-forecast-sharp-contractions.html

28. 'U.S. Warns of Stinger Missile and 155mm Ammunition Shortage in Its Stockpiles,' *Defense Express*, 18 November 2022, https://en.defence-ua.com/industries/us_warns_of_stinger_missile_and_155mm_ammunition_shortage_in_its_stockpiles-4879.html

29. Jen Judson, 'Lockheed Gets HIMARS Contract to Replenish Stock Sent to Ukraine,' *Defense News*, 2 December 2022, https://www.defensenews. com/industry/2022/12/02/lockheed-gets-himars-contract-to-replenish-stock-sent-to-ukraine/

30. Rupert Smith, *The Utility of Force: The Art of War in the Modern World* (New York: Vintage Books, 2007).

31. Smith, *Utility of Force*, 4–5; Thomas Kuhn, *The Structure of Scientific Revolutions* (Chicago: University of Chicago Press, 1962).

32. Barbara Salazar Torreon and Sofia Plagakis, 'Instances of Use of United States

357

NOTES pp. [38–45]

Armed Forces Abroad, 1798–2022, Updated March 8, 2022 (R42738),' *Congressional Research Service*, 8 March 2022, https://crsreports.congress. gov/product/details?prodcode=R42738

33. Jan Angstrom, 'Contribution Warfare: Sweden's Lessons of the War in Afghanistan,' *Parameters* 50, no. 4, (2020): 61–72.

34. 'UN Says 11 Million Have Fled Homes in Ukraine,' *Associated Press*, 5 April 2022, https://www.aol.com/news/live-updates-ukraine-reports-russian-063640006-122004857.html

2. THE FUTURES OF WAR

1. Eliot Cohen, 'A Revolution in Warfare', *Foreign Affairs* 75, no. 2 (1996): 39.

2. Ibid.; Keith Shimko, *The Iraq Wars and America's Military Revolution* (Cambridge: Cambridge University Press, 2010): 1–88; Williamson Murray, 'Thinking About Revolutions in Military Affairs', *Joint Force Quarterly* 16 (1997): 69–76.

3. Stephen Biddle, 'Victory Misunderstood: What the Gulf War Tells Us about the Future of Conflict', *International Security* 21, no. 2 (1996): 139–79.

4. Paul Norwood et al., 'Capturing the Character of Future War', *Parameters* 46, no. 2 (2016): 87.

5. Frans Osinga, 'The Rise of Transformation', in *A Transformation Gap? American Innovations and European Military Change*, eds. Terry Terriff, Frans Osinga and Theo Farrell (Stanford, CA: Stanford University Press, 2010): 14–34; David Alberts et al., *Network Centric Warfare* (Washington, DC: CCRP, 1999).

6. John Arquilla and David Ronfeldt, *Swarming and the Future of Conflict* (Santa Monica, CA: RAND, 2000).

7. Max Boot, 'The New American Way of War', *Foreign Affairs* 82, no. 4 (2003): 41–50; Michael O'Hanlon, 'A Flawed Masterpiece', *Foreign Affairs* 81, no. 3 (2002): 47–63.

8. See Colin Gray, *Strategy for Chaos, Revolutions in Military Affairs and the Evidence of History* (London: Frank Cass, 2002).

9. John Mueller, *Retreat from Doomsday: The Obsolescence of Major War* (New York: Basic Books, 1989).

10. Ibid.

11. Charles Moskos, John Allan Williams and David Segal (eds), *The Postmodern Military, Armed Forces After the Cold War* (Oxford: Oxford University Press, 2000).

12. Mary Kaldor and Andrew Salmon, 'Military Force and European Strategy', *Survival* 48, no. 1 (2006): 19–34; Adrian Hyde-Price, 'European Security, Strategic Culture, and the Use of Force', *European Security* 13, no. 4 (2004): 323–43.

13. Rupert Smith, *The Utility of Force* (New York: Knopf, 2007).

14. Ibid., 1.

pp. [45–48] NOTES

15. Alex Bellamy and Paul Williams, *Understanding Peacekeeping*, 2nd edition (Cambridge: Polity Press, 2010): 13–41.

16. Frank Harvey, 'Deterrence and Ethnic Conflict, The Case of Bosnia Herzegowina, 1993-1994', *Security Studies* 6, no. 3 (1997): 180–210.

17. Robert Owen, 'Operation Deliberate Force', in *A History of Air Warfare*, ed. John Olson (Washington, DC: Potomac Books, 2010): 201–24.

18. Benjamin Lambeth, *NATO's Air War for Kosovo* (Santa Monica, CA: RAND, 2001); Ivo Daalder and Michael O'Hanlon, *Winning Ugly, NATO's War to Save Kosovo* (Washington, DC: Brookings, 2004): 8.

19. Lawrence Freedman (ed.), *Strategic Coercion* (Oxford: Oxford University Press, 1998); Peter Viggo Jakobsen, *Western Use of Coercive Diplomacy After the Cold War* (London: MacMillan, 1998): 182–228; Daniel Byman and Matthew Waxman, *The Dynamics of Coercion* (Cambridge: Cambridge University Press, 2001).

20. Karl Mueller, 'Strategies of Coercion: Denial, Punishment and the Future of Air Power', *Security Studies* 7, no. 2 (1998): 182–228.

21. Max Boot, *War Made New* (New York: Gotham Books, 2005); Eliot Cohen, 'A Revolution'.

22. Rasmussen Mikkel Vedby, *The Risk Society at War: Terror, Technology and Strategy in the Twenty-First Century* (Cambridge: Cambridge University Press, 2006).

23. Christopher Coker, *Humane Warfare* (Abingdon: Routledge, 2001): 1–23.

24. Theo Farrell, *The Norms of War, Cultural Beliefs and Modern Conflict* (London: Lynn Riener, 2005); Edward Luttwak, 'Toward Post-Heroic Warfare', *Foreign Affairs* 74, no. 3 (1995): 109–22.

25. Christopher Coker, *Globalisation and Insecurity in the 21st Century: NATO and the Management of Risk*, (Oxford: Oxford University Press, 2002).

26. Martin Shaw, *The New Western Way of War* (Cambridge: Polity, 2005); Colin McInnes, *Spectator Sport Warfare* (London: Lynne Rienner, 2002).

27. Robert Kagan, *Of Paradise and Power: America and Europe in the New World Order* (New York: Alfred A. Knopf: Distributed by Random House, 2003): 3.

28. Colin S. Gray, *Another Bloody Century: Future Warfare* (London: Weidenfeld and Nicolson, 2005).

29. Martin van Creveld, *The Transformation of War* (New York: The Free Press, 1991); Mary Kaldor, *New and Old Wars, Organized Violence in a Global Era* (Cambridge: Polity Press, 1999).

30. Van Creveld, *The Transformation of War*; Kaldor, *New and Old Wars*; see also Klejda Mulaj (ed.), *Violent Non-State Actors in World Politics* (New York: Columbia University Press, 2010); Robert J. Bunker (ed.), *Networks, Terrorism and Global Insurgency* (Abingdon: Routledge, 2005).

31. Stuart Kaufman, 'Symbolic Politics or Rational Choice? Testing Theories of Extreme Ethnic Violence', *International Security* 30, no. 4 (2006): 45–86; Richard Jackson and Helen Dexter, 'The Social Construction of Organized

NOTES pp. [48–51]

Political Violence: An Analytical Framework', *Civil Wars* 16, no. 1 (2014): 1–23.

32. Chris Hedges, *War Is a Force That Gives Us Meaning* (New York: Public Affairs, 2002).

33. Van Creveld, *The Transformation*; Bill Lind et al., 'The Changing Face of War, into the Fourth Generation', *Marine Corps Gazette* 85, no. 11 (1989): 22–6; Charles Dunlap, 'Lawfare: A Decisive Element of 21st-Century Conflicts?', *JFQ* 54, no. 3 (2009): 34–9.

34. Frans Osinga, 'De Strategische Pauze Voorbij: Veiligheid, Defensie en de Westerse Krijgsmacht na 9/11', *Militaire Spectator* 180, no. 9 (2011): 398–415.

35. Francis Fukuyama, *The End of History and the Last Man* (New York: The Free Press, 1992).

36. Samuel P. Huntington, *The Clash of Civilizations and the Remaking of World Order* (New York: Simon & Schuster, 1996).

37. Antonio Giustozzi (ed.), *Decoding the Taliban* (New York: Columbia University Press, 2009); August Norton, *Hezbollah* (Princeton, NJ: Princeton University Press, 2007); Jessica Stern and J.M. Berger, *ISIS, the State of Terror* (New York: HarperCollins, 2015).

38. Jessica Stern, *Terror in the Name of God* (New York: Ecco, 2003); D.C. Rapoport, 'The Fourth Wave: September 11 in the History of Terrorism', *Current History* 100, no. 650 (2001): 419–24; M. Sageman, *Understanding Terror Networks* (Philadelphia, PA: University of Pennsylvania Press, 2004); Mark Juergensmeyer, *Terror in the Mind of God* (Berkeley, CA: University of California Press, 2003).

39. Francis Fukuyama, 'The Imperative of State-building', *Journal of Democracy* 15, no. 2 (2004): 17–31.

40. Mats Berdal and Domink Zaum, *Political Economy of Statebuilding* (Abingdon: Routledge, 2013).

41. Tobias Debiel and Patricia Rinck, 'Statebuilding', in *Routledge Handbook of Security Studies*, 2nd edition, eds. Myriam Dunn Cavelly and Thierry Balzacq (Abingdon: Routledge, 2017): 404–12; Roland Paris and Timothy Sisk, *The Dilemmas of Statebuilding* (London: Routledge, 2009).

42. Cf. Nicolas Lemay-Hébert, 'Statebuilding without Nation-building? Legitimacy, State Failure and the Limits of the Institutionalist Approach', *Journal of Intervention and Statebuilding* 3, no. 1 (2009): 21–45; Roland Paris, 'Saving Liberal Peacebuilding', *Review of International Studies* 36, no. 2 (2010): 337–65; Roger Mac Ginty and Oliver P. Richmond, 'The Local Turn in Peace Building: A Critical Agenda for Peace', *Third World Quarterly* 34, no. 5 (2013): 763–83.

43. See Paul Rich and Isabelle Duyvesteyn, *The Routledge Handbook of Insurgency and Counterinsurgency* (Abingdon: Routledge, 2012).

pp. [51–53]   NOTES

44. See Antonio Giustozzi, 'Hearts and Minds, and the Barrel of a Gun: The Taliban's Shadow Government', *Prism* 3, no. 2/3 (2012): 71–80.

45. See David Kilcullen, *Out of the Mountains: The Coming of Age of the Urban Guerilla* (Oxford: Oxford University Press, 2013): Chapter 3.

46. David Kilcullen, 'Counter-insurgency Redux', *Survival* 48, no. 4 (2006): 111–30; David Kilcullen, *Accidental Guerrilla* (Oxford: Oxford University Press, 2009); Frank Hoffman, 'Neo-Classical Counterinsurgency?', *Parameters* 37, no. 2 (2007): 71–87; Christopher Paul, Colin P. Clarke, Beth Grill and Molly Dunigan, *Paths to Victory: Lessons From Modern Insurgencies* (Santa Monica, CA: RAND, 2013); Kilcullen, *Out of the Mountains*.

47. T.X. Hammes, *The Sling and the Stone* (New York: Motorbooks International, 2006); Frank Hoffman, *Conflict in the 21st Century: The Rise of Hybrid Wars* (Arlington, VA: The Potomac Institute of Policy Studies, 2007).

48. David Betz, 'The Virtual Dimension of Contemporary Insurgency and Counterinsurgency', *Small Wars & Insurgencies* 19, no. 4 (2008): 510–40.

49. Carl J. Ciovacco, 'The Contours of Al Qaeda's Media Strategy', *Studies in Conflict & Terrorism*, 32, no. 10 (2009): 853–75; Stuart J. Kaufman, 'Narratives and Symbols in Violent Mobilization: The Palestinian-Israeli Case', *Security Studies* 18, no. 3 (2009): 400–34; James P. Farwell, 'Jihadi Video in the "War of Ideas"', *Survival* 52, no. 6 (2010): 127–50.

50. Lisa Blaker, 'The Islamic State's Use of Online Social Media', *Military Cyber Affairs* 1, no. 1 (2015): 1–9; James P. Farwell, 'The Media Strategy of ISIS', *Survival* 56, no. 6 (2014): 49–55.

51. Gideon Avidor and Russell W. Glenn, 'Information and Warfare: The Israeli Case', *Parameters* 46, no. 3 (2016): 99–107.

52. Eric V. Larson et al., *Foundations of Effective Influence Operations* (Santa Monica, CA: RAND, 2009); US Army Doctrine Manual, *FM 3-13, Inform and Influence Activities* (Washington, DC: Department of the Army, 2013).

53. Colin F. Jackson, 'Information Is Not a Weapons System', *Journal of Strategic Studies* 39, no. 5/6 (2016): 820–46.

54. Arturo Munoz, *U.S. Military Information Operations in Afghanistan: Effectiveness of Psychological Operations 2001–2010* (Santa Monica, CA: RAND, 2012); George Dimitriu and Beatrice De Graaf, 'Fighting the War at Home: Strategic Narratives, Elite Responsiveness, and the Dutch Mission in Afghanistan, 2006–2010', *Foreign Policy Analysis* 12, no. 1 (2016): 2–23.

55. Douglas Porch, 'The Dangerous Myths and Dubious Promise of COIN', *Small Wars & Insurgencies* 22, no. 2 (2011): 239–57; David Ucko, 'Whither Counterinsurgency', in *The Routledge Handbook of Insurgency and Counterinsurgency*, eds. Paul Rich and Isabelle Duyvesteyn (London: Routledge, 2012): 67–79.

56. A. Mack, 'Why Big Nations Lose Small Wars', in *Strategic Studies, A Reader*, eds. Thomas Mahnken and Joseph Maiolo (Abingdon: Routledge,

361

NOTES

2008): 308–25; Gil Merom, *How Democracies Lose Small Wars* (Cambridge: Cambridge University Press, 2003).

57. Stephen Watts et al., *Limited Intervention: Evaluating the Effectiveness of Limited Stabilization, Limited Strike, and Containment Operations* (Santa Monica, CA: RAND, 2017).

58. Benjamin Lambeth, 'Operation Enduring Freedom 2001', in *A History of Air Warfare*, ed. John Olson (Washington, DC: Potomac Books, 2010): 255–77; O'Hanlon, 'A Flawed Masterpiece', 47–63.

59. Stephen Biddle, 'Afghanistan and the Future of Warfare', *Foreign Affairs* 82, no. 2 (2003): 31–48.

60. Richard Andres, Craig Wills and Thomas E. Griffith, Jr., 'Winning with Allies: The Strategic Value of the Afghan Model', *International Security* 30, no. 3 (2005): 124–60.

61. Joseph L. Votel and Eero R. Keravuori, 'The By-With-Through Operational Approach', *Joint Force Quarterly* 89, no. 2 (2018): 40–7; Patrick Work, 'Fighting the Islamic State By, With, and Through: How Mattered as Much as What', *Joint Force Quarterly* 89, no. 2 (2018): 56–62.

62. Amos Fox, 'Conflict and the Need for a Theory of Proxy Warfare', *Journal of Strategic Security* 12, no. 1 (2019): 44–71.

63. Dmitry Dima Adamsky, 'From Israel with Deterrence: Strategic Culture, Intra-war Coercion and Brute Force', *Security Studies* 26, no. 1 (2017): 157–84; Alex S. Wilner, 'Targeted Killings in Afghanistan: Measuring Coercion and Deterrence in Counterterrorism and Counterinsurgency', *Studies in Conflict & Terrorism* 33, no. 4 (2010): 307–29.

64. Oliver Kessler and Wouter Werner, 'Extrajudicial Killing as Risk Management', *Security Dialogue* 39, no. 203 (2008): 298–308; Mathias Grossklaus, 'Friction, not Erosion: Assassination Norms at the Fault Line Between Sovereignty and Liberal Values', *Contemporary Security Policy* 38, no. 2 (2017): 260–80; Derek Gregory, 'The Everywhere War', *The Geographical Journal* 177, no. 3 (2011): 238–50; Peter Bergen and Daniel Rothenberg, *Drone Wars, Transforming Conflict, Law and Policy* (Cambridge: Cambridge University Press, 2015); Laurie Calhoon, *We Kill Because We Can, From Soldiering to Assassination in the Drone Age* (London: Zed Books, 2015); Ian G.R. Shaw, 'Predator Empire: The Geopolitics of US Drone Warfare', *Geopolitics* 18, no. 3 (2013): 1–24.

65. Price Bryan, 'Targeting Top Terrorists: How Leadership Decapitation Contributes to Counterterrorism', *International Security* 36, no. 4 (2012): 9–46; Patrick Johnston, 'Does Decapitation Work? Assessing the Effectiveness of Leadership Targeting in Counterinsurgency Campaigns', *International Security* 36, no. 4 (2012): 47–79.

66. Sam Jones, 'Dmitry Medvedev Warns of "new cold war"', *Financial Times*, 13 February 2016, http://www.ft.com/cms/s/0/a14e8900-d259-11e5-829b-8564e7528e54.html#axzz4IiVoaRgz

pp. [54–57] NOTES

67. Matthew Kroenig, 'Facing Reality: Getting NATO Ready for a New Cold War', *Survival* 57, no. 1 (2015): 49–70; Alexander Lanoszka, 'Russian Hybrid Warfare and Extended Deterrence in Eastern Europe', *International Affairs* 92, no. 1 (2016): 175–95.

68. Heidi Reisinger and Alexander Golts, 'Russia's Hybrid Warfare: Waging War Below the Radar of Traditional Collective Defense', in *NATO's Response to Hybrid Threats*, eds. Guillaume Lasconjarias and Jeffrey Larsen (Rome: NATO Defense College, 2015): 113–36.

69. Luke Harding, 'We Should Beware Russia's Links with Europe's Right', *The Guardian*, 8 December 2014; Peter Foster and Matthew Holehouse, 'Russia Accused of Clandestine Funding of European Parties as US Conducts Major Review of Vladimir Putin's Strategy', *The Telegraph*, 16 January 2016; NATO Parliamentary Assembly Report, *The Battle for the Hearts and Minds: Countering Propaganda Attacks Against the Euro-Atlantic Community*, NATO, http://www.nato-pa.int/default.asp?SHORTCUT=4012 (last accessed 28 September 2019).

70. Todd C. Helmus et al., *Russian Social Media Influence; Understanding Russian Propaganda in Eastern Europe* (Santa Monica, CA: RAND, 2018).

71. Mark Galeotti, 'Hybrid, Ambiguous, and Non-linear? How New is Russia's "new way of war"?', *Small Wars & Insurgencies* 27, no. 2 (2016): 282–301; Sandor Fabian, 'The Russian Hybrid Warfare Strategy – Neither Russian nor Strategy', *Defense and Security Analysis* 35, no. 3 (2019): 308–25.

72. Qiao Liang and Wang Xiangsui, *Unrestricted Warfare* (Beijing: PLA Literature and Arts Publishing House, 1999).

73. Andrew Monaghan, 'The "war" in Russia's Hybrid Warfare', *Parameters* 45, no. 4 (2015): 66–74.

74. Dmitri Trenin, 'The Revival of the Russian Military; How Moscow Reloaded', *Foreign Affairs* 95, no. 3 (2016): 23–9; Polina Sinoviets and Bettina Renz, 'Russia's 2014 Military Doctrine and Beyond: Threat Perceptions, Capabilities and Ambitions', in *NATO's Response to Hybrid Threats*, eds. Lasconjarias and Larsen, 73–91.

75. Jay Ballard, 'What is Past is Prologue, Why the Golden Age of Rapid Air Superiority is at an End', *Joint Air and Space Power Journal* 22 (2016): 56–64.

76. Elbridge Colby and Jonathan Solomon, 'Facing Russia: Conventional Defence and Deterrence in Europe', *Survival* 57, no. 6 (2015): 21–50; Martin Zapfe and Michael Carl Haas, 'Access for Allies?', *The RUSI Journal* 161, no. 3 (2016): 34–41; Stephan Frühling and Guillaume Lasconjarias, 'NATO, A2/AD and the Kaliningrad Challenge', *Survival* 58, no. 2 (2016): 95–116; Alexander Lanoszka and Michael A Hunzeker, 'Confronting the Anti-Access/Area Denial and Precision Strike Challenge in the Baltic Region', *The RUSI Journal* 161, no. 5 (2016): 12–18.

77. Dmitry Dima Adamsky, 'From Moscow with Coercion: Russian Deterrence

363

Theory and Strategic Culture', *Journal of Strategic Studies* 41, no. 1/2 (2017): 1–25.

78. AIV, *The Future of NATO and The Security of Europe* (The Hague: AIV, 2017).

79. Alex Wilner, 'Cyber Deterrence and Critical-Infrastructure Protection: Expectation, Application, and Limitation', *Comparative Strategy* 36, no. 4 (2017): 309–18.

80. Joshua Davis, 'Hackers Take Down the Most Wired Country in Europe', *Wired*, 21 August 2007, http://www.wired.com/politics/security/magazine/15-09/ff_estonia

81. Jon R. Lindsay, 'Stuxnet and the Limits of Cyber Warfare', *Security Studies* 22, no. 3 (2013): 365–404.

82. Richard A. Clarke and Robert K. Knake, *CyberWar, The Next Threat to National Security and What to Do About It* (New York: Ecco/HarperCollins, 2011).

83. General David Richards, *Taking Command: The Autobiography* (London: Headline, 2014).

84. Joseph S. Nye Jr., *The Future of Power* (New York: Public Affairs, 2011).

85. Thomas Rid, *Cyber War Will Not Take Place* (Oxford: Oxford University Press, 2013); Martin Libicki, *Cyberdeterrence and Cyberwar* (Santa Monica, CA: RAND, 2009); Timothy J. Junio, 'How Probable is Cyber War? Bringing IR Theory Back into the Cyber Conflict Debate', *Journal of Strategic Studies* 36, no. 1 (2013): 125–33; Erik Gartzke, 'The Myth of Cyberwar Bringing War in Cyberspace Back Down to Earth', *International Security* 38, no. 2 (2013): 41–73; Lucas Kello, 'The Meaning of the Cyber Revolution, Perils to Theory and Statecraft', *International Security* 38, no. 2 (2013): 7–40.

86. Joseph S. Nye Jr., 'Nuclear Lessons for Cyber Security?', *Strategic Studies Quarterly* 5, no. 4 (2011): 18–38; Lucas Kello, *The Virtual Weapon and International Order* (Hartford, CT: Yale University Press, 2017); Joseph S. Nye Jr., 'Deterrence and Dissuasion in Cyberspace', *International Security* 41, no. 3 (2016): 44–71; Stephen J. Cimbala, 'Nuclear Deterrence and Cyber Warfare: Coexistence or Competition?', *Defense & Security Analysis* 33, no. 3 (2017): 193–208.

87. Michael P. Fischerkeller and Richard J. Harknett, 'Deterrence is Not a Credible Strategy for Cyberspace', *Orbis* 61, no. 3 (2017): 381–93; Erica D. Borghard and Shawn W. Lonergan, 'The Logic of Coercion in Cyberspace', *Security Studies* 26, no. 3 (2017): 452–81; Max Smeets, 'Integrating Offensive Cyber Capabilities: Meaning, Dilemmas, and Assessment', *Defence Studies* 18, no. 4 (2018): 395–410.

88. John Stone, 'Cyber War Will Take Place!', *Journal of Strategic Studies* 36, no. 1 (2013): 101–8.

89. Kenneth Payne, 'Artificial Intelligence: A Revolution in Strategic Affairs?', *Survival* 60, no. 5 (2018): 7–32; Mark Fitzpatrick, 'Artificial Intelligence and Nuclear Command and Control', *Survival* 61, no. 3 (2019): 81–92; Jürgen

pp. [60–61] NOTES

Altmann and Frank Sauer, 'Autonomous Weapon Systems and Strategic Stability', *Survival* 59, no. 5 (2017): 117–42; Michael C. Horowitz, 'When speed kills: Lethal autonomous weapon systems, deterrence and stability', *Journal of Strategic Studies* 42, no. 6 (2019): 764–88; Todd S. Sechser, Neil Narang and Caitlin Talmadge, 'Emerging technologies and strategic stability in peacetime, crisis, and war', *Journal of Strategic Studies* 42, no. 6 (2019): 727–35; Jesse T. Wasson and Christopher E. Bluesteen, 'Taking the archers for granted: emerging threats to nuclear weapon delivery systems', *Defence Studies* 18, no. 4 (2018): 433–53.

90. Mary Kaldor, *Global Security Cultures* (Cambridge: Polity Press, 2018).

91. Kilcullen, *Out of the Mountains*; Sean McFate, *The New Rules of War: Victory in the Age of Durable Disorder* (New York: William Morrow, 2019).

92. Cf. Robert Kaplan, *The Coming Anarchy: Shattering the Dreams of the Post Cold War* (New York: Vintage Books, 2001).

93. Seth Jones, 'The Future of Warfare is Irregular', *The National Interest*, 26 August 2018.

94. Daniel Byman, 'Why States are Turning to Proxy War', *The National Interest*, 26 August 2018.

95. Andreas Krieg and Jean-Marc Rickli, 'Surrogate Warfare: The Art of War in the 21st Century?', *Defence Studies* 18, no. 2 (2018): 113–30; Thomas Waldman, 'Strategic Narratives and US Surrogate Warfare', *Survival* 61, no. 1 (2019): 161–78.

96. Michael Mazarr et al., *What Deters and Why* (Santa Monica, CA: RAND, 2018); Michael C. McCarthy, Matthew A. Moyer and Brett H. Venable, *Deterring Russia In The Gray Zone* (US Army SSI, March 2019); Gregory F. Treverton, Andrew Thvedt, Alicia R. Chen, Kathy Lee and Madeline McCue, *Addressing Hybrid Threats* (Stockholm: Swedish Defence University, 2018); Alina Polyakova and Spencer P. Boyer, *The Future of Political Warfare: Russia, The West, and the Coming Age of Global Digital Competition* (Washington, DC: Brookings Institute, 2017).

97. David Rothkopf, 'The Cool War', *Foreign Policy*, 20 February 2013; Noah Feldman, *Cool War: The Future of Global Competition* (New York: Random House, 2013); Michael Gross and Tamar Meisels (eds), *Soft War, The Ethics of Unarmed Conflict* (Cambridge: Cambridge University Press, 2017).

98. Todd C. Helmus et al., *Media Influence: Understanding Russian Propaganda in Eastern Europe* (Santa Monica, CA: RAND, 2018); Michael J. Mazarr et al., *Hostile Social Manipulation: Present Realities and Emerging Trends* (Santa Monica, CA: RAND, 2019); Ricardo A. Crespo, 'Currency Warfare and Cyber Warfare: The Emerging Currency Battlefield of the 21st Century', *Comparative Strategy* 37, no. 3 (2018): 235–50.

99. Peter Singer and Emerson T. Brooking, *LikeWar: The Weaponization of Social Media* (Boston, MA: Eamon Dolan, 2018).

100. Mark Galeotti, *The Weaponisation of Everything, A Field Guide to the New Way*

NOTES pp. [61–62]

of War (New Haven, NJ: Yale University Press, 2023); Elie Perot, 'The Blurring of War and Peace', *Survival* 61, no. 2 (2019): 101–10.

101. Sean Monaghan (ed.), *Countering Hybrid Warfare* (Shrivenham: DCDC, 2018); Lyle J. Morris et al., *Gaining Competitive Advantage in the Gray Zone: Response Options for Coercive Aggression Below the Threshold of Major War* (Santa Monica, CA: RAND, 2019); Linda Robinson et al., *Modern Political Warfare: Current Practices and Possible Responses* (Santa Monica, CA: RAND, 2018); Thomas G. Mahnken, Ross Babbage and Toshi Yoshihara, *Countering Comprehensive Coercion: Competitive Strategies Against Authoritarian Political Warfare* (Washington, DC: CSBA, 2018); Elizabeth G. Troeder, *A Whole-Of-Government Approach To Gray Zone Warfare* (Carlisle Barracks: US Army SSI, 2019).

102. Michael Mandelbaum, 'Is Major War Still Obsolete?', *Survival* 61, no. 5 (2019): 65–71; Michael Mandelbaum, *The Rise and Fall of Peace on Earth* (Oxford: Oxford University Press, 2019).

103. Richard Haass, 'How a World Order Ends: And What Comes in Its Wake', *Foreign Affairs* 98, no. 1 (2019): 22; Edward Luce, *The Retreat of Western Liberalism* (New York: Atlantic Monthly Press, 2017); John Peterson, 'Present at the Destruction? The Liberal Order in the Trump Era', *The International Spectator* 53, no. 1 (2018): 28–44.

104. Ronald Inglehart, 'The Age of Insecurity: Can Democracy Save Itself?', *Foreign Affairs* 97, no. 3 (2018): 20–8; Cas Mudde, 'Europe's Populist Surge, A Long Time in the Making', *Foreign Affairs* 95, no. 6 (2016): 25–30; Fareed Zakaria, 'Populism is on the March, Why the West is in Trouble', *Foreign Affairs* 95, no. 6 (2016): 9–17; Hal Brands, 'Democracy vs Authoritarianism: How Ideology Shapes Great-Power Conflict', *Survival* 60, no. 5 (2016): 61–114.

105. Mandelbaum, 'Is Major War Still Obsolete?', 65–71; Mandelbaum, *The Rise and Fall of Peace*; David Kilcullen, *The Dragons and the Snakes: How the Rest Learned to Fight the West* (New York: Oxford University Press, 2020).

106. Kenneth Payne, *Strategy, Evolution and War, From Apes to Artificial Intelligence* (Washington, DC: Georgetown University Press, 2018); Paul Scharre, 'The Real Danger of an AI Arms Race', *Foreign Affairs* 98 no. 3 (2019): 135–44.

107. Christian Brose, 'The New Revolution in Military Affairs, War's New Sci-Fi Future', *Foreign Affairs* 98, no. 3 (2019): 122–34; Robert Latiffe, *Future War: Preparing for the New Global Battlefield* (New York: Vintage Books, 2017).

108. Michael Raska, 'The Sixth RMA Wave: Disruption in Military Affairs?', *Journal of Strategic Studies* 44, no. 4 (2021): 456–79.

109. Daniel Fiott, 'A Revolution Too Far? US Defence Innovation, Europe and NATO's Military-Technological Gap', *Journal of Strategic Studies* 40, no. 3 (2017): 417–37; Kevin M. Woods and Thomas C. Greenwood, 'Multidomain Battle Time for a Campaign of Joint Experimentation', *JFQ* 88, no. 1 (2018): 14–21; US Army TRADOC, *Multi-Domain Battle: Evolution of Combined Arms*

pp. [63–71] NOTES

for the 21st Century (Carlisle Barracks, 2018); King Mallory, *New Challenges in Cross-Domain Deterrence* (Santa Monica, CA: RAND, 2018).

110. This reconstruction gratefully draws on one of the scarce reports on the air war: Justin Bronk, Nick Reynolds and Jack Watling, *The Russian Air War and Ukraine Requirements for Air Defense* (London: RUSI, 2022).

111. Peter Singer, 'One Year In: What Are the Lessons from Ukraine for the Future of War?', *New American Century*, 13 March 2023, https://www.newamerica.org/international-security/blog/one-year-in-what-are-the-lessons-from-the-war-in-ukraine-for-the-future-of-war/

112. Mykhaylo Zabrodskyi, Jack Watling, Oleksandr V. Danylyuk and Nick Reynolds, *Preliminary Lessons in Conventional Warfighting from Russia's Invasion of Ukraine: February–July 2022* (London: RUSI, 2022). Cf. Mick Ryan, 'A Year of War, Part I', *Substack*, 20 February 2023, https://mickryan.substack.com/p/a-year-of-war-part-i

113. @DefenceHQ, 'Latest Defence Intelligence update on the situation in Ukraine', *Twitter*, 17 February 2023, https://twitter.com/DefenceHQ/status/1626472945089486848

114. Cf. David Petraeus for CNN: Peter Bergen, 'Gen. David Petraeus: How the War in Ukraine Will End', *CNN*, 14 February 2023, https://edition.cnn.com/2023/02/14/opinions/petraeus-how-ukraine-war-ends-bergen-ctpr/index.html; Dara Massicot, 'What Russia Got Wrong, Can Moscow Learn From Its Failures in Ukraine?', *Foreign Affairs* 102, no. 3 (2023): 78–93; Rob Johnson, 'Dysfunctional Warfare: The Russian Invasion of Ukraine', *Parameters* 52, no. 2 (2022): 5–20.

115. Stephen Covington, *The Culture of Strategic Thought Behind Russia's Modern Approaches to Warfare* (Cambridge, MA: Belfer Center for Science and International Affairs, Harvard Kennedy School, 2016); Alex Vershinin, 'The Return of Industrial Warfare', *RUSI*, 17 June 2022, https://www.rusi.org/explore-our-research/publications/commentary/return-industrial-warfare; Michael Howard, 'The Forgotten Dimensions of Strategy', *Foreign Affairs* 57, no. 5 (1979): 975–86.

116. Cf. Frans Osinga, 'Boyd, Bin Laden and Fourth Generation Warfare as String Theory', in *New Wars and New Theories, Problems and Prospects*, ed. John Olson (Oslo: Norwegian Institute for Defence Studies, 2007): 168–98.

117. Frank Hoffman, 'The Future is Plural, Multiple Futures for Tomorrow's Joint Force', *JFQ* 88, no. 1 (2018): 4–13; Robert A. Johnson, 'Predicting Future War', *Parameters* 44, no. 1 (2014): 65–76.

3. AFTER RUSSIA'S INVASION OF UKRAINE

1. John Mueller, *Retreat from Doomsday: The Obsolescence of Major War* (New York: Basic Books, 1989); Azar Gat, *War in Human Civilization* (Oxford: Oxford University Press, 2006); idem., *The Causes of War and the Spread of Peace:*

NOTES pp. [73–82]

But Will War Rebound? (Oxford: Oxford University Press, 2017); Steven Pinker, *The Better Angels of Our Nature: Why Violence Has Declined* (New York: Penguin, 2011); Joshua Goldstein, *Winning the War on War: The Decline of Armed Conflict Worldwide* (New York: Penguin, 2011); Ian Morris, *War: What is It Good For?* (New York: Farrar, Straus, and Giroux, 2014).

2. Arie Kacowicz, *Zones of Peace: South America and West Africa in Comparative Perspective* (New York: SUNY, 1998).

3. John Lewis Gaddis, *The Long Peace: Inquiries Into the History of the Cold War* (Oxford: Oxford University Press, 1989).

4. See the data in Quincy Wright, *A Study of War* (Chicago: University of Chicago, 1965): 653; which is skewed against colonial powers, such as Britain and France, by the inclusion of the European powers' 'small wars' against minor rivals abroad. Also, the data in Evan Luard, *War in International Society* (London: Tauris, 1986): 23–82 (especially 24–5, 35, 45, 53, and appendices 1–4, 421–41), which is careful to distinguish between the two types of war.

5. Cf. Pitirim Sorokin, *Social and Cultural Dynamics, vol. 3: Fluctuation of Social Relationships, War, and Revolution* (New York: The Bedminster Press, 1962 [1937]); Wright, *A Study of War*; Jack S. Levy, *War in the Modern Great Power System, 1495–1975* (Lexington, KY: Kentucky University Press, 1983); Evan Luard, *War in International Society*; Kalevi Holsti, *Peace and War: Armed Conflicts and International Order 1648–1989* (Cambridge: Cambridge University Press, 1991); Jack Levy and William Thompson, *The Arc of War: Origins, Escalation, and Transformation* (Chicago: University of Chicago Press, 2011), extrapolates from the size of a few arbitrarily selected historical battles and their death toll, *inter alia* failing to calibrate for the size of the populations involved.

6. Victor Hanson, *A War Like No Other: How the Athenians and Spartans Fought the Peloponnesian War* (New York: Random House, 2005): 10–11, 79–80, 82, 264, 296.

7. Peter Brunt, *Italian Manpower 225 B.C.–A.D.14* (Oxford: Oxford University Press, 1971).

8. Frederick Mote, 'Chinese Society under Mongol Rule,' in *The Cambridge History of China: Alien Regimes and Border States, 907–1368*, eds. H. Franke and D. Twitchett (Cambridge: Cambridge University Press, 1994): 618–22.

9. Peter Wilson, *The Thirty Years War: Europe's Tragedy* (Cambridge, MA: Harvard University Press, 2009): 786–8.

10. Graham Allison, *Destined for War: Can America and China Escape Thucydides's Trap?* (Boston, MA: Houghton Mifflin Harcourt, 2017).

11. See Azar Gat, 'The Return of Authoritarian Great Powers,' *Foreign Affairs* 86, no. 4 (2007): 59–69. For the ensuing debate see idem., 'Which Way is History Marching: Debating the Authoritarian Revival,' *Foreign Affairs* 88, no. 4 (2009): 150–5.

368

pp. [82–86]   NOTES

12. Azar Gat, 'The Return of Authoritarian Great Powers,' *Foreign Affairs* 86, no. 4 (2007): 59–69.
13. Ibid.

4. THE NEXT WAR WOULD BE A CYBERWAR, RIGHT?

1. John Arquilla and David Ronfeldt, 'Cyberwar Is Coming', in *In Athena's Camp: Preparing for Conflict in the Information Age*, eds. John Arquilla and David Ronfeldt (Santa Monica, CA: RAND, 1997): 79 89.
2. Thomas Rid, 'Cyber War Will Not Take Place', *Journal of Strategic Studies* 35, no. 1 (2012): 5–32.
3. John Stone, 'Cyber War Will Take Place!', *Journal of Strategic Studies* 36, no. 1 (2013): 101–8.
4. Cf. Michael N. Schmitt, *Tallinn Manual 2.0 on the International Law Applicable to Cyber Operations*, 2nd edition (Cambridge: Cambridge University Press, 2017).
5. David Cattler and Daniel Black, 'The Myth of the Missing Cyberwar', *Foreign Affairs*, 6 April 2022.
6. Lennart Maschmeyer, 'The Subversive Trilemma: Why Cyber Operations Fall Short of Expectations', *International Security* 46, no. 2 (2021): 53–4.
7. Max Smeets, 'The Strategic Promise of Offensive Cyber Operations', *Strategic Studies Quarterly* 12, no. 3 (2018): 90–113.
8. Kim Zetter, *Countdown to Zero: Stuxnet and the Launch of the World's First Digital Weapon* (New York: Crown Publishers, 2015); James Long, 'Stuxnet: A Digital Staff Ride', *Modern War Institute*, 3 August 2019, https://mwi.usma.edu/stuxnet-digital-staff-ride/
9. Robert Lee, Michael Assante, and Tim Conway, *Analysis of the Cyber Attack on the Ukrainian Power Grid* (Washington, DC: SANS Industrial Control Systems Security, 2016); Kim Zetter, 'Inside the Cunning, Unprecedented Hack of Ukraine's Power Grid', *Wired*, 3 March 2016.
10. Special Council Robert S. Mueller, *Report On The Investigation Into Russian Interference in The 2016 Presidential Election, vol. I and II* (Washington, DC: US Department of Justice, 2019); Renee DiResta et al., *The Tactics and Tropes of the Internet Research Agency* (U.S. Senate Documents—Congress of the United States, October 2018): 101; Fergus Hanson et al., 'Hacking Democracies: Cataloguing Cyber-Enabled Attacks on Elections', *ASPI Policy Brief Report,* no. 16 (2019): 2–30.
11. Max Smeets, 'Building a Cyber Force Is Even Harder Than You Thought', *War On The Rocks*, 12 May 2022, https://warontherocks.com/2022/05/building-a-cyber-force-is-even-harder-than-you-thought/
12. North Atlantic Treaty Organization, 'Warsaw Summit Communiqué', *NATO*, 9 July 2016.

NOTES pp. [86–88]

13. William Courtney and Peter A. Wilson, 'If Russia Invaded Ukraine', *Rand*, 8 December 2021, https://www.rand.org/blog/2021/12/expect-shock-and-awe-if-russia-invades-ukraine.html; Andy Greenberg, 'The WIRED Guide to Cyberwar', *Wired*, 23 August 2019; Alina Polyakova and Spencer P. Boyer, *The Future of Political Warfare: Russia, the West, and the Coming Age of Global Digital Competition the New Geopolitics* (Washington, DC: Brookings—Robert Bosch Foundation, 2018): 1.

14. Referring *inter alia* to the framing used in the Rid–Stone discourse. See Mykhaylo Zabrodskyi et al., *Preliminary Lessons in Conventional Warfighting from Russia's Invasion of Ukraine: February–July 2022* (London: RUSI, 2022); Rob Lee, 'The Tank Is Not Obsolete, and Other Observations about the Future of Combat', *War on the Rocks*, 6 September 2022, https://warontherocks.com/2022/09/the-tank-is-not-obsolete-and-other-observations-about-the-future-of-combat/

15. Timothy Thomas, 'Russia's Information Warfare Strategy: Can the Nation Cope in Future Conflicts?', in *The Transformation of Russia's Armed Forces: Twenty Lost Years*, ed. Roger McDermott (New York: Taylor and Francis, 2016): 130; Jason Healey, 'Preparing for Inevitable Cyber Surprise', *War On The Rocks*, 12 January 2022, https://warontherocks.com/2022/01/preparing-for-inevitable-cyber-surprise/

16. Antonius J.C. Selhorst, 'Russia's Perception Warfare', *Militaire Spectator*, no. 4 (2016): 148–64.

17. Dea Bankova et al., 'Six Months of the War in Ukraine', *Reuters*, 24 August 2022, https://www.reuters.com/graphics/UKRAINE-CRISIS/jnvwenoqdvw/; Kateryna Stepanenko and Mason Clark, 'Russian Offensive Campaign, August 17', *Institute for the Study of War*, 17 August 2022, https://www.understandingwar.org/backgrounder/russian-offensive-campaign-assessment-august-17

18. Nick Beecroft, 'Evaluating the International Support to Ukrainian Cyber Defense', *Carnegie Endowment for International Peace*, 3 November 2022, https://carnegieendowment.org/2022/11/03/evaluating-international-support-to-ukrainian-cyber-defense-pub-88322

19. At the moment of writing, the Russo-Ukrainian War, starting as of 24 February 2022, had been ongoing for almost a year. The analysis of this chapter does not reach beyond February 2023.

20. Armed conflict is here used as the technical state of affairs described in Common Article 2 to the Geneva Conventions of 1949 ('international armed conflict').

21. Erik Gartzke and Jon R. Lindsay, 'Weaving Tangled Webs: Offense, Defense, and Deception in Cyberspace', *Security Studies* 24, no. 2 (2015): 316–48.

22. Russell Buchan and Inaki Navarrete, 'Cyber Espionage and International Law', in *Research Handbook on International Law and Cyberspace*, eds. Nicholas

pp. [88–89] NOTES

Tsagourias and Russell Buchan, 2nd edition (Cheltenham: Edward Elgar, 2021): 231–52.

23. These cyber-related intelligence operations are not the focus of this chapter. For more information on this, see Parts 1 and 2 of Neveen Shaaban Abdalla et al., 'Intelligence and the War in Ukraine', *War On The Rocks*, 11 May 2022, https://warontherocks.com/2022/05/intelligence-and-the-war-in-ukraine-part-1/

24. Cyber operations make use of the electromagnetic spectrum, and the latter can also be used as a 'standalone' or supplementary capability to generate effects in electronic warfare (EW). EW makes use of energy whereas cyberspace operations use a binary code and are often Internet-based. Both aspects can be combined if a malign virus (binary software) is sent via an EW asset, e.g. during the 2007 Operation Orchard. See Stone, 'Cyber War Will Take Place!', 16–17.

25. James Shires, 'The Simulation of Scandal: Hack-and-Leak Operations, the Gulf States, and U.S. Politics', *Texas National Security Review* 3, no. 4. (2020): 11–28.

26. Timothy L. Thomas, 'Russia's Reflexive Control Theory and the Military', *The Journal of Slavic Military Studies* 17, no. 2 (2004): 237–43. See also: Media Ajir and Bethany Vailliant, 'Russian Information Warfare: Implications for Deterrence Theory', *Strategic Studies Quarterly* 12, no. 3 (2018): 72–3; Keir Giles, 'Handbook of Russian Information Warfare', *NATO Defence College* 9, November (2016): 19.

27. Martin Kragh and Sebastian Åsberg, 'Russia's Strategy for Influence through Public Diplomacy and Active Measures: The Swedish Case', *Journal of Strategic Studies* 40, no. 6 (2017): 790–7.

28. Andrew Radin, Alyssa Demus and Krystyna Marcinek, *Understanding Russian Subversion: Patterns, Threats, and Responses* (Santa Monica, CA: RAND, 2020): 2–3.

29. United States Senate Committee on Intelligence, *Report on Russian Active Measures Campaigns and Interference in the 2016 U.S. Election—Volume 2: Russia's Use of Social Media* (Washington, DC: United States Senate, 2019): 12–13; *EU vs Disinformation, Election Meddling and Pro-Kremlin Disinformation: What You Need to Know* (EUvsDisinfo, 2019): 4; William Aceves, 'Virtual Hatred: How Russia Tried to Start a Race War in the United States', *Michigan Journal of Race and Law* 24, no. 2 (2019): 185–9.

30. Peter B.M.J. Pijpers and Eric H. Pouw, 'Sovereignty in Cyberspace: Lessons from the Ukrainian Case', *Atlantisch Perspectief* 46, no. 3 (2022): 36–41; Albert J.H. Bouwmeester, 'Krym Nash: An Analysis of Modern Russian Deception Warfare', PhD diss. (Utrecht University, 2020).

31. M.C. Libicki, 'Correlations Between Cyberspace Attacks and Kinetic Attacks', *2020 12th International Conference on Cyber Conflict (CyCon)* (2020): 199–213.

371

NOTES

32. Brendan Chrzanowski, 'An Episode of Existential Uncertainty: The Ontological Security Origins of the War in the Donbas', *Texas National Security Review* 4, no. 3 (2021): 11–32.

33. Holger Möldera and Vladimir Sazonov, 'Information Warfare as the Hobbesian Concept of Modern Times—the Principles, Techniques, and Tools of Russian Information Operations in the Donbass', *Journal of Slavic Military Studies* 31, no. 3 (2018): 322–5.

34. David E. Sanger, *The Perfect Weapon: War, Sabotage, and Fear in the Cyber Age* (New York: Scribe, 2018): 90–8.

35. Lee, Assante and Conway, 'Analysis of the Cyber Attack on the Ukrainian Power Grid'.

36. Kaspersky Lab, 'Newly Discovered BlackEnergy Spear-Phishing Campaign Targets Ukrainian Entities', *Kaspersky*, 28 January 2016.

37. Sanger, *The Perfect Weapon*, 90; Kim Zetter, 'The Ukrainian Power Grid Was Hacked Again', *Motherboard, Tech by Vice*, 10 Janaury 2017.

38. See 'Tactical Unit Combat Control System "Kropyva"', *Logika*, 8 June 2023, https://logika.ua/en/automation-systems/

39. Crowdstrike Global Intelligence Team, 'Use of Fancy Bear Android Malware in tracking of Ukrainian Field Artillery Units', *CrowdStrike*, 22 December 2016 (updated 23 March 2017).

40. Polyakova and Boyer, 'The Future of Political Warfare', 14.

41. US Department of Justice, 'The NotPetya Cyber Attacks', *United States District Court Western District of Pennsylvania*, Indictment No. 20-316, 15 October 2020, 16–23.

42. Dutch Safety Board, *Crash of Malaysia Airlines flight MH17, Hrabove, Ukraine, 17 July 2014* (The Hague: Dutch Safety Board, 2015): 253.

43. 'Увага: жодної загрози для коштів вкладників «Приватбанку» немає' ['Attention: there is no threat to the funds of "Privatbank" depositors'], *Ukrainian Centre for Strategic Communication*, 15 February 2022, https://spravdi.gov.ua/uvaga-zhodnoyi-zagrozy-dlya-koshtiv-vkladnykiv-pryvatbanku-nemaye/

44. Security Service of Ukraine, 'Cyber Attacks on Government Websites', *Security Services of Ukraine*, 14 January 2022, https://ssu.gov.ua/en/novyny/shchodo-aktak-na-saity-derzhavnykh-orhaniv

45. Nadiya Kostyuk and Aaron Brantly, 'War in the Borderland through Cyberspace: Limits of Defending Ukraine through Interstate Cooperation,' *Contemporary Security Policy* 43, no. 3 (2022): 498.

46. Microsoft Threat Intelligence Centre (MSTIC), 'Destructive malware targeting Ukrainian organizations', *Microsoft*, 15 January 2022.

47. Brad Smith, 'Defending Ukraine: Early Lessons from the Cyber War', *Microsoft*, 22 June 2022, https://blogs.microsoft.com/on-the-issues/2022/06/22/defending-ukraine-early-lessons-from-the-cyber-war/

pp. [91–92] NOTES

48. Davey Alba, 'Russia Has Been Laying Groundwork Online for a "False Flag" Operation', *The New York Times*, 19 February 2022; Lucy Fisher, Danielle Sheridan, and Josie Ensor, 'Russia Begins "False Flag" Attacks against Ukraine', *The Telegraph*, 18 February 2022.

49. Lally Weymouth, 'Volodymyr Zelensky : "Everyone Will Lose" If Russia Invades Ukraine', *The Washington Post*, 20 January 2022; Paul Sonne, Missy Ryan and John Hudson, 'Russia Planning Potential Sabotage Operations in Ukraine, U.S. Says', *The Washington Post*, 14 January 2022.

50. Alexander Martin, 'US Military Hackers Conducting Offensive Operations in Support of Ukraine, Says Head of Cyber Command', *Sky News*, 1 June 2022, https://news.sky.com/story/us-military-hackers-conducting-offensive-operations-in-support-of-ukraine-says-head-of-cyber-command-12625139; Gordon Corera, 'Inside a US Military Cyber Team's Defence of Ukraine', *BBC News*, 30 October 2022, https://www.bbc.com/news/uk-63328398

51. Joe Tidy, 'Ukraine: EU Deploys Cyber Rapid-Response Team', *BBC News*, 22 February 2022, https://www.bbc.com/news/technology-60484979. Contributing states are Lithuania, Poland, Croatia, Estonia, Romania and the Netherlands.

52. A.J. Vicens, 'Details Emerge on Hack of Belarusian Railways and the Group behind It', *Cyberscoop*, 26 January 2022, https://cyberscoop.com/cyber-partisans-belarus-ukraine-russia/; Peter Dickinson, 'Cyber Partisans Target Russian Army in Belarus amid Ukraine War Fears', *Atlantic Council*, 26 January 2022, https://www.atlanticcouncil.org/blogs/belarusalert/cyber-partisans-target-russian-army-in-belarus-amid-ukraine-war-fears/

53. Symantec Threat Hunter Team, 'Ukraine: Disk-wiping Attacks precede Russian Invasion', *Symantec*, 24 February 2022, https://symantec-enterprise-blogs.security.com/blogs/threat-intelligence/ukraine-wiper-malware-russia

54. Jon Bateman, *Russia's Wartime Cyber Operations in Ukraine: Military Impacts, Influences, and Implications* (Washington, DC: Carnegie Endowment for International Peace, 2022): 10.

55. Ukraine's Victor Zhora, Chief Digital Transformation Officer, State Service of Special Communication and Information Protection of Ukraine, initially said it caused 'a really huge loss in communications in the very beginning of war'; see Raphael Satter, 'Satellite Outage Caused "Huge Loss in Communications" at War's Outset—Ukrainian Official', *Reuters*, 15 March 2022, https://www.reuters.com/world/satellite-outage-caused-huge-loss-communications-wars-outset-ukrainian-official-2022-03-15/. Later on, Zhora argued that there was 'no information that [the hack] worsened communications within Ukraine's military'; see Kim Zetter, 'Viasat Hack "Did Not" Have Huge Impact on Ukrainian Military Communications,

373

NOTES pp. [92–93]

Official Says', *Zero Day*, 26 September 2022, https://zetter.substack. com/p/viasat-hack-did-not-have-huge-impact

56. Jason Blessing, 'Revisiting the Russian Viasat Hack: Four Lessons About Cyber on the Battlefield', *American Enterprise Institute*, 2 September 2022, https://www.aei.org/foreign-and-defense-policy/revisiting-the-russian-viasat-hack-four-lessons-about-cyber-on-the-battlefield/

57. 'Satellite Outage Knocks out Thousands of Enercon's Wind Turbines', *Reuters*, 28 February 2022, https://www.reuters.com/business/energy/satellite-outage-knocks-out-control-enercon-wind-turbines-2022-02-28/

58. @elonmusk, 'Starlink service is now active in Ukraine', *Twitter*, 26 February 2022, https://twitter.com/elonmusk/status/1497701484003213317?lang=en

59. Hyunjoo Jin, 'Musk Says Starlink Active in Ukraine as Russian Invasion Disrupts Internet', *Reuters*, 26 February 2022, https://www.reuters.com/technology/musk-says-starlink-active-ukraine-russian-invasion-disrupts-internet-2022-02-27/

60. Stephan Sestanovich, Thomas Graham and Charles A. Kupchan, 'Conflict in Ukraine', *Council on Foreign Relations*, 2022, https://www.cfr.org/global-conflict-tracker/conflict/conflict-ukraine (last updated 15 August 2023).

61. @YourAnonOne, 'The Anonymous collective is officially in cyber war against the Russian government', *Twitter*, 24 February 2022, https://twitter.com/YourAnonOne/status/1496965766435926039

62. Emma Vail, 'Russia or Ukraine: Hacking groups take sides', *The Record*, 25 February 2022, https://therecord.media/russia-or-ukraine-hacking-groups-take-sides (last updated 3 March 2022).

63. Bateman, 'Russia's Wartime Cyber Operations in Ukraine', 2.

64. Ibid., 14.

65. Cathal McDaid and Rowland Corr, 'The Mobile Network Battlefield in Ukraine—Part 3', *Adaptive Mobile Security*, 25 April 2022, https://blog.adaptivemobile.com/the-mobile-network-battlefield-in-ukraine-part-3

66. Alexander Martin, 'US Military Hackers Conducting Offensive Operations in Support of Ukraine, Says Head of Cyber Command', *Sky News*, 1 June 2022, https://news.sky.com/story/us-military-hackers-conducting-offensive-operations-in-support-of-ukraine-says-head-of-cyber-command-12625139

67. Cyberknow, 'Update 21. 2022 Russia-Ukraine War: Cyber Group Tracker', *Medium*, 19 December 2022, https://cyberknow.medium.com/update-21-2022-russia-ukraine-war-cyber-group-tracker-december-19-24c61f3349e3

68. 'Website of NAFO | OFAN', *NAFO |OFAN*, https://nafo-ofan.org (last accessed 7 February 2023).

69. @FedorovMykhailo, 'We are creating an IT army', *Twitter*, 26 February 2022, https://twitter.com/FedorovMykhailo/status/1497642156076511233.

pp. [93–95] NOTES

See the analysis by Stefan Soesanto, *The IT Army of Ukraine Structure, Tasking, and Ecosystem* (Zurich: CSS Cyberdefense Report, 2022).

70. Soesanto, *IT Army of Ukraine Structure*, 28. The IT-Army initially appeared to be government-controlled, but closer research alluded to an affiliation with the Anonymous group.

71. Ibid., 4.

72. Cyberknow, 'Update 21. 2022 Russia-Ukraine War: Cyber Group Tracker'; Andrew S. Bowen, 'Russian Cyber Units', *Congressional Research Service*, updated 2 February 2022, https://crsreports.congress.gov/

73. Andrei Soldatov and Irina Borogan, 'Russian Cyberwarfare: Unpacking the Kremlin's Capabilities', (Washington, DC: Centre for European Policy Analysis, 2022).

74. Vail, 'Russia or Ukraine'; Peter B.M.J. Pijpers, 'Exploiting Cyberspace: International Legal Challenges and the New Tropes, Techniques and Tactics in the Russo-Ukraine War', *Hybrid CoE* 15, October (2022): 8–9.

75. Smith, 'Defending Ukraine'.

76. Ibid,. 3. Microsoft, however, seems to build this conclusion on correlation rather than on causality.

77. ESET Research, 'Ukraine hit by destructive attacks before and during the Russian invasion with HermeticWiper and IsaacWiper', *ESET*, 1 March 2022, https://www.eset.com/int/about/newsroom/press-releases/research/eset-research-ukraine-hit-by-destructive-attacks-before-and-during-the-russian-invasion-with-hermet/

78. Symantec Threat Hunter Team, 'Ukraine: Disk-wiping Attacks Precede Russian Invasion'.

79. See *Financial Times*, describing the neutralization of malware in Ukraine's railways systems: Mehul Srivastava, Madhumita Murgia and Hannah Murphy, 'The Secret US Mission to Bolster Ukraine's Cyber Defences Ahead of Russia's Invasion', *Financial Times*, 9 March 2022.

80. Smith, 'Defending Ukraine', 8.

81. ESET Research, 'Industroyer2: Industroyer reloaded', *ESET*, 12 April 2022, https://www.welivesecurity.com/2022/04/12/industroyer2-industroyer-reloaded/

82. 'Чотири місяці війни: статистика кібератак' ['Four months of war: statistics of cyberattacks'], State Service of Special Communications and Information Protection, 30 June 2022, https://cip.gov.ua/ua/news/chotiri-misyaci-viini-statistika-kiberatak (last accessed 8 September 2022).

83. Theodore W. Kleisner and Trevor T. Garmey, 'Tactical TikTok for Great Power Competition—Applying the Lessons of Ukraine's IO Campaign to Future Large-Scale Conventional Operations', *Military Review Online Exclusive*, April (2022): 12–14.

84. Alden Wahlstrom et al., 'The IO Offensive : Information Operations

375

Surrounding the Russian Invasion of Ukraine', *Mandiant*, 19 May 2022, https://www.mandiant.com/resources/blog/information-operations-surrounding-ukraine

85. Marco Longobardo, 'Legal Perspectives on the Role of the Notion of "Denazification" in the Russian Invasion of Ukraine under Jus Contra Bellum and Jus in Bello', *Revue Belge de Droit International* (Forthcoming).

86. Dennis Lichtenstein et al., 'Framing the Ukraine Crisis: A Comparison between Talk Show Debates in Russian and German Television', *The International Communication Gazette* 81, no. 1 (2019): 66–88.

87. Mario Baumann, '"Propaganda Fights" and "Disinformation Campaigns": The Discourse on Information Warfare in Russia-West Relations', *Contemporary Politics* 26, no. 3 (2020): 295–6.

88. Justin Ling, 'How U.S. Bioweapons in Ukraine Became Russia's New Big Lie', *Foreign Policy*, 10 March 2022; EU vs Disinformation, 'Weapons of Mass Delusion', 10 March 2022, https://euvsdisinfo.eu/weapons-of-mass-delusion/

89. Dalia Al-Aqidi, 'How Zelensky Used Social Media to His Advantage', *Center for Security Policy*, 6 March 2022, https://www.arabnews.com/node/2037236

90. Paul Adams, 'How Ukraine Is Winning the Social Media War', *BBC*, 16 October 2022, https://www.bbc.com/news/world-europe-63272202

91. Sharon Braithwaite, 'Zelensky Refuses US Offer to Evacuate, Saying "I Need Ammunition, Not a Ride"', *CNN*, 26 February 2022, https://edition.cnn.com/2022/02/26/europe/ukraine-zelensky-evacuation-intl/index.html

92. Matthew Ford, 'The Smartphone as Weapon—Part 2: The Targeting Cycle in Ukraine', *Academia*, 29 April 2022, https://www.academia.edu/76011845/The_Smartphone_as_Weapon_part_2_the_targeting_cycle_in_Ukraine

93. Air Force Command of the Ukrainian Forces, 'ПРОСИМО НЕ НАПОВНЮВАТИ ІНФОПРОСТІР ФЕЙКАМИ' ['PLEASE DO NOT FILL THE INFOSPACE WITH FAKES!'], *Facebook*, 30 April 2022, https://www.facebook.com/kpszsu/posts/363834939117794

94. The actual phrase was (translated into English) 'go f**k yourself'. See '"Go fuck yourself", Ukrainian soldiers on Snake Island tell Russian ship – audio', *The Guardian*, 25 February 2022, https://www.theguardian.com/world/video/2022/feb/25/go-fuck-yourself-ukrainian-soldiers-snake-island-russian-ship-before-being-killed-audio

95. Digital Forensic Research Lab, 'Russian War Report: Competing Narratives about the Sinking of Russia's Moskva Warship', *Atlantic Council* (blog), 15 April 2022, https://www.atlanticcouncil.org/blogs/new-atlanticist/russian-war-report-competing-narratives-about-the-sinking-of-russias-moskva-warship/

pp. [96–98] NOTES

96. Jenny Hill, 'Russian Warship: Moskva Sinks in Black Sea', *BBC News*, 15 April 2022, https://www.bbc.com/news/world-europe-61114843

97. 'З початку війни СБУ ліквідувала 5 ворожих ботоферм потужністю понад 100 тис. фейкових акаунтів' ['Since the beginning of the war, the SBU has eliminated 5 enemy bot farms with a capacity of more than 100,000 fake accounts'], *Security Service of Ukraine*, https://ssu.gov.ua/novyny/z-pochatku-viiny-sbu-likviduvala-5-vorozhykh-botoferm-potuzhnistiu-ponad-100-tys-feikovykh-akauntiv (last accessed 8 September 2022).

98. Niels Nagelhus Schia, 'The Digital Battlefield', *Prio Bolgs*, 16 March 2022, https://blogs.prio.org/2022/03/the-digital-battlefield/

99. Jelena Vicic and Rupal N. Netha, 'Why Russian Cyber Dogs Have Mostly Failed to Bark', *War On The Rocks*, 14 March 2022, https://warontherocks.com/2022/03/why-cyber-dogs-have-mostly-failed-to-bark/; Nadiya Kostyuk and Erik Gartzke, 'Why Cyber Dogs Have Yet to Bark Loudly in Russia's Invasion of Ukraine', *Texas National Security Review* 5, no. 3 (2022): 113–26.

100. Andy Greenberg, 'Russia's Sandworm Hackers Attempted a Third Blackout in Ukraine', *Wired*, 12 April 2022.

101. Bobby Allyn, 'Deepfake Video of Zelenskyy Could Be "Tip of the Iceberg" in Info War, Experts Warn', *NPR*, 16 March 2022, https://www.npr.org/2022/03/16/1087062648/deepfake-video-zelenskyy-experts-war-manipulation-ukraine-russia

102. Stephanie Burnett, 'Fact check: The deepfakes in the disinformation war between Russia and Ukraine', *DW*, 18 March 2022, https://www.dw.com/en/fact-check-the-deepfakes-in-the-disinformation-war-between-russia-and-ukraine/a-61166433

103. Contrary to the more common involvement of proxy actors such as APTs or Glavsent (Internet Research Agency).

104. Vail, 'Russia or Ukraine'.

105. Anh V. Vu et al., 'Getting Bored of Cyberwar: Exploring the Role of the Cybercrime Underground in the Russia-Ukraine Conflict', *Arxiv*, 3 December 2022, https://arxiv.org/abs/2208.10629

106. Tom Burt, 'The Hybrid War in Ukraine', *Microsoft*, 27 April 2022, https://blogs.microsoft.com/on-the-issues/2022/04/27/hybrid-war-ukraine-russia-cyberattacks/

107. Russ Mitchell, 'How Amazon Put Ukraine's "Government in a Box"—and Saved Its Economy from Russia', *Los Angeles Times*, 15 December 2022.

108. McDaid and Corr, 'The Mobile Network Battlefield in Ukraine—Part 3'.

109. Burt, 'The Hybrid War in Ukraine'; Morgan Meaker, 'Russia Blocks Facebook and Twitter in a Propaganda Standoff', *Wired*, 4 March 2022.

110. Musk, 'Starlink Service Is Now Active in Ukraine'; Jin, 'Musk Says Starlink Active in Ukraine as Russian Invasion Disrupts Internet'.

377

NOTES

pp. [98–101]

111. Shubham Verma, 'ICANN Rejects Ukraine Request to Kick Russia off the Internet', *India Today*, 4 March 2022.

112. Laurens Cerulus, 'Kyiv's Hackers Seize Their Wartime Moment', *Politico*, 10 March 2022.

113. Based on threat intelligence advances, supported by artificial intelligence, and Internet-connected endpoint protection. See: Smith, 'Defending Ukraine', 2.

114. Nick Beecroft, 'Evaluating the International Support'.

115. Nadiya Kostyuk and Aaron Brantly, 'War in the Borderland through Cyberspace: Limits of Defending Ukraine through Interstate Cooperation', *Contemporary Security Policy* 43, no. 3 (2022): 509–10.

116. Joseph Marks and Aaron Schaffer, '11 Reasons We Haven't Seen Big Russian Cyberattacks yet', *The Washington Post*, 3 March 2022.

117. Jamey Keaten, 'NATO Chief Fears Ukraine War Could Become a Wider Conflict', *AP News*, 9 December 2022, https://apnews.com/article/russia-ukraine-jens-stoltenberg-government-f01121d32693881920b1fe5e8c75b22d

118. A 'zero-day' software vulnerability is a vulnerability that is not yet known (hence, 'zero-days' known) to the creator of that software, or for which no adequate patch has yet been developed.

119. G.D. Vynck, R.L.C.. Zakrzewski and C. Zakrzewski, 'How Ukraine's Internet Still Works despite Russian Bombs, Cyberattacks', *The Washington Post*, 29 March 2022; Cairan Martin, 'Cyber Realism in a Time of War', *Lawfare*, 2 March 2022, https://www.lawfareblog.com/cyber-realism-time-war

120. In contrast to kinetic weapons, the effects of hard-cyber weapons can be hardly visible with the human eye, or sometimes only be discovered months after their deployment.

121. Maschmeyer, 'The Subversive Trilemma'; Gavin Wilde, 'Assess Russia's Cyber Performance Without Repeating Its Past Mistakes', *War On The Rocks*, 21 July 2022, https://warontherocks.com/2022/07/assess-russias-cyber-performance-without-repeating-its-past-mistakes/

122. Herbert S. Lin, 'Russian Cyber Operations in the Invasion of Ukraine', *The Cyber Defense Review* 7, no. 4 (2022): 36–8.

123. An uncontrolled, self-propagating worm in combination with a global IT vulnerability.

124. A cyberattack on any member of the Alliance could trigger Article 5.

125. Nadiya Kostyuk and Yuri M. Zhukov, 'Invisible Digital Front: Can Cyber Attacks Shape Battlefield Events?', *Journal of Conflict Resolution* 63, no. 2 (2019): 333.

126. Seth G. Jones, Philip G. Wasielewski and Joseph S. Bermudez, *Russia's Losing Hand in Ukraine* (Washington, DC: CSIS Briefs, 2022); Peter A Mattsson,

378

'Russian Military Thinking—A New Generation of Warfare', *Journal on Baltic Security* 1, no. 1 (2015): 62–3.

127. Antony Beevor, 'Putin Doesn't Realize How Much Warfare Has Changed', *The Atlantic*, 24 March 2022.

128. The instruments of power include diplomacy, the military, the economy or information; the domains are land, sea, air and cyberspace; and the dimensions are physical, virtual and cognitive.

129. Chris Stokel-Walker, 'Ukraine's Army of Hackers Failed to Thwart Russia and Quickly Gave up Group Subscriptions', *New Scientist*, 1 September 2022.

130. Martin, 'Cyber Realism in a Time of War'.

131. Michael P. Fischerkeller and Richard J. Harknett, 'Persistent Engagement, Agreed Competition, and Cyberspace Interaction Dynamics and Escalation', *The Cyber Defense Review Special Edition* (2019): 267–87.

5. FOUR FACES OF WAR

1. Brent L. Sterling, *Other People's Wars: The US Military and the Challenge of Learning from Foreign Conflicts* (Washington, DC: Georgetown University Press, 2021).

2. Lotje Boswinkel and Tim Sweijs, *Wars to Come, Europeans to Act, A Multimethod Foresight Study into Europe's Military Future* (The Hague: Hague Centre for Strategic Studies, 2022): 1.

3. Lawrence Freedman, *Future War, A History* (New York: Public Affairs, 2017); for an extensive discussion on trends see Azar Gat, *The Causes of War and the Spread of Peace: But Will War Rebound?* (Oxford: Oxford University Press, 2017).

4. Mick Ryan, *War Transformed: The Future of Twenty-First Century Great Power Competition and Conflict* (Annapolis, MD: Naval Institute Press, 2022).

5. On the use of air power to coerce civilians, see Robert A. Pape, *Bombing to Win: Air Power and Coercion in War* (Ithaca, NY: Cornell University Press, 1997).

6. On the history of cyber vulnerabilities, see Fred Kaplan, *Dark Territory: The Secret History of Cyber War* (New York: Simon & Schuster, 2017); David Sanger, *The Perfect Weapon, War, Sabotage, and Fear in the Cyber Age* (New York: Crown, 2018); Richard Clarke and Roger Knake, *The Fifth Domain: Defending Our Country, Our Companies, and Ourselves in the Age of Cyber Threats* (New York: Penguin, 2019).

7. Jack Dutton, 'Russia-Linked Cyberattack on JBS Meats Hits 10,000 Jobs,' *Newsweek*, 2 June 2021, https://www.newsweek.com/russia-linked-cyber-attack-jbs-meats-hits-7000-jobs-1596713; 'JBS: Cyber-attack Hits World's Largest Meat Supplier,' *BBC News*, 2 June 2021, https://www.bbc.co.uk/news/world-us-canada-57318965

8. David Sanger, Clifford Krauss and Nicole Perlroth, 'Major Pipeline Forced to Close by Cyberattack,' *New York Times*, 8 May 2021, A1.

NOTES pp. [110–112]

9. Melissa Berry, 'Ransomware Attacks Against Healthcare Organizations Nearly Doubled in 2021,' *Thomson Reuters*, 5 July 2022, https://www.thomsonreuters.com/en-us/posts/investigation-fraud-and-risk/ransomware-attacks-against-healthcare/; Steve Adler, 'Healthcare Ransomware Attacks Increased by 94% in 2021,' *HIPPA Journal*, 6 June 2022, https://www.hipaajournal.com/healthcare-ransomware-attacks-increased-by-94-in-2021/

10. Benjamin Jensen, Brian Valierno and Ryan Maness, 'Fancy Bears and Digital Trolls: Cyber Strategy with a Russian Twist,' *Journal of Strategic Studies* 42, no. 2 (2019): 212–34.

11. See the U.S. government's overview: 'Russian State-Sponsored and Criminal Cyber Threats to Critical Infrastructure,' *U.S. Cybersecurity and Infrastructure Security Agency*, 9 May 2022, https://www.cisa.gov/news-events/cybersecurity-advisories/aa22-110a

12. Nicole Perlroth, 'Microsoft Says Russian Hackers Viewed Some of its Source Code,' *New York Times*, 31 December 2020, https://www.nytimes.com/2020/12/31/technology/microsoft-russia-hack.html

13. See 'Advanced Persistent Threat Compromise of Government Agencies, Critical Infrastructure, and Private Sector Organizations,' *Cyberstructure & Infrastructure Agency*, 17 December 2020, https://www.cisa.gov/uscert/ncas/alerts/aa20-352a

14. Ellen Nakashima, 'Russian Military was Behind "NotPetya" Cyberattack in Ukraine, CIA Concludes,' *The Washington Post*, 12 January 2018.

15. Andy Greenberg, 'The Untold Story of NotPetya, The Most Devastating Cyberattack in History,' *Wired*, 22 August 2018, https://www.wired.com/story/notpetya-cyberattack-ukraine-russia-code-crashed-the-world/; Cyberspace Solarium Commission, *Cyberspace Solarium Commission Final Report* (Washington, DC: Cyberspace Solarium Commission, 2020): 9.

16. Office of the US Director of National Intelligence, *2021 Annual Threat Assessment* (Washington, DC: Office of the Director of National Intelligence, 2021), https://www.dni.gov/files/ODNI/documents/assessments/ATA-2021-Unclassified-Report.pdf

17. David Cattler and Daniel Black, 'The Myth of the Missing Cyberwar,' *Foreign Affairs*, 6 April 2022, https://www.foreignaffairs.com/articles/ukraine/2022-04-06/myth-missing-cyberwar

18. Yurii Shchyol, 'Russia's Cyberwar against Ukraine Offers Vital Lessons for the West,' *Atlantic Council*, 31 January 2023, https://www.atlanticcouncil.org/blogs/ukrainealert/russias-cyberwar-against-ukraine-offers-vital-lessons-for-the-west/

19. See the cybersecurity alert, 'Cyber-Attack Against Ukrainian Critical Infrastructure,' *Cybersecurity and Infrastructure Security Agency*, 20 July 2021, https://www.cisa.gov/uscert/ics/alerts/IR-ALERT-H-16-056-01;

pp. [112–114]   NOTES

Catherine Stupp, 'Russian Cyberattacks Increase on Ukraine's Critical Infrastructure,' *Wall Street Journal*, 5 April 2022, https://www.wsj.com/articles/russian-cyberattacks-increase-on-ukraines-critical-infrastructure-report-11649186873

20. Shane Huntley, 'Fog of War: How the Ukraine Conflict Transformed the Cyber Threat Landscape,' *Google Threat Analysis Group*, 16 February 2023, https://blog.google/threat-analysis-group/fog-of-war-how-the-ukraine-conflict-transformed-the-cyber-threat-landscape/

21. The FBI Director, for instance, issued a blunt statement; see Christopher Wray, 'The Threat Posed by the Chinese Government and the Chinese Communist Party to the Economic and National Security of the United States,' *Hudson Institute*, 7 July 2020, https://www.fbi.gov/news/speeches/the-threat-posed-by-the-chinese-government-and-the-chinese-communist-party-to-the-economic-and-national-security-of-the-united-states

22. Tom Burt, 'New Nation-State Cyberattacks,' *Microsoft*, 2 March 2021, https://blogs.microsoft.com/on-the-issues/2021/03/02/new-nation-state-cyberattacks/; Zolan Kanno-Youngs and David Sanger, 'U.S. and Key Allies Accuse China in String of Global Cyberattacks,' *New York Times*, 19 July 2021, A1.

23. US Director of National Intelligence, *Annual Threat Assessment of the U.S. Intelligence Community* (Washington, DC: Office of the Director of National Intelligence, February 2023): 10.

24. Richard J. Harknett and Max Smeets, 'Cyber Campaigns and Strategic Outcomes,' *Journal of Strategic Studies* 45, no. 4 (2022): 534–67.

25. Mathew Ingram, 'What Should we Do About the Algorithmic Amplification of Disinformation?' *Columbia Journalism Review*, 11 March 2021, https://www.cjr.org/the_media_today/what-should-we-do-about-the-algorithmic-amplification-of-disinformation.php

26. The weaponization of social media is captured in Peter Singer and Emerson T. Brooking, *Like War: The Weaponization of Social Media* (New York: Houghton Mifflin Harcourt, 2018); as well as David Patrikarakos, *War in 140 Characters, How Social Media is Reshaping Conflict in the Twenty-First Century* (New York: Basic Books, 2017).

27. Bernard Claverie and François du Cluzel, *The Cognitive Warfare Concept* (NATO Innovation Hub, 2020), https://www.innovationhub-act.org/sites/default/files/2022-02/CW%20article%20Claverie%20du%20Cluzel%20final_0.pdf

28. Mike Mazarr et al., *The Emerging Risk of Virtual Societal Warfare* (Santa Monica, CA: RAND, 2019).

29. *Disinformation: A Primer in Russian Active Measures and Influence Campaigns, Statement prepared for the Senate Select Committee on Intelligence*, 115th Congress (Clint Watts, Robert A. Fox, Foreign Policy Research Institute, 2017);

NOTES pp. [114–115]

Disinformation: A Primer in Russian Active Measures and Influence Campaigns Panel II, Statement prepared for the Senate Select Committee on Intelligence, 115th Congress (Thomas Rid, King's College, London, 2017); Cyber-enabled Information Operations, Statement prepared for the SASC Subcommittee on Cybersecurity, 115th Congress (Clint Watts, Robert A. Fox, Foreign Policy Research Institute, 2017); Extremist Content and Russian Disinformation Online, Statement prepared for the Senate Judiciary Committee, Subcommittee on Crime and Terrorism, 115th Congress (Clint Watts, Robert A. Fox, Foreign Policy Research Institute, 2017).

30. Timothy Thomas, 'Russia's Information Warfare Strategy: Can the Nation Cope in Future Conflicts?' Journal of Slavic Military Studies 27, no. 1 (2014): 101–30; Martin Kragh and Sebastian Asberg, 'Russia's Strategy for Influence through Public Diplomacy and Active Measures: The Swedish Case,' Journal of Strategic Studies 40, no. 6 (2017): 773–816. For a comprehensive study on Russian activity, see Thomas Rid, Active Measures: The Secret History of Disinformation and Political Warfare (New York: Farrar, Strauss & Giroux, 2020).

31. T.S. Allen and A.J. Moore, 'Victory without Casualties: Russia's Information Operations,' Parameters 48, no. 1 (2018): 59–71; Buddhika B. Jayamahaa and Jahara Matisek, 'Social Media Warriors: Leveraging a New Battlespace,' Parameters 48, no. 4 (2018): 11–23.

32. Robert Mueller, Report on the Investigation into Russian Interference in the 2016 Presidential Election (Washington, DC: US Department of Justice, 2019); Oliver Backes and Andrew Swab, Cognitive Warfare: The Russian Threat to Election Integrity in the Baltic States (Cambridge, MA: Belfer Center for Science and International Affairs, 2019); US Senate Select Committee on Intelligence, Report of the U.S. Senate Select Committee on Intelligence, On Russian Active Measures Campaigns and Interference in the 2016 U.S. Election (Washington, DC, 2020), www.intelligence.senate.gov/sites/default/files/documents/report_volume5.pdf

33. Vladislav Surkov quoted in Heather Conley et al., Playbook 2: The Enablers (Lanhan, MD: Center for Strategic and International Studies, 2019): v.

34. Heather Conley, James Minz, Ruslan Stefanov and Martim Vladimirov, The Kremlin Playbook: Understanding Russian Influence in Central and Eastern Europe (Washington, DC: Center for Strategic and International Studies, 2016).

35. Conley et al., Kremlin Playbook, vi.

36. Koichiro Takagi, 'The Future of China's Cognitive Warfare, Lessons from the War in Ukraine,' War on the Rocks, 22 July 2022, https://warontherocks.com/2022/07/the-future-of-chinas-cognitive-warfare-lessons-from-the-war-in-ukraine/

37. For a clear background on the CCP's United Front Work Department, see Alexander Bowe, China's Overseas United Front Work Background and Implications

NOTES

for the United States (Washington, DC: U.S.-China Economic and Security Review Commission, 2018).

38. Sangkuk Lee, 'China's "Three Warfares": Origins, Applications, and Organizations,' *Journal of Strategic Studies* 37, no. 2 (2014): 198–221; Peter Mattis, 'China's Three Warfares in Perspective,' *War on the Rocks*, 30 January 2018, https://warontherocks.com/2018/01/chinas-three-warfares-perspective/

39. US Defense Department, *Military and Security Developments Involving the People's Republic of China* (Washington, DC: US Defense Department, 2022): 160–2.

40. Nathan Beauchamp-Mustafaga, 'Cognitive Domain Operations: The PLA's New Holistic Concept for Influence Operations,' *Jamestown Foundation China Brief* 19, no. 16 (2019); Tzu-Chieh Hung and Tzu-Wei Hung, 'How China's Cognitive Warfare Works: A Frontline Perspective of Taiwan's Anti-Disinformation Wars,' *Journal of Global Security Studies* 7, no. 4 (2020): 1–18; Yamaguchi Shinji, Yatsuzuka Masaaki and Momma Rira, *China Security Report 2023: China's Quest for Control of the Cognitive Domain and Gray Zone Situations* (Tokyo: National Institute for Defense Studies, 2023): 28–52.

41. Michael J. Mazarr et al., *Hostile Social Manipulation: Present Realities and Emerging Trends* (Santa Monica, CA: RAND, 2019); Scott W. Harold, Nathan Beauchamp-Mustafaga and Jeffrey W. Hornung, *Chinese Disinformation Efforts on Social Media* (Santa Monica, CA: RAND, 2020).

42. Tiffany Hsu, 'Influence Networks in Russia Misled European Users, TikTok Says,' *New York Times*, 9 February 2023, https://www.nytimes.com/2023/02/09/business/media/tiktok-misinformation-russia-ukraine.html

43. On deepfakes, see 'Deep Fake It Till You Make It: Pro-Chinese Actors Promote AI-Generated Video Footage of Fictitious People in Online Influence Operation,' *Graphika*, 7 February 2023, https://www.graphika.com/reports/deepfake-it-till-you-make-it

44. Alina Polyakova and Spencer Phipps Boyer, *The Future of Political Warfare: Russia, the West, and the Coming Age of Global Digital Competition* (Washington, DC: Brookings, March 2018); Todd C. Helmus et al., *Russian Social Media Influence: Understanding Russian Propaganda in Eastern Europe* (Santa Monica, CA: RAND, 2018).

45. Sally Adee, 'What Are Deepfakes and How Are They Created?', *IEEE Spectrum*, 29 April 2020, https://spectrum.ieee.org/what-is-deepfake; adam satariano and paul mozur, 'How Deepfake Videos are Used to Spread Disinformation', *New York Times*, 7 February 2023, https://www.nytimes.com/2023/02/07/technology/artificial-intelligence-training-deepfake.html; Tiffany Hsu and Stuart A. Thompson, 'A.I.'s Ease at Spinning Deception Raises Alarm', *New York Times*, 9 February 2023, A1, A22.

NOTES

pp. [116–117]

46. Singer and Brooking, *LikeWar*, 261.
47. Dominic Tierney, 'The Future of Sino-U.S. Proxy War', *Texas National Security Review* 4, no. 2 (2021): 50-73, 67.
48. Austin Carson, *Secret Wars: Covert Conflict in International Politics* (Princeton, NJ: Princeton University Press, 2018).
49. Andrew Mumford, *Proxy Warfare* (London: Polity, 2013): 111.
50. Aaron Stein and Ryan Fishel, 'Syria, Airpower and the Future of Great-Power War', *War on the Rocks*, 13 August 2021, https://warontherocks.com/2021/08/syria-airpower-and-the-future-of-great-power-war/; Frank Hoffman and Andrew Orner, 'The Return of Great-Power Proxy Wars', *War on the Rocks*, 2 September 2021, https://warontherocks.com/2021/09/the-return-of-great-power-proxy-wars/
51. See Andrew Mumford, 'Proxy Warfare and the Future of Conflict', *RUSI Journal* 158, no. 2 (2013): 40–6; Candance Rondeaux and David Sterman, *Twenty-first Century Proxy Warfare* (Washington, DC: New America, 2019); Vladimir Rauta, 'Proxy Warfare and the Future of Conflict: Take Two', *RUSI Journal* 165, no. 2 (2020): 1–10.
52. Meia Nouwens and Helen Lengrada, 'Guarding the Silk Road, How China's Private Security Companies are Going Global', *World Economic Forum*, 24 October, 2018, at https://weforum.org/agenda/2018/guarding-the-silk-road-how-chinas-private-security-companies-are-going-global; Alessandro Arduino, 'China's Private Security Companies: The Evolution of a New Security Actor', *NBR Special Report* no. 80 (Washington, DC: The National Bureau of Asian Research, September 2019); Fatoumata Dialio, *Private Security Companies: The New Notch in Beijing's Belt and Road Initiative?* (Stockholm: Institute for Security and Development Policy, June 2018); Sergey Sukhankin, 'Chinese Private Security Contractors: New Trends and Future Prospects', *China Brief* 20, no. 9 (2020).
53. Åse G. Østensen and Tor Bukkvoll, *Russian Use of Private Military and Security Companies-Implications for European and Norwegian Security* (Oslo: Norwegian Defence Research Establishment, 2018); Christopher Spearin, 'NATO, Russian and Private Military and Security Companies', *RUSI Journal* 163, no. 3, (2018): 66–72; Kimberly Marten, 'Russia's Use of Semi-State Forces: The Case of the Wagner Group', *Post-Soviet Affairs* 35, no. 3 (2019): 181–204.
54. Neil Hauer, 'Russia's Favorite Mercenaries', *The Atlantic*, 27 August 2018, https://www.theatlantic.com/international/archive/2018/08/russian-mercenaries-wagner-africa/568435/
55. Justin Bristow, *Russian Private Military Companies: An Evolving Set of Tools in Russian Military Strategy* (Fort Leavenworth, KS: Foreign Military Studies Office, 2019); Seth Jones et al., *Russia's Corporate Soldiers: The Global Expansion of Russia's Private Military Companies* (Washington, DC: Center for Strategic and International Studies, 2021).

pp. [117–119] NOTES

56. As evidenced by Chechen units in Ukraine since 2015, see Nicholas Waller, 'A Chechen War by Proxy', *Foreign Affairs*, 1 April 2015, https://www.foreignaffairs.com/articles/eastern-europe-caucasus/2015-04-01/chechen-war-proxy

57. Brian Katz, 'What the U.S. Can Learn from Iranian Warfare', *The Atlantic*, 19 October, 2019, https://www.theatlantic.com/politics/archive/2019/10/what-us-can-learn-iranian-warfare/600082/; Ariane M. Tabatabai and Colin P. Clarke, 'Iran's Proxies are More Powerful Than Ever', *Rand.blog*, 16 October 2019; *Iran's Networks of Influence in the Middle East* (Washington, DC: International Institute for Strategic Studies, 2019), https://www.iiss.org/publications/strategic-dossiers/iran-dossier/iran-19-03-ch-1-tehrans-strategic-intent

58. Afshon Ostovar, 'The Grand Strategy of Militant Clients: Iran's Way of War', *Security Studies* 28, no. 1 (2019): 158–88.

59. Michael Eisenstadt, *Operating in the Gray Zone: Countering Iran's Asymmetric Way of War* (Washington, DC: Washington Institute for Near East Policy, January 2020).

60. Daniel L. Byman, 'Why Engage in Proxy War? A State's Perspective', *Brookings*, 21 May, 2018, https://www.brookings.edu/blog/order-from-chaos/2018/05/21/why-engage-in-proxy-war-a-states-perspective/

61. Mumford, 'Proxy Warfare and the Future of Conflict', 44.

62. For a current assessment, see Assaf Moghadam, Vladimir Rauta and Michael Wyss (eds), *Routledge Handbook of Proxy Warfare* (Oxon: Routledge, 2023).

63. Aaron Stein and Ryan Fishel, 'Syria, Airpower and the Future of Great-Power War', *War on the Rocks*, 13 August 2021, https://warontherocks.com/2021/08/syria-airpower-and-the-future-of-great-power-war/; Emmanuel Karagiannis, 'Russian Surrogate Warfare in Ukraine and Syria: Understanding the Utility of Militias and Private Military Companies', *Journal on Balkan and Near Easter Studies* 23, no. 4 (2021): 549–65.

64. Michael Noonan, *Irregular Soldiers and Rebellious States* (Lanham, MD: Roman & Littlefield, 2021): 146.

65. Graham Allison and Jonah Glick-Unterman, *The Great Military Rivalry: China vs the U.S.* (Cambridge, MA: Belfer Center, Harvard University, 2021); Hal Brands, *Getting Ready for a Long War with China: Dynamics of Protracted Conflict in the Western Pacific* (Washington, DC: American Enterprise Institute, 2022): Timothy Heath, Kristen Gunness and Tristan Finazzo, *The Return of Great Power War* (Santa Monica, CA: RAND, 2022).

66. Rob Johnson, Martijn Kitzen and Tim Sweijs, *The Conduct of War in the 21st Century: Kinetic, Connected and Synthetic* (Oxon: Routledge, 2021): 3–14.

67. For a balanced assessment of this emerging revolution, see Paul Scharre, *Four Battlegrounds: Power in the Age of Artificial Intelligence* (New York: Norton, 2023).

NOTES

pp. [119–121]

68. William O. Odom and Christopher D. Hayes, 'Cross-Domain Synergy: Advancing Jointness', *Joint Force Quarterly* 73 (2014): 123.

69. US Army, *TRADOC Pamphlet 525-3-1: The U.S. Army in Multi-Domain Operations 2028* (Army Training and Doctrine Command, 2018): 20. Multi-Domain Operations were officially enshrined in US Army Doctrine in Field Manual 3-0, *Operations* (1 October 2022): 3-1 to 3-2.

70. Davis Ellison, 'Mastering the Fundamentals – Developing the Alliance for the Future Battlefield', *NATO Joint Warfare Centre, Three Swords* 37, no. 3 (2021): 12–18; Jack Watling and Dan Roper, *European Allies in US Multi-Domain Operations* (London: Royal United Services Institute, 2019).

71. Headquarters Allied Command Transformation, 'NATO's Warfighting Capstone Concept', *NATO,* last modified 6 June 2023, https://www.act.nato.int/nwcc

72. Ellison, 'Mastering the Fundamentals', 17.

73. On how the Chinese look at multidomain challenges, see Edward Burke, Kristen Gunness, Cortez Cooper III and Mark Cozad, *People Liberation Army Operational Concepts* (Santa Monica, CA: RAND, 2020); Phillip C. Saunders, 'China's Approach to Multi-Domain Conflict', in *Getting the Multi-Domain Challenge Right*, ed. Brad Roberts (Livermore: Lawrence Livermore National Laboratory, 2021).

74. Mark Cozad et al., *Gaining Victory in Systems Warfare China's Perspective on the U.S.-China Military Balance* (Santa Monica, CA: RAND, 2022): 70–6; Timothy Thomas, *The Chinese Way of War: How Has it Changed?* (McLean, VA: MITRE, 2020).

75. Michael Kofman et al., *Russian Military Strategy: Core Tenets and Operational Concepts* (Alexandria, VA: CNA, 2021); V.B. Zarudnitskiy, 'Character and Content of Armed Conflicts in Modern Conditions and Near-term Perspective', *Voennaya Mysl'* no. 1 (2021): 34–44.

76. Jeffrey Engstrom, *Systems Confrontation and System Destruction Warfare: How the Chinese People's Liberation Army Seeks to Wage Modern Warfare* (Santa Monica, CA: RAND, 2016),): 15.

77. See the 2019 Chinese Defense White Paper, *China's National Defense in a New Era*, http://www.xinhuanet.com/english/download/whitepaperon nationaldefenseinnewera.doc; Michael Losacco, 'Blinding the Elephant: Assessing PLA Systems Confrontation and the Fight for Information Dominance', *Georgetown Security Studies Review blog*, 25 February 2022, https://georgetownsecuritystudiesreview.org/2022/02/25/blinding-the-elephant-assessing-pla-systems-confrontation-and-the-fight-for-information-dominance/

78. Quote in Jeffrey Engstrom, 'China Has Big Plans to Win the Next War It Fights', *National Interest*, 9 February 2018, https://nationalinterest.org/blog/the-buzz/china-has-big-plans-win-the-next-war-it-fights-24449?nopaging=1

pp. [121–126] NOTES

79. Michael Horowitz and Laura Kahn, 'DoD's 2021 China Military Power Report: How Advances in AI and Emerging Technologies Will Shape China's Military', *Council on Foreign Relations*, 4 November 2021, https://www.cfr.org/blog/dods-2021-china-military-power-report-how-advances-ai-and-emerging-technologies-will-shape

80. Michael Dahm, 'Chinese Debates on the Military Utility of Artificial Intelligence', *War on the Rocks*, 5 June 2020, https://warontherocks.com/2020/06/chinese-debates-on-the-military-utility-of-artificial-intelligence/

81. Cozad, *Systems of Systems Warfare*, 71.

82. Kofman et al., *Russian Military Strategy*, 76. See also Yu. E. Donskov, A.L. Moraresku, and V.V. Panasuyk, 'On the Issue of Disorganizing C2 of Troops (forces) and Arms', *Voennaya Mysl'* no. 8 (2017): 19–25; A.N. Klyushin, D.V. Holuenko and V.A. Anokhin, 'On the Elements of the Theory of Disorganization of C2 of Troops', *Voennaya Mysl'* no. 9 (2017): 65–9; V.A. Anokhin and D.V. Kholuenko, 'Methodological Foundations of the Assessment of Effectiveness of Disorganization of Network-Centric Information-Command Systems', *Voennaya Mysl'* no. 12 (2020): 92–7.

83. Boswinkel and Sweijs, *Wars to Come, Europeans to Act*, 47–50.

84. Krieg and Rickli, *Surrogate Warfare*, 56–82; Sean McFate, *New Rules of War: Victory in the Age of Durable Disorder* (New York: William Morrow, 2020).

85. Charles Bartles, "Getting Gerasimov Right", *Military Review* 96, no. 1 (2016): 30–8.

86. Valery Gerasimov, 'The Value of Science Is in the Foresight New Challenges Demand Rethinking the Forms and Methods of Carrying out Combat Operations,' *Military Review* 96, no. 1 (2016): 25.

87. Oscar Jonsson and Robert Seely, 'Russian Full-Spectrum Conflict: An Appraisal After Ukraine,' *Journal of Slavic Military Studies* 28, no. 1 (2015): 1–22.

6. PEOPLE'S WAR VS PROFESSIONAL WAR

1. Carl von Clausewitz, 'Bekenntnissdenkschrift', in idem., *Schriften, Aufsätze, Studien, Briefe*, ed. Werner Hahlweg, vol. 1 (Göttingen: Vandenhoeck & Ruprecht, 1966): 681–750.

2. Peter Paret, 'Responses and Reform', in idem., *The Cognitive Challenge of War: Prussia 1806* (Princeton, NJ: Princeton University Press, 2009): 93–9; Karen Hagemann, *Revisiting Prussia's War against Napoleon: History, Culture and Memory* (Cambridge: Cambridge University Press, 2015): 47–60; Vanya Eftimova Bellinger, 'Educating Clausewitz: Gerhard von Scharnhorst's Influence on Carl von Clausewitz's Life and Thought', PhD diss., King's College London, 2023.

387

NOTES   pp. [126–133]

3. Carl von Clausewitz, *Hinterlassene Werke über Krieg und Kriegführung, Vol. 3:Vom Kriege* (Berlin: Dümmler, 1834): Bookk VIII, Chapter 3B, 116.

4. Historian of Germany James Sheehan was one of the few who noted how remarkable this break with history was: James Sheehan, *Where Have All the Soldiers Gone? The Transformation of Modern Europe* (Boston, MA: Houghton Mifflin, 2008).

5. Clausewitz, *Vol. 1: Vom Kriege*, Book I, Chapter 1, §28, 31.

6. Ibid. Italics in original.

7. Cf. Marc Boone, 'The Dutch Revolt and the Medieval Tradition of Urban Dissent', *Journal of Early Modern History* 11, no. 4/5 (2007): 351–75.

8. For figures, see William Serman and Jean-Paul Bertaud, *Nouvelle histoire militaire de la France, 1789–1919* (Paris: Fayard, 1998): 48–9, 54–9.

9. Heinz Stübig, 'Die Wehrverfassung Preußens in der Reformzeit', in *Die Wehrpflicht, Entstehung, Erscheinungsformen und politisch-militärische Wirkung*, ed. Roland G. Foerster (Munich: R. Oldenbourg Verlag, 1994): 40.

10. I am following François Furet's interpretative model, which in turn followed De Tocqueville's, by putting forward democracy, equality and liberty as the three key elements of the French Revolution, of which Napoleon retained the first two, but not liberty, whereas the monarchies only offered a form of equality: François Furet, *La Revolution Française, Vol. I: De Turgot à Napoléon, 1770–1814* (Paris: Fayard, 1988); Alexis de Tocqueville, *Democracy in America*, vol. 2, Pt. 4, trans./eds. Harvey C. Mansfield and Delba Winthrop (London: University of Chicago Press, 2002): 640–3.

11. Hagemann, *Revisiting Prussia's War*, 40.

12. J.F. Verbruggen, 'De militaire dienst in het graafschap Vlaanderen', *Tijdschrift voor Rechtsgeschiedenis* 26, no. 4 (1958): 446–53.

13. Cf. R.R. Palmer, 'Frederick the Great, Guibert, Bülow: From Dynastic to National War', in *Makers of Modern Strategy: Military Thought from Machiavelli to Hitler*, ed. Edward Meade Earle (Princeton, NJ: Princeton University Press, 1943): 49–74; Ute Planert, 'Innovation or Evolution? The French Wars in Military History', in *War in an Age of Revolution 1775–1815*, eds. Roger Chickering and Stig Förster (Cambridge: Cambridge University Press): 83.

14. Convention Nationale, 'Rapport et décret, du 23 août 1793, d l'an II de la République, sur la réquisition civique des jeunes citoyens pour la défense de la patrie', in *Les archives de la Révolution française; 11.1a.109*, Bibliothèque nationale de France, 15 October 2007, https://gallica.bnf.fr/ark:/12148/bpt6k43937b/f15.item.texteImage

15. This is a complex and under-researched field which I analyse in a forthcoming article: Jan Willem Honig, 'Discourses of Calculated Brutality: The Regularity of Irregular Warfare in the Dutch Revolt' (forthcoming).

16. For the astonishing number of riots, revolts and revolutions the French army faced between 1815 and 1870, see Jean Delmas, 'Armée, Garde nationale et

388

pp. [134–139] NOTES

maintain de l'ordre', in *Histoire militaire de la France, Vol. 2: De 1715 à 1871*, ed. idem. (Paris: Presses Universitaires de France, 1992): 536–48.

17. Dierk Walter, *Preußische Heeresreformen, 1807–1870, Militärische Innovation und der Mythos der 'Roonschen Reform'* (Paderborn: Ferdinand Schöning, 2003), 93.

18. Edward M. Spiers, *The Army and Society, 1815–1914* (London: Longman, 1980): 243–4; see also 254 5, 258–9, 266–71, 274–81.

19. Quoted in ibid., 258.

20. Richard D. Challener, *The French Theory of the Nation in Arms, 1866 1939* (New York: Columbia University Press, 1965).

21. William Serman, 'French Mobilization in 1870', in *On the Road to Total War: The American Civil War and the German Wars of Unification, 1861–1871*, eds. Stig Förster and Jörg Nagler (Cambridge: Cambridge University Press, 1997): 283–94.

22. Gerhard Ritter, quoted approvingly in Walter, *Heeresreformen*, 51.

23. Ibid., 472–3. By the end of the century, the rest of Europe had caught up, and the French Republic even managed 82 per cent by 1914.

24. Serman and Bertaud, *Nouvelle histoire*, 455–81.

25. The then legality of reprisals, however deplorable in retrospect, is often overlooked in modern histories: cf. Alan Kramer, *Dynamics of Destruction: Culture and Mass Killing in the First World War* (Oxford: Oxford University Press, 2007); Hamburg Institute for Social Research (ed.), *The German Army and Genocide: Crimes Against War Prisoners, Jews, and Other Civilians* (New York: The New Press, 1999). The latter is the catalogue of a New York exhibition that was pulled at the last minute because of its failure to point this out.

26. Michael Geyer, 'Insurrectionary Warfare: The German Debate about a *Levée en masse* in October 1918', *Journal of Modern History* 73, no. 3 (2001): 459–527.

27. Ibid., 503 and 493–4. In his English translations, Geyer is not always careful to note the politically informed French-German language dichotomy.

28. Léon Daudet, *La guerre totale* (Paris: Nouvelle Librairie Nationale, 1918): 8–9. See also Jan Willem Honig, 'The Idea of Total War: From Clausewitz to Ludendorff', in *The Pacific War as Total War: Proceedings of the 2011 International Forum on War History* (Tokyo: National Institute for Defence Studies, 2012): 29–41.

29. Cf. Ofer Fridman, *Russian Hybrid Warfare: Resurgence and Politicization* (London: Oxford University Press, 2018); Oscar Jonsson, *The Russian Understanding of War: Blurring the Lines between Peace and War* (Washington, DC: Georgetown University Press, 2019).

30. Amanda Alexander, 'The Genesis of the Civilian', *Leiden Journal of International Law* 20, no. 2 (2007): 359–76.

31. Frank Reichherzer, *'Alles ist Front!' Wehrwissenschaften in Deutschland und*

NOTES pp. [139–148]

die Bellifizierung der Gesellschaft vom Ersten Weltkrieg bis in den Kalten Krieg (Paderborn: Ferdinand Schöningh, 2012); Talbot C. Imlay, *Facing the Second World War: Strategy, Politics, and Economics in Britain and France, 1938–1940* (Oxford: Oxford University Press, 2003); R. Elberton Smith, *The Army and Economic Mobilization, The United States Army in World War II: The War Department* (Washington, DC: Center of Military History United States Army, 1959); Harold W. Thatcher, *Planning for Industrial Mobilization, 1920–1940* (Washington, DC: Historical Section General Administrative Services Division Office of the Quartermaster General, 1943).

32. Erich Ludendorff, *Der totale Krieg* (Munich: Ludendorffs Verlag, 1935). See also Hans Speier, 'Ludendorff: The German Concept of Total War', in *Makers of Modern Strategy*, ed. Earle, 306–21.

33. Jan-Werner Müller, *Contesting Democracy: Political Ideas in Twentieth-Century Europe* (New Haven, CT: Yale University Press, 2011): esp. 49–124.

34. Bernhard R. Kroener, 'Wollt ihr den totalen Krieg …? Die Angst vor dem totalen Krieg', in *La guerre totale, La défense totale, 1789–2000, Actes XXVIe Congrès International d'Histoire Militaire* (Stockholm: IKO, 2001), 120–9.

35. Ilmari Käihkö, *'Slava Ukraini!' Strategy and the Spirit of Ukrainian Resistance, 2014–2022* (Helsinki: Helsinki University Press, 2023).

36. Geyer, 'Insurrectionary Warfare', 504.

37. Cf. Matthew Ford and Andrew Hoskins, *Radical War: Data, Attention and Control in the 21st Century* (London: Oxford University Press, 2022).

38. Ibid.

7. URBICIDE AND THE FUTURE OF CIVIL WAR

1. Notably Barbara F. Walter, *How Civil Wars Start and How to Stop Them* (New York: Crown, 2022). In defining civil war, one may be a 'lumper' or a 'splitter', depending upon how intent one is on separating various types of internal conflict—rebellion, insurrection, terrorism and so on—from one another. I am a lumper. In terms of (a) internality, (b) non-state character, (c) anti-status quo orientation, and (d) scale of greater than 1,000 annual deaths, my conception of what is coming would meet all conventional understandings of civil war. For more, see Bill Kassane, *Nations Torn Asunder: The Challenge of Civil War* (Oxford: Oxford University Press, 2016): 1–65.

2. In particular, see Peter Turchin, *Age of Discord: A Structural Demographic Analysis of America History* (Chaplin, CT: Beresta Books, 2016). See also, Peter Turchin, *War and Peace and War: The Rise and Fall of Empires* (London: Penguin, 2007).

3. See Alexander Adams, *Culture War: Art, Identity, Politics and Cultural Entryism* (Exeter: Societas, 2019).

4. Henry Foy, 'A Year of War in Ukraine has Left Europe's Armouries

390

pp. [148–150] NOTES

Dry', *Financial Times*, 15 February 2023, https://www.ft.com/content/661a7278-3dc4-4afd-b6b4-3f0e8eb4477c

5. Walter, *How Civil Wars Start*, 198.

6. H.G. Wells, *Anticipations* (London: Chapman and Hall, 1901): 13. I have been influenced also by Matt Carr, 'Slouching Towards Dystopia: The New Military Futurism', *Race and Class* 51, no. 3 (2010): 13–32; also the very sensible discussion of the limitations of prediction in Lars-Erik Cederman and Nils Weismann, 'Predicting Armed Conflict: Time to Adjust Our Expectations?', *Science* 355, no. 6324 (2017): 474–6.

7. For more on the method and difficulty of such studies, see Andrea Nagle, *Kill All Normies: Online Culture Wars from 4Chan and Tumblr to Trump and The Alt-Right* (Ropley: John Hunt Publishing, 2017); and Thomas Colley and Martin Moore, 'The Challenges of Studying 4chan and the Alt-Right: "Come on in the Water's Fine"', *New Media & Society* 24, no. 1 (2022): 5–30. I participated in and helped to organize the 2019 workshop on which the latter work is based.

8. See Martin van Creveld, *The Transformation of War* (London: The Free Press, 1991); Rupert Smith, *The Utility of Force* (London: Penguin, 2006); Kalevi Holsti, *The State, War, and the State of War* (Cambridge: Cambridge University Press, 2010).

9. 'Edelman Trust Barometer 2022', *Edelman*, https://www.edelman.com/trust/2022-trust-barometer (last updated 24 January 2022).

10. Lee Rainie et al., 'Trust in America', *Pew Research Centre*, 22 July 2019, https://www.pewresearch.org/politics/2019/07/22/trust-and-distrust-in-america/

11. Barbara Walters, '"These Are Conditions Ripe for Political Violence": How Close is the US to Civil War?', *The Guardian*, 6 November 2022, https://www.theguardian.com/us-news/2022/nov/06/how-close-is-the-us-to-civil-war-barbara-f-walter-stephen-march-christopher-parker

12. Robert Putnam, *Bowling Alone: The Collapse and Revival of American Community* (New York: Simon and Schuster, 2001).

13. Robert Putnam, 'E Pluribus Unum: Diversity and Community in the Twenty-first Century: The 2006 Johan Skytte Prize Lecture', *Scandinavian Political Studies* 30, no. 2 (2007): 137–74.

14. See Matthew Weaver, 'Angela Merkel: German Multiculturalism has "Utterly Failed"', *The Guardian*, 17 October 2010, https://www.theguardian.com/world/2010/oct/17/angela-merkel-german-multiculturalism-failed; Jamie Doward, 'David Cameron's Attack on Multiculturalism Divides the Coalition', *The Guardian*, 6 February 2011, https://www.theguardian.com/politics/2011/feb/05/david-cameron-attack-multiculturalism-coalition

NOTES pp. [150–153]

15. Alicia Wanless and Michael Buerk, 'Participatory Propaganda', in *Social Media and Social Order*, eds. David Herbert and Stefan Fisher-Hoyrem (Warsaw: De Gruyter, 2021): 111–32.

16. See also Pankaj Mishra, *The Age of Anger* (New York: Farrar, Straus, and Giroux, 2017).

17. Alastair Walton, 'The Elephant in the Room: The US Fiscal Deficit & Debt Outlook Over the Next 30 Years' (Lecture, Australian National University, College of Business and Economics, 10 September 2019).

18. For more on the relationship of economics and causes of civil war, see Eok Leong Swee, 'Economics of Civil War', *Australian Economic Review* 49, no. 1 (2016): 105–11.

19. Megan Henney, 'US National Debt on Pace to be 225% of GDP By 2050, Penn Wharton Says', *Fox Business*, 19 December 2022, https://www.foxbusiness.com/economy/us-national-debt-pace-gdp-2050-penn-wharton-says

20. Correspondence by author with Alastair Walton (1 February 2023). See also Ray Dalio, *Principles for Dealing with the Changing World Order: Why Nations Succeed or Fail* (New York: Simon and Schuster, 2021).

21. Economist Intelligence Unit, *Europe Outlook 2023: The Threats to Europe's Industrial Competitiveness* (London: Economist Intelligence, 2022), https://pages.eiu.com/rs/753-RIQ-438/images/europe-outlook-2023.pdf

22. See the exchange between James Chowning Davies, 'The J-Curve and Power Struggle Theories of Collective Violence', *American Sociological Review* 39, no. 4 (1974): 607–10; and David Snyder and Charles Tilly, 'On Debating and Falsifying Theories of Collective Violence', *American Sociological Review* 39, no. 4 (1974): 610–13.

23. Charles Davis, 'Weapons in Ukraine Aren't Flooding Europe's Black Markets, But That Could Change', *Business Insider*, 27 October 2022, https://www.businessinsider.com/no-sign-of-mass-arms-trafficking-from-ukraine-authorities-say-2022-10

24. The strategy is not altogether dissimilar to the Chinese communist Lin Biao's argument for encircling cities in the chapter 'Rely on the Peasants and Establish Rural Base Areas', in *Long Live the Victory of People's War!* (1965), https://www.marxists.org/reference/archive/lin-biao/1965/09/peoples_war/index.htm (last accessed 10 May 2023).

25. Murray Bookchin, *The Limits of the City* (New York: Harper Torchbooks, 1974): 3–4.

26. David Kilcullen, *Out of the Mountains: The Coming Age of the Urban Guerrilla* (London: Hurst, 2013): 76.

27. Martin Coward, *Urbicide: The Politics of Urban Destruction* (Abingdon: Routledge, 2007): 178.

28. Jessica Murray, Aina J Khan and Rajeev Syal, '"It Feels Like People Want to Fight": How Communal Unrest Flared in Leicester', *The Guardian*,

pp. [153–161] NOTES

23 September 2022, https://www.theguardian.com/uk-news/2022/sep/23/how-communal-unrest-flared-leicester-muslim-hindu-tensions

29. See Andrew Hussey, *The French Intifada: The Long War Between France and its Arabs* (New York: Faber and Faber, 2014).

30. See Phil Shea, 'Mapping Major Accident Hazard Pipelines for Land Use Planning Decision Making', *UK Onshore Pipeline Operators' Association* (undated), https://www.ukopa.co.uk/mapping-major-accident-hazard-pipelines-for-land-use-planning-decision-making/ (last accessed 10 May 2023).

31. See 'Ghislenghien Pipeline Explosion 2004', *Process, Safety, Integrity* (undated), https://processsafetyintegrity.com/events/2004-07-30_ghislenghien/ (last accessed 10 May 2023).

32. George Legg, 'Security Experiments: London, Belfast, and the Ring of Steel', *Divided Society—Northern Ireland 1990-1998*, https://www.dividedsociety.org/essays/security-experiments-london-belfast-and-ring-steel (last accessed 10 May 2023).

33. Department for Environment, Food, and Rural Affairs, *UK Food Security Report 2021* (London: Department for Environment, Food, and Rural Affairs, 2021): 149–206.

34. Peter Baudains, Alex Braithwaite and Shane D. Johnson, 'Target Choice during Extreme Events: A Discrete Spatial Choice Model of the 2011 London Riots', *Criminology* 51, no. 2 (2013): 251–86.

35. Paul Kirby, 'Germany Arrests 25 Accused of Plotting Coup', *BBC News*, 7 December 2022, https://www.bbc.co.uk/news/world-europe-63885028

36. Zoltan Barany, *How Armies Respond to Revolutions and Why* (Princeton, NJ: Princeton University Press, 2016).

37. Marek N. Posard, Leslie Adrienne Payne, and Laura L. Miller, *Reducing the Risk of Extremist Activity in the U.S. Military* (Santa Monica, CA: RAND, 2021).

38. See Barany, *How Armies Respond to Revolutions*.

39. Mark Henderson and Natasha Livingstone, 'Want to Trace the UK's New Transatlantic Internet Cable, Mr Putin? Just Watch This Google Video', *Daily Mail*, 23 January 2022.

40. See Duncan Campbell, *War Plan UK* (London: Paladin, 1983).

41. As recounted in William Prochnau's, *Once Upon a Distant War* (New York: Vintage Books, 1996): 162.

42. Sheldon Wolin, *Democracy Incorporated: Managed Democracy and the Spectre of Inverted Totalitarianism* (Princeton, NJ: Princeton University Press, 2004); Noam Chomsky, 'The Carter Administration: Myth and Reality', 1981, https://chomsky.info/priorities01/

43. Pentti Linkola, *Can Life Prevail?* (Budapest: Arktos Media, 2021).

44. Peter Laurie, *Beneath the City Streets: A Private Inquiry into the Government Preparations for National Emergency* (London: Penguin, 1970): 102.

393

NOTES

8. LIVING WITH DENIAL?

1. Keith Crane, Olga Oliker and Brian Nichiporuk, *Trends in Russia's Armed Forces: An Overview of Budgets and Capabilities* (Santa Monica, CA: RAND Corporation, 2019).

2. Paul Van Hooft, Nora Nijboer and Tim Sweijs, *Raising the Costs of Access: Active Denial Strategies by Small and Middle Powers against Revisionist Aggression* (The Hague: The Hague Centre for Strategic Studies, 2021), https://hcss.nl/report/deterrence-raising-the-costs-of-access/

3. Sam J. Tangredi, *Anti-Access Warfare. Countering A2/D2 Strategies* (Annapolis, MD): Naval Institute Press, 2013); Evan Braden Montgomery, 'Contested Primacy in the Western Pacific: China's Rise and the Future of US Power Projection,' *International Security* 38, no. 4 (2014): 115–49; Stephen Biddle and Ivan Oelrich, 'Future Warfare in the Western Pacific: Chinese Antiacess/area Denial, US Airsea Battle, and Command of the Commons in East Asia,' *International Security* 41, no. 1 (2016): 7–48; Robert Dalsjö, Christofer Berglund and Michael Jonsson, *Bursting the Bubble? Russian A2/AD in the Baltic Sea Region: Capabilities, Countermeasures, and Implications* (Stockholm: Swedish Defence Research Agency, 2019): 11.

4. Jonathan D. Caverley and Peter Dombrowski, 'Cruising for a Bruising: Maritime Competition in an Anti-Access Age,' *Security Studies* 29, no. 4 (2020): 671–700.

5. Barry Posen applied the concept to the areas of the globe that all actors make use of but that only the United States can deny to others, which specifically refers to the maritime commons, but also air above 15,000 feet, space, and cyber and informational space; see Barry R. Posen, 'Command of the Commons: The Military Foundation of US Hegemony,' *International Security* 28, no. 1 (2003): 5–46; Barry R. Posen, *Restraint: A New Foundation for US Grand Strategy* (Ithaca, NY, & London: Cornell University Press, 2014).

6. Shaan Shaikh and Wes Rumbaugh, *The Air and Missile War in Nagorno-Karabakh: Lessons for the Future of Strike and Defense* (Washington, DC: Center for Strategic and International Studies, 2020), https://www.csis.org/analysis/air-and-missile-war-nagorno-karabakh-lessons-future-strike-and-defense

7. Audrey Kurth Cronin, *Power to the People: How Open Technological Innovation Is Arming Tomorrow's Terrorists* (New York: Oxford University Press, 2019); Thomas X. Hammes, 'Technologies Converge and Power Diffuses,' *Cato Institute Policy Analysis*, no. 786 (2016): 1–14.

8. Any conclusions from the Russo-Ukrainian War must take into account the uncertainty about contingency (meaning poor Russian preparations and decisions in the opening stages) and structural features (meaning assessments of, for example, the future of air power); see Justin Bronk, Nick Reynolds, and Jack Watling, *The Russian Air War and Ukrainian Requirements for Air Defence* (London: Royal United Services Institute, 2022).

pp. [164–167]   NOTES

9. Paul van Hooft and Lotje Boswinkel, *Surviving the Deadly Skies: Integrated Air and Missile Defence 2021-2035* (The Hague: The Hague Centre for Strategic Studies, 2021).

10. Robert O. Keohane and Joseph S. Nye, 'Globalization: What's New? What's Not? (And so What?),' in *Making Policy Happen*, eds. Leslie Budd, Julie Charlesworth and Rob Paton (New York: Routledge, 2020): 105–13; 'Economic Growth in a Shrinking World: The IMF and Globalization, Address by Anne Krueger, Acting Managing Director, IMF,' *IMF*, https://www.imf.org/en/News/Articles/2015/09/28/04/53/sp060204 (last accessed 19 February 2023).

11. 'Trade and Globalization,' *Our World in Data*, https://ourworldindata.org/trade-and-globalization (last accessed 26 September 2022).

12. 'World Seaborne Trade by Cargo Type,' *Statista*, https://www.statista.com/statistics/1277810/international-seaborne-trade-breakdown-by-cargo/ (last accessed 23 April 2023).

13. 'Trade and Globalization,' *Our World in Data*.

14. Chris Miller, *Chip War: The Fight for the World's Most Critical Technology* (New York: Simon and Schuster, 2022); Saif M. Khan, Alexander Mann and Dahlia Peterson, *The Semiconductor Supply Chain* (Washington, DC: Center for Security and Emerging Technology, 2021): 5.

15. Robert M. Farley and Davida H. Isaacs, *Patents for Power: Intellectual Property Law and the Diffusion of Military Technology* (Chicago: University of Chicago Press, 2020): 104–6; Jonathan D. Caverley, 'United States Hegemony and the New Economics of Defense,' *Security Studies* 16, no. 4 (2007): 598–614; Stephen G. Brooks, *Producing Security: Multinational Corporations, Globalization, and the Changing Calculus of Conflict* (Princeton, NJ, and Oxford: Princeton University Press, 2011).

16. Douglas W. Skinner, *AirLand Battle Doctrine* (Alexandria, VA: Center for Naval Analysis, September 1988).

17. Bronk, Reynolds and Watling, *The Russian Air War and Ukrainian Requirements for Air Defence*; Antonio Calcara et al., 'Why Drones Have Not Revolutionized War: The Enduring Hider-Finder Competition in Air Warfare,' *International Security* 46, no. 4 (2022): 130–71.

18. Henry Farrell and Abraham L. Newman, 'Weaponized Interdependence: How Global Economic Networks Shape State Coercion,' *International Security* 44, no. 1 (2019): 42–79; Daniel W. Drezner, Henry Farrell and Abraham L. Newman (eds), *The Uses and Abuses of Weaponized Interdependence* (Washington, DC: Brookings Institution Press, 2021).

19. Adam Segal, 'Huawei, 5G, and Weaponized Interdependence', in *The Uses and Abuses of Weaponized Interdependence*, eds. Drezner, Farrell and Newman, 149–68.

20. Andre Barbe and Will Hunt, *Preserving the Chokepoints—Center for Security and*

NOTES pp. [167–170]

Emerging Technology (Washington, DC: Center for Security and Emerging Technology, 2022), https://cset.georgetown.edu/publication/preserving-the-chokepoints/ (last accessed 19 February 2023); Mathieu Duchâtel, 'Great Power Chokepoints: China's Semiconductor Industry in Search of Breakthroughs | Institut Montaigne,' *Institut Montaigne,* https://www.institutmontaigne.org/en/analysis/great-power-chokepoints-chinas-semiconductor-industry-search-breakthroughs (last accessed 19 February 2023).

21. Miller, *Chip War*; Christian Brose, *The Kill Chain: Defending America in the Future of High-Tech Warfare* (London: Hachette UK, 2020); Elsa B. Kania, *Battlefield Singularity Artificial Intelligence, Military Revolution, and China's Future Military Power* (Washington, DC: Center for a New American Security, 2017).

22. Middle East Eye, 'Iranian Press Review: Revolutionary Guard Equips Speed Boats with Suicide Drones,' *Middle East Eye,* http://www.middleeasteye.net/news/iranian-press-review-revolutionary-guard-equips-speed-boats-suicide-drones (last accessed 15 April 2021).

23. Paul Iddon, 'Turkey and Iran Show Off Their Homegrown Air Defense Systems,' *Forbes,* 2 January 2023, https://www.forbes.com/sites/pauliddon/2023/01/02/turkey-and-iran-show-off-their-homegrown-air-defense-systems/

24. Bronk, Reynolds and Watling, *The Russian Air War and Ukrainian Requirements for Air Defence.*

25. Maya Carlin, 'How the Turkish-Made TB2 Drone Gave Ukraine an Edge against Russia,' *Business Insider,* 18 September 2022, https://www.businessinsider.com/how-turkish-baykar-tb2-drone-gave-ukraine-edge-against-russia-2022-9

26. 'Submarine Detection and Monitoring: Open-Source Tools and Technologies,' *The Nuclear Threat Initiative* (blog), 2 March 2021, https://www.nti.org/analysis/articles/submarine-detection-and-monitoring-open-source-tools-and-technologies/; Sergio Silva, *Advances in Sonar Technology* (Vienna: IntechOpen, 2009); Justin Bronk, *Modern Russian and Chinese Integrated Air Defence Systems* (London: Royal United Services Institute (RUSI), 2020), https://rusi.org/sites/default/files/20191118_iads_bronk_web_final.pdf; Justin Bronk, 'Disruptive Trends in Long-Range Precision Strike, ISR, and Defensive Systems,' *The Nonproliferation Review* 27, no. 1–3 (2020): 39–47.

27. Alex Hollings, 'What Sets a '5th-Generation' Fighter Jet Apart from a '4th-Generation' Fighter Jet,' *Business Insider,* 11 February 2021, https://www.businessinsider.com/differences-between-fifth-and-fourth-generation-fighter-jets-2020-4

28. Owen R. Cote, 'Invisible Nuclear-Armed Submarines, or Transparent

pp. [170–173] NOTES

Oceans? Are Ballistic Missile Submarines Still the Best Deterrent for the United States?' *Bulletin of the Atomic Scientists* 75, no. 1 (2019): 30–5; Owen R. Cote, *Assessing the Undersea Balance between the US and China* (Cambridge, MA: MIT Center for International Studies, 2011); Keir A. Lieber and Daryl G. Press, 'The New Era of Counterforce: Technological Change and the Future of Nuclear Deterrence,' *International Security* 41, no. 4 (2017): 9–49.

29. 'World Missiles,' *Missile Threat*, https://missilethrcat.csis.org/missile/ (last accessed 1 October 2022).

30. Andrea Gilli and Mauro Gilli, 'Why China Has Not Caught Up Yet: Military-Technological Superiority and the Limits of Imitation, Reverse Engineering, and Cyber Espionage,' *International Security* 43, no. 3 (2019): 141–89.

31. Farley and Isaacs, *Patents for Power*; Caverley, 'United States Hegemony and the New Economics of Defense.'

32. Ronen Bergman and Farnaz Fassihi, 'The Scientist and the A.I.-Assisted, Remote-Control Killing Machine,' *The New York Times*, 18 September 2021, https://www.nytimes.com/2021/09/18/world/middleeast/iran-nuclear-fakhrizadeh-assassination-israel.html

33. Jon R. Lindsay, 'Stuxnet and the Limits of Cyber Warfare,' *Security Studies* 22, no. 3 (2013): 365–404.

34. Brose, *The Kill Chain*.

35. Emily Meierding, 'Weaponizing Energy Interdependence,' in *The Uses and Abuses of Weaponized Interdependence*, eds. Drezner, Farrell, and Newman, 169–84.

36. Phillips Payson O'Brien, *How the War Was Won* (Cambridge: Cambridge University Press, 2015).

37. Parisa Hafezi and Phil Stewart, 'Israel Appears to Have Been behind Drone Strike on Iran Factory, U.S. Official Says,' *Reuters*, 29 October 2023, https://www.reuters.com/world/middle-east/blast-heard-military-plant-irans-central-city-isfahan-state-media-2023-01-28/

38. O'Brien, *How the War Was Won*.

39. Jeffrey Sonnenfeld et al., *Business Retreats and Sanctions Are Crippling the Russian Economy* (Social Science Research Network, July 2022).

40. David Axe, 'What's Perfectly Round, Made of Metal, and Keeping Russia from Replacing the 2,000 Tanks It's Lost in Ukraine?' *Forbes*, 19 April 2023, https://www.forbes.com/sites/davidaxe/2023/04/19/whats-perfectly-round-made-of-metal-and-keeping-russia-from-replacing-the-2000-tanks-its-lost-in-ukraine/

41. 'Russia's War on Ukraine: A Sanctions Timeline,' *Peterson Institute for International Economics*, 14 March 2022, https://www.piie.com/blogs/realtime-economics/russias-war-ukraine-sanctions-timeline

42. Drezner, Farrell and Newman, *The Uses and Abuses of Weaponized Interdependence*, 185–202.

NOTES pp. [173–175]

43. Howard Altman, 'Ukraine Strikes Back at Bomber Base Deep Inside Russia,' *The Drive*, 26 December 2022, https://www.thedrive.com/the-war-zone/ukraine-strikes-back-at-bomber-base-deep-inside-russia

44. Stephen D. Biddle, *Military Power: Explaining Victory and Defeat in Modern Battle* (Princeton, NJ, and Oxford: Princeton University Press, 2004).

45. Michael Beckley, 'The Emerging Military Balance in East Asia: How China's Neighbors Can Check Chinese Naval Expansion,' *International Security* 42, no. 2 (2017): 78–119; Eric Heginbotham and Richard J. Samuels, 'Active Denial: Redesigning Japan's Response to China's Military Challenge,' *International Security* 42, no. 4 (2018): 128–69; Paul van Hooft, 'Don't Knock Yourself Out: How America Can Turn the Tables on China by Giving up the Fight for Command of the Seas,' *War on the Rocks*, 23 February 2021, https://warontherocks.com/2021/02/dont-knock-yourself-out-how-america-can-turn-the-tables-on-china-by-giving-up-the-fight-for-command-of-the-seas/

46. Fiona S. Cunningham and M. Taylor Fravel, 'Assuring Assured Retaliation: China's Nuclear Posture and US-China Strategic Stability,' *International Security* 40, no. 2 (2015): 7–50; Caitlin Talmadge, 'Would China Go Nuclear? Assessing the Risk of Chinese Nuclear Escalation in a Conventional War with the United States,' *International Security* 41, no. 4 (2017): 50–92; Paul van Hooft, 'All-In or All-Out: Why Insularity Pushes and Pulls American Grand Strategy to Extremes,' *Security Studies* 29, no. 4 (2020): 701–29.

47. Toby Sterling, Karen Freifeld and Alexandra Alper, 'Dutch to Restrict Semiconductor Tech Exports to China, Joining US Effort | Reuters,' *Reuters*, 8 March 2023, https://www.reuters.com/technology/dutch-responds-us-china-policy-with-plan-curb-semiconductor-tech-exports-2023-03-08/

48. The United States and its allies should control, with presumptive denial of licenses, advanced SME (especially extreme ultraviolet photolithography and argon fluoride immersion photolithography tools), advanced materials (photomasks and photoresists), and the software necessary for China to build and use advanced chip factories. As a backup option, they can consider limited controls on electronic design automation software (for chip design) and intellectual property to slow improvement in China's chip design capabilities. These controls would ensure China's dependence on imports for advanced chips; see Saif Khan, *Securing Semiconductor Supply Chains* (Washington, DC: Center for Security and Emerging Technology, 2021).

49. James Johnson, 'Washington's Perceptions and Misperceptions of Beijing's Anti-Access Area-Denial (A2-AD) 'Strategy': Implications for Military Escalation Control and Strategic Stability,' *The Pacific Review* 30, no. 3 (2017): 271–88.

50. Congressional Research Service, 'China Naval Modernization: Implications for U.S. Navy Capabilities—Background and Issues for Congress'

pp. [175–184] NOTES

(Washington, DC: Congressional Research Service, 2021), https://fas.org/sgp/crs/row/RL33153.pdf; Kirsten Huang, 'China Fires 'Aircraft-Carrier Killer' Missile in 'Warning to US'', *South China Morning Post*, 26 August 2020, https://www.scmp.com/news/china/military/article/3098972/chinese-military-launches-two-missiles-south-china-sea-warning

51. Bronk, 'Modern Russian and Chinese Integrated Air Defence Systems.'

52. As Josh Rovner notes, the resulting war is likely to be both intense and enduring. Joshua Rovner, 'Two Kinds of Catastrophe: Nuclear Escalation and Protracted War in Asia,' *Journal of Strategic Studies* 40, no. 5 (2017): 696–730.

53. Posen, 'Command of the Commons,' 7, 22–4.

9. MILITARY-TECHNOLOGICAL INNOVATION IN THE DIGITAL AGE

1. The text is based on my keynote address at the *Future of War Conference* hosted by the War Studies Research Centre, delivered in Amsterdam on 6 October 2022.

2. Adam Grissom, 'The Future of Military Innovation Studies,' *The Journal of Strategic Studies* 29, no. 5 (2006): 905–34.

3. Samuel Huntington, *The Soldier and the State* (Cambridge, MA: Harvard University Press, 1981).

4. Williamson R. Murray and Allan R. Millett (eds), *Military Innovation in the Interwar Period* (Cambridge: Cambridge University Press, 1998).

5. Ibid.

6. William A. Owens and Edward Offley, *Lifting the Fog of War* (New York: Farrar, Straus, and Giroux, 2000).

7. Stephen A. Rosen, *Innovation and the Modern Military: Winning the Next War* (New York: Cornell University Press, 1994).

8. Barry Posen, *The Sources of Military Doctrine: France, Britain, and Germany Between the World Wars* (Ithaca, NY: Cornell University Press, 1984).

9. Robert O. Work and Shawn Brimley, *20YY: Preparing for War in the Robotic Age* (Washington, DC: Center for a New American Century, 2014).

10. Eliot Cohen, 'Change and Transformation in Military Affairs,' *Journal of Strategic Studies* 27, no. 3 (2004): 404.

11. Winston S. Churchill, Speech to Royal College of Physicians, 2 March 1944.

12. David Johnson, *Fast Tanks and Heavy Bombers: Innovation in the U.S. Army, 1917–1945* (Ithaca, NY, & London: Cornell University Press, 1998).

13. Stephen Rosen, *Innovation and the Modern Military: Winning the Next War* (Ithaca, NY, & London: Cornell University Press, 1991).

14. Harvey M. Sopolski, *The Polaris System Development: Bureaucratic and Programmatic Success in Government* (Cambridge, MA: Harvard University Press, 1972).

NOTES pp. [184–191]

15. Shawn Brimley, 'Offset Strategies and Warfighting Regimes,' *War on the Rocks*, 15 October 2014, https://warontherocks.com/2014/10/offset-strategies-warfighting-regimes/

16. Audrey Kurth Cronin, *Power to the People: How Open Technological Innovation Is Arming Tomorrow's Terrorists* (New York: Oxford University Press, 2022).

17. Sebastian Mallaby, *The Power Law: Venture Capital and the Making of the New Future* (New York: Penguin Press, 2022).

18. For more on how to adapt to open innovation, see my *Power to the People* and Audrey Kurth Cronin, 'Technology and Strategic Surprise: Adapting to an Era of Open Innovation,' *Parameters* 50, no. 3 (2020): 71–84.

19. Special Competitive Studies Project, 'Chapter 5: The Future of Conflict and the New Requirements of Defense,' in *Mid-Decade Challenges to National Competitiveness* (Arlington, VA: Special Competitive Studies Project, 2022): 122–44.

20. Gian Gentile, Michael Shurkin, Alexandra T. Evans, Michelle Grise, Mark Hvizda and Rebecca Jensen, *A History of the Third Offset, 2014–2018* (Santa Monica, CA: The RAND Corporation, 2021), https://www.rand.org/pubs/research_reports/RRA454-1.html

21. Graham Allison and Jonah Glick-Unterman, *The Great Military Rivalry: China vs the U.S.* (Cambridge, MA: The Belfer Center, 2021): 18, https://www.belfercenter.org/sites/default/files/GreatMilitaryRivalry_ChinavsUS_211215.pdf

22. Graham Allison, Kevin Klyman, Karina Barbesino and Hugo Yen, *The Great Tech Rivalry: China vs. the US,* (Cambridge, MA: The Belfer Center, 2021): 2, https://www.belfercenter.org/sites/default/files/GreatTechRivalry_ChinavsUS_211207.pdf

23. Robert O. Work and Greg Grant, *Beating the Americans at Their Own Game: An Offset Strategy with Chinese Characteristics* (Washington, DC: Center for a New American Security, 2019), https://www.cnas.org/publications/reports/beating-the-americans-at-their-own-game

24. J. Michael Lane and Lila Summer, 'Smallpox as a Weapon for Bioterrorism,' in *Bioterrorism and Infectious Agents: A New Dilemma for the 21st Century*, eds. I. W. Fong and Kenneth Alibek (New York: Springer, 2009), 147–67.

25. Tiago Sousa et al., 'Generative Deep Learning for Targeted Compound Design,' *Journal of Chemical Information and Modeling* 61, no. 11 (2021): 5343–61.

26. Charles Edward White, *The Enlightened Soldier: Scharnhorst and the Militärische Gesellschaft in Berlin, 1801–1805* (New York: Praeger, 1989).

27. Carl von Clausewitz, *On War*, Book 2, ed. and trans. by Michael Howard and Peter Paret (Princeton, NJ: Princeton University Press, 1976): 127. Also cited by Michael Handel, 'Clausewitz in the Age of Technology,' *The Journal of Strategic Studies* 9, no. 2-3 (1986): 51–92.

pp. [191–196]   NOTES

28. Michael Howard, 'The Wars of the Nations,' in *War in European History* (Oxford: Oxford University Press, 2001): 102–5.

29. The Special Competitive Studies Project argues that promotion within the senior executive service ranks of the intelligence community should require executive education on artificial intelligence and emerging technologies. I agree; but it is just as essential to mid- and senior-level professional military and other DoD policymakers' education. Special Competitive Studies Project, *Mid-Decade Challenges to National Competitiveness*, September 2022, 152, https://www.scsp.ai/reports/mid-decade-challenges-for national-competitiveness/.pdf

30. Joint Chiefs of Staff, *Summary of the 2018 National Defense Strategy of the United States of America* (Washington, DC: Joint Chiefs of Staff, 2018): 8, https://dod.defense.gov/Portals/1/Documents/pubs/2018-National-Defense-Strategy-Summary.pdf

31. Memorandum by the Under Secretary of Defense for Research and Engineering, *Technology Vision for an Era of Competition* (Washington, DC: Pentagon, 2022), https://www.cto.mil/wp-content/uploads/ 2022/02/usdre_strategic_vision_critical_tech_areas.pdf

32. *U.S. Senate, Senate Armed Services Committee, Subcommittee on Emerging Threats and Capabilities, Hearing on Accelerating Innovation for the Warfighter*, 118th Cong. 4 (2022) (Statement of Michael Brown, Director, Defense Innovation Unit).

33. Yanek Mieczkowski, *Eisenhower's Sputnik Moment: The Race for Space and World Prestige* (Ithaca, NY: Cornell University Press, 2013).

34. *U.S. Senate, Senate Armed Services Committee*, 5.

35. Mike Brown, interviewed by Martijn Rasser, *CNAS Live*, 22 June 2022, https://www.cnas.org/events/virtual-event-conversation-with-michael-brown-director-of-the-defense-innovation-unit

36. Mallaby, *The Power Law*.

37. Special Competitive Studies Project, *Mid-Decade Challenges*, 140.

38. Charlie Parker, 'Specialist Ukrainian Drone Unit Picks off Invading Russian Forces as They Sleep,' *The Times*, 18 March 2022, https://www.thetimes.co.uk/article/specialist-drone-unit-picks-off-invading-forces-as-they-sleep-zlx3dj7bb. At the time of writing, Starlink President Gwynne Shotwell has threatened to put in place measures to limit the use of the technology for offensive targeting. Joey Roulette, 'SpaceX Curbed Ukraine's Use of Starlink Internet for Drones – Company President,' *Reuters*, 9 February 2023, https://www.reuters.com/business/aerospace-defense/spacex-curbed-ukraines-use-starlink-internet-drones-company-president-2023-02-09/

39. Cristiano Lima, 'U.S. Quietly Paying Millions to Send Starlink Terminals to Ukraine, Contrary to SpaceX Claims', *The Washington Post*, 8 April 2022,

401

https://www.washingtonpost.com/politics/2022/04/08/us-quietly-paying-millions-send-starlink-terminals-ukraine-contrary-spacexs-claims/

40. Amazon Staff, 'Safeguarding Ukraine's Data to Preserve its Present and Build its Future,' *News AWS*, 9 June 2022, https://www.aboutamazon.com/news/aws/safeguarding-ukraines-data-to-preserve-its-present-and-build-its-future

41. Frank G. Hoffman, *Mars Adapting: Military Change during War* (Annapolis, MD: Naval Institute Press, 2021). Notably, the US military adapted well under fire in Iraq and Afghanistan, but fell short in both wars for strategic reasons.

42. Alexander Kersten and Jillian Cota, 'The Russian Invasion Must Not Halt Ukraine's High-Tech Ambitions,' *Center for Strategic and International Studies*, 22 April 2022, https://www.csis.org/blogs/perspectives-innovation/russian-invasion-must-not-halt-ukraines-high-tech-ambitions

43. Drew Harrell, 'Instead of Consumer Software, Ukraine's Tech Workers Build Apps of War,' *The Washington Post*, 24 March 2022.

44. Jonathan Guyer, 'The West is Testing Out a Lot of Shiny New Military Tech in Ukraine,' *Vox*, 21 September 2022, https://www.vox.com/2022/9/21/23356800/us-testing-tech-ukraine-russia-war

45. Gregory C. Allen, 'Across Drones, AI, and Space, Commercial Tech is Flexing Military Muscle in Ukraine,' *CSIS*, 13 May 2022, https://www.csis.org/analysis/across-drones-ai-and-space-commercial-tech-flexing-military-muscle-ukraine

46. Peter Beaumont, 'Russia Escalating Use of Iranian 'Kamikaze' Drones in Ukraine,' *The Guardian*, 29 September 2022, https://www.theguardian.com/world/2022/sep/29/russia-escalating-use-of-iranian-kamikaze-drones-ukraine-reports-say

47. Ibid.

10. THE RISING DOMINANCE OF THE TACTICAL DEFENSE?

1. Military revolutions are much rarer and have a much greater impact than revolutions in military affairs (RMAs). See MacGregor Knox and Williamson Murray (eds), *The Dynamics of Military Revolutions, 1300–2050* (Cambridge: Cambridge University Press, 2001).

2. For a history of how tactical defense become dominant near the end of the American Civil War, see Bruce Catton, *Never Call Retreat* (New York: Doubleday & Company, 1961). To understand how the Germans restored offensive movement near the end of the First World War, see Bruce I. Gudmundsson, *Stormtroop Tactics: Innovation in the German Army, 1914–1918* (Westport, CT: Praeger, 1995).

pp. [204–206] NOTES

3. 'Planet Imagery and Archive,' *Planet Lab*, https://www.planet.com/markets/defense-and-intelligence/ (last accessed 8 May 2023).

4. 'Capella SAR-X,' *eoPortal*, 19 December 2018, https://directory.eoportal.org/web/eoportal/satellite-missions/c-missions/capella-x-sar

5. 'Mission Space,' *Hawkeye 360*, https://www.he360.com/products/mission-space-commercial-rf-analysis-platform/ (last accessed 29 October 2022); Aria Alamalhodaci, 'HawkEye 360 Raises $145M to Scale Space-based Radio Frequency Data and Analytics,' *TechCrunch*, 8 November 2021, https://techcrunch.com/2021/11/08/hawkeye-360-raises-145m-to-scale-space-based-radio-frequency-data-and-analytics/

6. Andy, 'Satellites Orbiting the Earth in 2022,' *Pixalytics*, 2 March 2022, https://www.pixalytics.com/satellites-in-2022/#:~:text=To%20the%20end%20of%20January,and%20even%20submarine%2Dbased%20launches

7. 'Manufacturing Revenues for Earth Observation to Grow to $76.1 Billion by 2030, Bolstered by Existing Government Programs, new Entrants and Diversified Commercial Constellations,' *Euroconsult*, 13 January 2022, https://www.euroconsult-ec.com/press-release/manufacturing-revenues-for-earth-observation-to-grow-to-76-1-billion-by-2030-bolstered-by-existing-government-programs-new-entrants-and-diversified-commercial-constellations/

8. Christian Davenport, 'Commercial Satellites Test the Rules of War in Russia-Ukraine Conflict,' *The Washington Post*, 10 March 2022, https://www.washingtonpost.com/technology/2022/03/10/commercial-satellites-ukraine-russia-intelligence/

9. David Ignatius, 'How the Algorithm Tipped the Balance in Ukraine,' *The Washington Post*, 20 December 2022, https://www.washingtonpost.com/opinions/2022/12/19/palantir-algorithm-data-ukraine-war/

10. 'Synthetic-aperture Radar is Making the Earth's Surface Watchable 24/7,' *The Economist*, 29 January 2022, https://www.economist.com/technology-quarterly/2022/01/27/synthetic-aperture-radar-is-making-the-earths-surface-watchable-24/7

11. VICE, 'Ukrainians Are Bombing Russians with Custom Drones,' *YouTube*, 17 June 2022, https://www.youtube.com/watch?v=VqOdnCEbDU4

12. Sam Schechner and Daniel Michaels, 'Ukraine Has Digitized Its Fighting Forces on a Shoestring,' *Wall Street Journal*, 3 January 2023, https://www.wsj.com/articles/ukraine-has-digitized-its-fighting-force-on-a-shoestring-11672741405

13. 'V-BAT 128 Unmanned Aircraft System (UAS), US,' *Air Force Technology*, 21 April 2021, https://www.airforce-technology.com/projects/v-bat-128/

14. 'Orion Unmanned Aircraft System (UAS),' *Air Force Technology*, 17 March

NOTES

2014, https://www.airforce-technology.com/projects/orion-unmanned-aircraft-system-uas/

15. Jordan Golson, 'A Military-Grade Drone That Can Be Printed Anywhere,' *Wired*, 16 September 2014, https://www.wired.com/2014/09/military-grade-drone-can-printed-anywhere/

16. Justin Bronk, Nick Reynolds and Jack Watling, *The Russian Air War and Ukrainian Requirements for Air Defence* (London: Royal United Services Institute, 2022): 2, https://rusi.org/explore-our-research/publications/special-resources/russian-air-war-and-ukrainian-requirements-air-defence

17. Ignatius, 'How the Algorithm Tipped the Balance in Ukraine.'

18. Wayne Hughes, Fleet Tactics & Naval Operations (Naval Institute Press, 2018), https://www.usni.org/press/books/fleet-tactics-and-naval-operations-third-edition

19. Maximilian K. Bremer and Kelly A. Grieco, 'In Denial about Denial: Why Ukraine's Air Success Should Worry the West,' *War on the Rocks*, 15 June 2022, https://warontherocks.com/2022/06/in-denial-about-denial-why-ukraines-air-success-should-worry-the-west/

20. Bronk, Reynolds and Watling, *The Russian Air War*.

21. Mia Jankowicz, 'Iranian-made Drones Cost as Little as $20,000 to Make But up to $500,000 to Shoot Down, a Growing Concern in Ukraine, Report Says,' *Business Insider*, 4 January 2023, https://www.businessinsider.com/suicide-drones-much-cheaper-launch-than-shoot-down-ukraine-nyt-2023-1

22. Ilan Berman, 'The Logic of Israel's Laser Wall,' *National Institute for Public Policy*, no. 526 (2022), https://www.realcleardefense.com/articles/2022/06/23/the_logic_of_israels_laser_wall_838770.html

23. Elisabeth Bumiller and Thomas Shanker, 'Panetta Warns of Possible 'Cyber-Pearl Harbor',' *New York Times*, 12 October 2012, https://www.nytimes.com/2012/10/12/world/panetta-warns-of-dire-threat-of-cyberattack.html

24. Brandon Valeriano, Benjamin Jensen and Ryan C. Maness, *Cyber Strategy: The Evolving Character of Power and Coercion* (New York: Oxford University Press, 2018).

25. Charles Smythe, 'Cult of the Cyber Offensive: Misperceptions of the Cyber Offense/Defense Balance,' *Yale Journal of International Affairs*, 10 June 2020, https://www.yalejournal.org/publications/cult-of-the-cyber-offensive-misperceptions-of-the-cyber-offensedefense-balance

26. US Cyber Command, *Achieve and Maintain Cyberspace Superiority: Command Vision for U.S. Cyber Command* (Washington, DC: United States Cyber Command, 2018): 6, https://www.cybercom.mil/Portals/56/Documents/USCYBERCOM%20Vision%20April%202018.pdf

pp. [215–225] NOTES

27. Marcus Willett, 'The Cyber Dimension of the Russia-Ukraine War,' *Survival* 64, no. 5, (2022): 12–15.

28. Kim Zetter, 'An Unprecedented Look at Stuxnet, the World's First Digital Weapon,' *Wired*, 3 November 2014, https://www.wired.com/2014/11/countdown-to-zero-day-stuxnet/

29. Theresa Hitchens, "Spectrum Superiority' Key to All Domain Operations: Gen. Hyten,' *Breaking Defense*, 20 January 2021, https://breakingdefense.com/2021/01/spectrum-superiority-key-to-all-domain-operations-gen-hyten/

30. Sandra Erwin, 'DoD Space Agency: Cyber Attacks, not Missiles, are the Most Worrisome Threat to Satellites,' *Space News*, 14 April 2021, https://spacenews.com/dod-space-agency-cyber-attacks-not-missiles-are-the-most-worrisome-threat-to-satellites

31. Alex J. Bellamy, 'Massacres and Morality: Mass Atrocities in an Age of Civilian Immunity,' in *Massacres and Morality: Mass Atrocities in an Age of Civilian Immunity* (Oxford: Oxford Academic, 2013), https://oxford.universitypressscholarship.com/view/10.1093/acprof:oso/9780199288427.001.0001/acprof-9780199288427-chapter-5

32. Tami Biddle, *Rhetoric and Reality in Air Warfare: The Evolution of British and American Ideas about Strategic Bombing, 1914–1945* (Princeton, NJ and Oxford: Princeton University Press, 2004); T.X. Hammes, 'Independent Long-Range Strike: A Failed Theory,' *War on the Rocks*, 8 June 2015, https://warontherocks.com/2015/06/independent-long-range-strike-a-failed-theory/

33. Stephen van Evera, 'Offense, Defense, and the Causes of War,' *International Security* 22, no. 4 (1998): 5–43.

11. ARTIFICIAL INTELLIGENCE AND THE NATURE OF WAR

1. Carl von Clausewitz, *On War*, 1st edition (New York: David Campbell Publishers Ltd., 1993).

2. Meta Fundamental AI Research Diplomacy Team (FAIR) et al., 'Human-Level Play in the Game of Diplomacy by Combining Language Models with Strategic Reasoning', *Science* 378, no. 6624 (2022): 1067–74.

3. Christopher D. Frith and D.M. Wolpert (eds), *The Neuroscience of Social Interaction: Decoding, Imitating, and Influencing the Actions of Others* (Oxford: Oxford University Press, 2004); Francesca Happé, 'Theory of Mind and the Self', *Annals of the New York Academy of Sciences* 1001, no. 1 (2003): 134–44; Stephen M. Fleming, *Know Thyself: The New Science of Self-Awareness* (London: John Murray, 2021).

4. Laurent Keller and Elisabeth Gordon, *The Lives of Ants*, 1st edition (Oxford & New York: Oxford University Press, 2009); Terrence P. McGlynn, 'Do Lanchester's Laws of Combat Describe Competition in Ants?', *Behavioral*

405

Ecology 11, no. 6 (2000): 686–90; Richard W. Wrangham and Luke Glowacki, 'Intergroup Aggression in Chimpanzees and War in Nomadic Hunter-Gatherers', *Human Nature* 23, no. 1 (2012): 5–29; L. David Mech and Luigi Boitani, *Wolves: Behavior, Ecology, and Conservation* (Chicago, IL & London: University of Chicago Press, 2003).

5. Kenneth Payne, *Strategy Evolves: From Apes to Artificial Intelligence* (Washington, DC: Georgetown University Press, 2017).

6. Frans De Waal, *Are We Smart Enough to Know How Smart Animals Are?* (London: Granta Books, 2017); Peter Godfrey-Smith, *Metazoa: Animal Minds and the Birth of Consciousness* (London: William Collins, 2021).

7. Robert Trivers, *Deceit and Self-Deception: Fooling Yourself the Better to Fool Others* (London: Penguin Books, 2013).

8. Robin Dunbar, *Grooming, Gossip and the Evolution of Language* (London: Faber & Faber, 2004); Sverker Johansson, *The Dawn of Language: How We Came to Talk* (London: MacLehose Press, 2021).

9. Cristina Moya and Joseph Henrich, 'Culture–Gene Coevolutionary Psychology: Cultural Learning, Language, and Ethnic Psychology', *Current Opinion in Psychology*, 8 (2016): 112–18.

10. Azar Gat, *War in Human Civilization* (Oxford: Oxford University Press, 2006); Lawrence H. Keeley, *War before Civilization* (New York & Oxford: Oxford University Press, 1996); Payne, *Strategy Evolves*.

11. Richard Dawkins, *The Selfish Gene*, 40th anniversary edition (Oxford: Oxford University Press, 2016); Daniel C. Dennett, *From Bacteria to Bach and Back: The Evolution of Minds*, 1st edition (London: Penguin, 2018).

12. Angela Saini, *Superior: The Return of Race Science*, 1st edition (London: Fourth Estate, 2020).

13. Rory Bowden et al., 'Genomic Tools for Evolution and Conservation in the Chimpanzee: Pan Troglodytes Ellioti is a Genetically Distinct Population', *PLOS Genetics* 8, no. 3 (2012): e1002504.

14. Joshua Greene, *Moral Tribes: Emotion, Reason and the Gap Between Us and Them* (New York: Atlantic Books, 2014).

15. Joseph Henrich, *The Secret of Our Success: How Culture Is Driving Human Evolution, Domesticating Our Species, and Making Us Smarter* (Princeton, NJ: Princeton University Press, 2015).

16. Batja Mesquita, *Between Us: How Cultures Create Emotions* (New York: W.W. Norton & Co, 2022); Lisa Feldman Barrett, *How Emotions Are Made: The Secret Life of the Brain* (London: Pan Books, 2018).

17. Jean-Francois Bonnefon, *The Car That Knew Too Much: Can a Machine Be Moral?* (Cambridge, MA: MIT Press, 2021); Edmond Awad et al., 'Universals and Variations in Moral Decisions Made in 42 Countries by 70,000 Participants', *Proceedings of the National Academy of Sciences* 117, no. 5 (2020): 2332–7.

pp. [230–237] NOTES

18. Kenneth Payne, 'Artificial Intelligence: A Revolution in Strategic Affairs?', *Survival* 60, no. 5 (2018): 7–32.

19. Kenneth Payne, *I, Warbot: The Dawn of Artificially Intelligent Conflict* (London: Hurst, 2021); Paul Scharre, *Army of None: Autonomous Weapons and the Future of War* (New York & London: W.W. Norton & Company, 2018); James Johnson, 'Artificial Intelligence & Future Warfare: Implications for International Security', *Defense & Security Analysis* 35, no. 2 (2019): 147–69.

20. MoD, 'Human-Machine Teaming (JCN 1/18)', *GOV.UK*, 18 May 2018, https://www.gov.uk/government/publications/human-machine-teaming-jcn-118

21. Nick Bostrom, *Superintelligence: Paths, Dangers, Strategies* (Oxford: Oxford University Press, 2014).

22. Michael Wooldridge, *The Road to Conscious Machines: The Story of AI* (London: Pelican, 2020); Gary Marcus and Ernest Davis, *Rebooting AI: Building Artificial Intelligence We Can Trust* (New York: Dutton / Signet, 2020).

23. William A. Owens, *Lifting the Fog of War* (New York: Farrar, Straus and Giroux, 2000).

24. Antoine J. Bousquet, *The Scientific Way of Warfare: Order and Chaos on the Battlefields of Modernity* (London: Hurst, 2022).

25. Michael S.A. Graziano, *Consciousness and the Social Brain* (New York: Oxford University Press, 2015).

26. @davidchalmers42, 'one criticism of large language models', *Twitter*, 20 September 2022, https://twitter.com/davidchalmers42/status/1572234123577532416?lang=en

27. @GaryMarcus, 'their inconsistency and persistent inability to maintain coherence', *Twitter*, 20 September 2022, https://twitter.com/GaryMarcus/status/1572249139790118914

28. Steven T. Piantadosi and Felix Hill, 'Meaning without Reference in Large Language Models', *ArXiv*, 12 August 2022, https://doi.org/10.48550/arXiv.2208.02957; Belinda Z. Li, Maxwell Nye and Jacob Andreas, 'Implicit Representations of Meaning in Neural Language Models', *ArXiv*, 1 June 2021, https://doi.org/10.48550/arXiv.2106.00737

29. Michal Kosinski, 'Theory of Mind May Have Spontaneously Emerged in Large Language Models', *ArXiv*, 29 August 2023, https://doi.org/10.48550/arXiv.2302.02083

30. Andy Zeng et al., 'Socratic Models: Composing Zero-Shot Multimodal Reasoning with Language', *ArXiv*, 27 May 2022, https://doi.org/10.48550/arXiv.2204.00598

31. Bernard J. Baars, *In the Theater of Consciousness: The Workspace of the Mind* (New York & Oxford: Oxford University Press, 1997).

32. Ted Chiang, 'Catching Crumbs from the Table', *Nature* 405, no. 6786 (2000): 517.

NOTES pp. [241–243]

12. ASSEMBLING THE FUTURE OF WARFARE

1. Tarak Barkawi and Shane Brighton, 'Powers of War: Fighting, Knowledge, and Critique,' *International Political Sociology* 5, no. 2 (2011): 137.
2. David Ignatius, 'How the Algorithm Tipped the Balance in Ukraine,' *The Washington Post*, 12 November 2020.
3. Ibid.
4. Robin Fontes and Jorrit Kamminga, 'Commentary: Ukraine a Living Lab for AI Warfare,' *National Defense*, 23 March 2023, nationaldefensemagazine. org/articles/2023/3/24/ukraine-a-living-lab-for-ai-warfare
5. On loitering munitions, see Peter Burt, 'Loitering Munitions, The Ukraine War, and the Drift Towards 'Killer Robots',' *Drone Wars*, 8 June 2022, dronewars.net/2022/06/08/loitering-munitions-the-ukraine-war-and-the-drift-towards-killer-robots/. On automated target recognition, see Jeffrey Dastin, 'Ukraine is Using Palantir's Software for 'Targeting', CEO Says,' *Reuters*, 2 February 2023, reuters.com/technology/ukraine-is-using-palantirs-software-targeting-ceosays2023-02-02/. On facial recognition technology in warfare, see James Clayton, 'How Facial Recognition is Identifying the Dead in Ukraine,' *BBC*, 13 April 2022, bbc.com/news/technology-61055319
6. Cf. Larry Lewis, 'Resolving the Battle over Artificial Intelligence in War,' *The RUSI Journal* 164, no. 5/6 (2019): 62–71.
7. For a critique on technological determinism in international relations and security studies, see Daniel McCarthy, *Technology and World Politics: An Introduction* (London: Routledge, 2017); Marijn Hoijtink and Matthias Leese (eds), *Technology and Agency in International Relations* (London & New York: Routledge, 2019).
8. For more on 'remote warfare' and its strategies, technologies and effects, see Alasdair McKay, Abigail Watson and Megan Karlshøj-Pedersen (eds), *Remote Warfare: Interdisciplinary Perspectives* (Bristol: E-International Relations, 2021).
9. Melissa Heikkilä, 'Why Business is Booming for Military AI Startups,' *MIT Technology Review*, 7 July 2022, technologyreview. com/2022/07/07/1055526/why-business-is-booming-for-military-ai-startups; 'Defensie koopt voor miljarden aan nieuwe schepen en raketsystemen,' *NOS,* 3 April 2023, nos.nl/artikel/2470027-defensie-koopt-voor-miljarden-aan-nieuwe-schepen-en-raketsystemen
10. Marina Favaro, Ulrich Kuhn and Neil Renic, 'Will DIANA – NATO's DARPA-style Innovation Hub – Improve or Degrade Global Stability?' *Bulletin of the Atomic Scientists*, 23 March 2023, thebulletin.org/2023/03/will-diana-natos-darpa-style-innovation-hub-improve-or-degrade-global-stability/; 'NATO Launches Innovation Fund,' *NATO*, 30 June 2022, nato. int/cps/en/natohq/news_197494.htm

pp. [243–246]

NOTES

11. Dastin, 'Ukraine is Using Palantir's Software.'

12. International Committee for the Red Cross, 'Artificial Intelligence and Machine Learning in Armed Conflict: A Human-Centered Approach,' *International Committee of the Red Cross*, 6 June 2019, icrc.org/en/document/artificial-intelligence-and-machine-learning-armed-conflict-human-centred-approach

13. Cf. Sarah Brayne, 'Big Data Surveillance: The Case of Policing,' *American Sociological Review*, 82, no. 5 (2017): 977–1008.

14. Cf. Linda Weiss, *America Inc.?* (Ithaca, NY: Cornell University Press, 2014).

15. William Lynn, "The End of the Military-Industrial Complex: How the Pentagon Is Adapting to Globalization,' *Foreign Affairs* 93, no. 6 (2014): 104–10; Maaike Verbruggen, 'The Role of Civilian Innovation in the Development of Lethal Autonomous Weapon Systems,' *Global Policy* 10, no. 3 (2019): 338–42.

16. Lewis, 'Resolving the Battle.'

17. Cf. Lauren Gould and Nora Stel, 'Strategic Ignorance and the Legitimation of Remote Warfare: The Hawija Bombardments,' *Security Dialogue* 53, no. 1 (2022): 57–74; Jennifer Gibson, 'Death by Data: Drones, Kill Lists and Algorithms,' in *Remote Warfare: Interdisciplinary Perspectives*, eds. McKay, Watson and Karlshøj-Pedersen, 187–98.

18. Cf. Ingvild Bode and Hendrik Huelss, *Autonomous Weapons Systems and International norms* (Montreal: McGill-Queen's Press-MQUP, 2022); Denise Garcia, 'Future Arms, Technologies, and International Law: Preventive Security Governance,' *European Journal of International Security* 1, no. 1 (2016): 94–111.

19. Eugene Kagan, Nir Shvalb and Irad Ben-Gal, *Autonomous Mobile Robots and Multi-Robot Systems: Motion-Planning, Communication, and Swarming* (Hoboken, NJ: Wiley, 2020): 12.

20. Maurits Korthal Altes, 'Realities of Algorithmic Defence Symposium' (Panel contribution, Utrecht University, Utrecht, 8 June 2022); David Kilcullen, *Out of the Mountains: The Coming Age of the Urban Guerilla* (Oxford: Oxford University Press, 2013).

21. See Patrick Tucker, 'The US military's Drone Swarm Strategy Just Passed a Key Test,' *DefenseOne*, 21 November 2018, defenseone.com/technology/2018/11/us-militarys-drone-swarm-strategy-just-passed-key-test/153007; Devin Coldewey, 'Swarming Drones Autonomously Navigate a Dense Forest (and Chase a Human),' *TechCrunch*, 4 May 2022, techcrunch.com/2022/05/04/swarming-drones-autonomously-navigate-a-dense-forest-and-chase-a-human; David Stern, 'Russia Attacks Kyiv Overnight with Swarm of Self-detonating Drones,' *The Washington Post*, 19 December 2022, washingtonpost.com/world/2022/12/19/kyiv-drones-attack-belarus-putin; Zak Kallenborn, 'Israel's Drone Swarm

409

Over Gaza Should Worry Everyone,' *DefenseOne*, 7 July 2021, defenseone. com/ideas/2021/07/israels-drone-swarm-over-gaza-should-worry-everyone/183156; Stijn Mitzer, 'Swarm-UAVs are Coming – And Turkey Just Called Dibs,' *Oryxspioenkop*, 11 January 2022, oryxspioenkop. com/2022/01/swarm-uavs-are-coming-and-turkey-just.html; Paul Scharre, *Army of None*, 22; David Hambling, 'What Are Drone Swarms and Why Does Every Military Suddenly Want One?' *Forbes*, 1 March 2021, forbes.com/sites/davidhambling/2021/03/01/what-are-drone-swarms-and-why-does-everyone-suddenly-want-one/?sh=129eaeb82f5c

22. Maria Cramer, 'A.I. Drone May Have Acted on its Own in Attacking Fighters, U.N. Says,' *The New York Times*, 4 June 2021, nytimes.com/2021/06/03/world/africa/libya-drone.html

23. Mitzer, 'Swarm-UAVs Are Coming.'

24. Kallenborn, 'Israel's Drone Swarm.'

25. Peter Huis in 't Veld, 'Delft Companies to Work on Swarm-based Recon,' *NIDV*, 29 July 2021, nidv.eu/en/news/delft-companies-to-work-on-swarm-based-recon/

26. Tania Li, 'Practices of Assemblage and Community Forest Management,' *Economy and Society* 36, no. 2 (2007): 263–93.

27. Li, 'Practices of Assemblage,' 264.

28. Li, 'Practices of Assemblage,' 263–93.

29. Ministerie van Defensie, 'Strategische Kennis- En Innovatieagenda (SKIA) 2021–2025' (The Hague, December 2020), 6, defensie. nl/downloads/publicaties/2020/11/25/strategische-kennis--en-innovatieagenda-2021-2025; 'Robotica en Autonome Systemen Technologie,' in *N PRGRSS: Samen Sneller Sterker*, eds. Natascha Koetsier-van der Werff et al. (The Hague: Ministerie van Defensie, 2021): 141–5, defensie.nl/downloads/publicaties/2021/12/17/magazine-n-prgrss

30. Arno Marchand, 'De Kunst van het Loslaten,' *Materieelgezien*, 18 December 2014, magazines.defensie.nl/materieelgezien/2014/ 09/ fokker-50; 'Damen & The Royal Netherlands Navy,' *Damen Naval*, nlnavy. damen.com (last accessed 26 July 2022).

31. NLR primarily works on air force-specific projects, MARIN on maritime problems, and TNO on projects throughout the entire organization. See 'Defensie en Kennisinstituten Gaan nog Nauwer Samenwerken,' *MARIN*, 29 April 2021, marin.nl/nl/news/defensie-en-kennisinstituten-gaan-nog-nauwer-samenwerken

32. *N PRGRSS*.

33. 'Startups verrassen Defensie nu vaker met innovaties,' *PIANOo*, July 2019, pianoo.nl/startups-verrassen-defensie-nu-vaker-met-innovaties

34. *N PRGRSS*.

35. Lynn, 'The End.'

pp. [249–252]   NOTES

36. 'Startups verrassen Defensie nu vaker met innovaties.'
37. Author interview (Den Haag, 25 April 2022).
38. Ministerie van Defensie, *Defensie Innovatie Strategie: Samen Sneller Innoveren* (The Hague, 2018), defensie.nl/downloads/publicaties/2018/11/16/defensie-innovatie-strategie-2018
39. Author interview (Den Haag, 25 April 2022).
40. Ministerie van Defensie and Ministerie van Economische Zaken en Klimaat, *Defensie Industrie Strategie* (The Hague, 2018), defensie.nl/downloads/beleidsnota-s/2018/11/15/defensie-industrie-strategie
41. Such as the Navy's Maritime Drone Team, the Royal Marechaussee's Deep Vision Team, and the Army's RAS unit. See Robert den Hartog, 'Snelle innovaties specialiteit van CD&E,' *Landmacht*, 3 September 2019, magazines. defensie.nl/landmacht/2019/07/09_snelle_innovaties_specialiteit_van_cde_07-2019
42. 'Chinese drones van Rijkswaterstaat en politie mogelijk onveilig,' *NOS*, 30 September 2021.
43. Charlotte Snel, 'Trouwe viervoeter met potentie,' *KMarMagazine*, 29 July 2021, magazines.defensie.nl/kmarmagazine/2021/07/01_robothond_spot_07-2021
44. Boston Dynamics, 'Spot: The Agile Mobile Robot,' *Boston Dynamics*, bostondynamics.com/products/spot (last accessed 27 July 2022).
45. Boston Dynamics, 'About,' *Boston Dynamics*, bostondynamics.com/about; @BostonDynamics, 'We Condemn the Portrayal of our Technology in any Way that Promotes Violence,' *Twitter*, 20 February 2021, t.co/8BjVXhfZjP (last accessed 27 July 2022).
46. This claim was made by a Navy commander during an interview; author interview (Den Helder, 3 May 2022).
47. 'Het binden van slimme koppen aan Defensie levert Ereteken in goud op,' *Defensie.nl*, 13 January 2023, opdefensie.nl/actueel/nieuws/2023/01/13/het-binden-van-slimme-koppen-aan-defensie-levert-ereteken-in-goud-op
48. Ministerie van Defensie, 'Méér dan een bijbaan: Defensity College werkstudent,' werkenbijdefensie.nl/interessegebieden/werkstudent (last accessed 13 December 2022).
49. 'Het binden van slimme koppen aan Defensie levert Ereteken in goud op.'
50. Jaap Wolting, 'Gevechtskracht Omhoog, Kosten Omlaag,' *Landmacht*, 7 September 2021, magazines.defensie.nl/landmacht/2021/07/06_cde-rubriek
51. *N PRGRSS.*
52. 'Robotica en Autonome Systemen Technologie,' 145.
53. Author interview (Den Haag, 25 April 2022).
54. Author interview (Den Haag, 25 April 2022).

NOTES

pp. [252–254]

55. Ministerie van Defensie, 'Aantallen Personeel,' *Defensie.nl*, 29 March 2019, defensie.nl/onderwerpen/overdefensie/het-verhaal-van-defensie/aantallen-personeel

56. To make his point, the Navy Commander emphasized the flexibility and adaptivity crucial for the operational success of small units: author interview (Den Helder, 3 May 2022).

57. Ministerie van Defensie, 'Frontdoor: De Toegangspoort voor Innovatieve Samenwerking met Defensie,' *Defensie.nl*, 19 April 2018, defensie.nl/actueel/videos/2018/04/19/frontdoor

58. 'Landmacht Zoekt Samenwerking met Maakindustrie Oor [*sic*] Aanschaf Robots,' *Link Magazine*, 31 October 2018, linkmagazine.nl/landmacht-zoekt-samenwerking-met-maakindustrie-oor-aanschaf-robots

59. See 'Innovatienetwerk Defensie,' *LinkedIn*, linkedin.com/company/innovatienetwerk-defensie (last accessed 27 July 2022).

60. Arno Marchand, 'Achter de Schermen bij Deep Vision,' *KMarMagazine*, 24 September 2020, magazines.defensie.nl/kmarmagazine/2020/08/07_achter-de-schermen_08-2020

61. Like Brainport near Eindhoven University of Technology, YES!Delft, located near Delft University of Technology, and Brightlands, in the Maastricht area: 'Wat is Brainport Eindhoven?' *Brainport Eindhoven*, brainporteindhoven.com/nl/ontdek/wat-is-brainport-eindhoven (last accessed 11 July 2022); 'We Are YES!Delft', *YES!Delft*, yesdelft.com/about-us (last accessed 11 July 2022); Ministerie van Defensie, 'Defensie en Brightlands Onderzoeken 3D-printing voor Militair gebruik,' *Defensie.nl*, 9 July 2020, defensie.nl/actueel/nieuws/2020/07/09/defensie-en-brightlands-onderzoeken-3d-printing-voor-militair-gebruik

62. Evert Brouwer, 'Vernieuwing in Oude Haven,' *Materieelgezien*, 2 July 2020, magazines.defensie.nl/materieelgezien/2020/05/09_mindbase

63. Linde Arentze, *DMO MIND Log* (Rotterdam, 29 March 2022).

64. Author interview (Den Helder, 3 May 2022).

65. Author interview (Den Haag, 25 April 2022).

66. Ibid.

67. Li, 'Practices of Assemblage,' 267.

68. Author interview (Den Haag, 25 April 2022).

69. Li, 'Practices of Assemblage,' 265.

70. Ibid., 280.

71. Nikolas Rose, *Powers of Freedom: Reframing Political Thought* (Cambridge: Cambridge University Press, 1999): 192.

72. Lennart Hofman, 'Robotsoldaten Zijn Allang geen Sciencefiction Meer (en dus Bouwt Nederland ze ok),' *De Correspondent*, 21 January 2020; Koninklijke Landmacht, '🎧 PODCAST #04 🧑‍🦰Gijs Tuinman & 🙂🗂

pp. [255–257]   NOTES

Martijn Hädicke | BOOTS on the GROUND | Mens vs Machine,' *YouTube*, 9 June 2022, youtube.com/watch?v=3gAxLlsH5wo; 'Visie op Robotica bij de Koninklijke Landmacht,' *Vision + Robotics*, 4 October 2018, visionenrobotics. nl/artikelen/visie-op-robotica-bij-de-koninklijke-landmacht

73. Snel, "Trouwe Viervoeter met Potentie'; KAP Jaap Wolting, 'Informatiegestuurd Optreden Dankzij Drones,' *Landmacht*, 6 September 2021, defensie.nl/landmacht/2021/07/06_cde-rubriek; Martijn Hädicke, 'Realities of Algorithmic Defence Symposium' (Panel contribution, Utrecht University, Utrecht, 8 June 2022).

74. Tweede Kamer der Staten-Generaal, *Verslag van een Algemeen Overleg: Onbemande Vliegtuigen (UAV)* (The Hague, 2 April 2014), tweedekamer.nl/ kamerstukken/detail?id=2013Z22754&did=2014D06913

75. Jip van Dort and Lauren Gould, 'Bewapende Drone is Helemaal Niet zo'n Superwapen,' *Het Parool*, 27 April 2022, parool.nl/columns-opinie/opinie-bewapende-drone-is-helemaal-niet-zo-n-superwapen~bffbd5b9

76. Something referred to as the 'slippery slope' in 'function creep'; see Tim Dekkers, 'Technology Driven Crimmigration? Function Creep and Mission Creep in Dutch Migration Control,' *Journal of Ethnic and Migration Studies* 46 no. 9 (2020): 1852.

77. Author interview (Den Haag, 25 April 2022).

78. Saheed A. Gbadegeshin et al., 'Overcoming the Valley of Death: A New Model for High Technology Startups,' *Sustainable Futures* 4 (2022): 1–15.

79. Jack Oosthoek, 'Het is Soms Moeilijk om Alles bij te Benen,' *Defensiekrant*, 28 February 2022, magazines.defensie.nl/defensiekrant/2022/07/06_ innovatie-competitie_07

80. Ministerie van Defensie, 'Defensie Innovatie Competitie (DIC),' *Defensie.nl*, 2022, defensie.nl/onderwerpen/innovatie/defensie-innovatie-competitie (last accessed 20 June 2022).

81. Author interview (Delft, 5 May 2022).

82. Oosthoek, 'Het is Soms Moeilijk om Alles bij te Benen.'

83. Arno Marchand, 'Innovatieprijs Naar Dronesysteem,' *Defensiekrant*, 6 March 2020, magazines.defensie.nl/defensiekrant/2020/09/03_ innovatiecompetitie_09

84. Author interview (Delft, 5 May 2022).

85. Li, 'Practices of Assemblage,' 265.

86. Author interview (Delft, 5 May 2022); Wiebe de Jager, 'Delftse Dronebouwers Gaan Autonome Dronezwerm Ontwikkelen voor Defensie,' 22 July 2021, dronewatch.nl/delftse-dronebouwers-gaan-autonome-dronezwerm-ontwikkelen-voor-defensie

87. Author interview (Delft, 5 May 2022).

88. Jeroen Lappenschaar, 'Afgelopen Weekend was Mijn Laatste Oefening,' *LinkedIn*, March 2023, linkedin.com/posts/jeroen-lappenschaar-

NOTES pp. [257–259]

68616130_afgelopen-weekend-was-mijn-laatste-oefening-activity-7046233177371090944-NEu1 (last accessed 27 April 2023); Maurits Korthals Altes, 'Afgestudeerd, Avalor AI opgericht, en Direct op Zoek Naar Nieuwe Collega's!,' *LinkedIn*, 2022, linkedin.com/ posts/mauritskorthalsaltes_ai-artificialintelligence-spear-activity-6825493547857399808-Gah_ (last accessed 27 April 2023).

89. Jaap Wolting, 'Informatiegestuurd Optreden Dankzij Drones,' *Landmacht*, 6 June 2021, magazines.defensie.nl/landmacht/2021/07/06_cde-rubriek

90. Evert Brouwer, 'Vernieuwing in Oude Haven,' *Materieelgezien*, 2 July 2020, magazines.defensie.nl/materieelgezien/2020/05/09_mindbase

91. Li, 'Practices of Assemblage,' 279.

92. Author interview (Den Helder, 3 May 2022); Arno Marchand, 'Zwerm Drones voor Red Force Tracking,' *Landmacht*, 6 May 2020, magazines.defensie. nl/landmacht/2020/04/06_zwerm_drones_voor_red_force_tracking

93. Ministerie van Defensie, *Defensie Innovatie Strategie*.

94. Jopke Rozenberg-van Lisdonk, 'Verandermanager aan de Slag met Digitale Transformatie,' *De Vliegende Hollander*, 26 October 2021, magazines. defensie.nl/vliegendehollander/2021/10/06_digitale-transformatie-clsk

95. Ministerie van Defensie, *Defensie Innovatie Strategie*; 'Proeftuin of Pilot Inrichten,' *PIANOo*, n.d., pianoo.nl/nl/themas/innovatiegericht-inkopen/aan-de-slag/proeftuin-pilot-inrichten (last accessed 20 June 2022).

96. Evert Brouwer, 'Nieuwe 'Robot' op Proef,' *Defensiekrant*, 13 November 2020, magazines.defensie.nl/defensiekrant/2020/45/03_mission-master_45

97. Marchand, 'Zwerm Drones voor Red Force Tracking'; Author interview (Delft, 5 May 2022); Author interview (Den Haag, 25 April 2022); Author interview (Den Helder, 3 May 2022).

98. For an in-depth analysis of experimentation and failure by Western militaries, see Marijn Hoijtink, '"Prototype Warfare": Innovation, Optimisation, and the Experimental Way of Warfare,' *European Journal of International Security* 7, no. 3 (2022): 322–36.

99. Ministerie van Defensie, *Defensie Innovatie Strategie*; Ministerie van Defensie, *Defensievisie 2035: Vechten Voor Een Veilige Toekomst* (The Hague, October 2020): 20; Ministerie van Defensie, *Strategische Kennis — En Innovatieagenda (SKIA) 2021–2025* (The Hague, November 2020); Ministerie van Defensie and Ministerie van Economische Zaken en Klimaat, *Defensie Industrie Strategie* (The Hague, November 2018).

100. Ministerie van Defensie, *Defensie Innovatie Strategie*.

101. 'Startups Verrassen Defensie nu Vaker met Innovaties.'

102. *N PRGRSS.*

103. Author interview (Delft, 5 May 2022).

104. Ibid.

pp. [261–272]   NOTES

105. Saskia Sassen, *Expulsions: Brutality and Complexity in the Global Economy* (Cambridge, MA: Harvard University Press, 2014).

106. Barkawi and Brighton, 'Powers of War,' 137.

13. THE PAST AS GUIDE TO THE FUTURE

1. Valery Gerasimov, 'The Value of Science Is in the Foresight: New Challenges Demand Rethinking the Forms and Methods of Carrying out Combat Operations', *Military Review* 96, no. 1 (Jan–Feb. 2016): 29.

2. Cf. David S. Yost, 'Political Philosophy and the Theory of International Relations', *International Affairs* 70, no. 2 (1994): 285.

3. Jane Kramer, *The Politics of Memory: Looking for Germany in the New Germany* (New York: Random House, 1996).

4. Robert M. Sapolsky, 'A Natural History of Peace', *Foreign Affairs* 85, no. 1 (2006): 104–20.

5. Office of the Director of National Intelligence, 'Anticipatory Intelligence', in *The National Intelligence Strategy of the USA* (Washington, DC: Office of the Director of National Intelligence, 2019): 9, https://www.dni.gov/files/ODNI/documents/National_Intelligence_Strategy_2019.pdf (last accessed 28 June 2022).

6. Jeannie Johnson and Marilyn Maines, 'The Cultural Topography Analytic Framework', in *Crossing Nuclear Thresholds: Leveraging Sociocultural Insights into Nuclear Decision-making*, eds. Jeannie Johnson, Kerry Kartchner and Marilyn Maines (London: Palgrave Macmillan, 2018): 29–60.

7. Cf. Beatrice Heuser and Cyril Buffet, 'Introduction', in *Haunted by History: Myths in International Relations,* eds. Cyril Buffet and Beatrice Heuser (Oxford: Berghahn, 1998): ix.

8. Thucydides, *History of the Peloponnesian War*, trans. Rex Warner (London: Penguin Books, 1972): II.34.

9. P. Connerton, *How Societies Remember* (Cambridge: Cambridge University Press, 1989); Elizabeth Hallam and Jenny Hockley, *Death, Memory and Material Culture* (Oxford: Berg, 2001).

10. T.G. Ashplant, Graham Dawson and Michael Roger (eds), *The Politics of War Memory and Commemoration* (London: Routledge, 2002).

11. Geoffrey Hosking and George Schöpflin (eds), *Myths and Nationhood* (London: Hurst, 1997).

12. Cf. Beatrice Heuser and Athena Leoussi (eds), *Famous Battles and How They Have Shaped the Modern World* (Barnsley: Pen & Sword, 2018).

13. A. Borg, *War Memorials* (London: Leo Cooper, 1991).

14. G. Kavanagh, *Museums and the First World War: A Social History* (Leicester: Leicester University Press, 1994); A. King, *Memorials of the Great War in Britain: The Symbolism and Politics of Remembrance* (Oxford: Berg, 1998); Annette

NOTES   pp. [273–281]

Becker, 'Le Culte du Souvenir Après la Grande Guerre: Les Monuments aux Morts', in *Reconstructions et Modernisation*, ed. J. Favier (Paris: Archives de France, 1991); Jay Winter, *Sites of Memory, Sites of Mourning: The Great War in European Cultural History* (Cambridge: Cambridge University Press, 1995); Jay Winter and Emmanuel Sivan (eds), *War and Remembrance in the Twentieth Century* (Cambridge: Cambridge University Press, 2000).

15. Brian Murdoch and William Kidd, *Memory and Memorials: The Commemorative Century* (London & New York: Ashgate Publishing, 2003).

16. We hesitate to give examples here. Our visit to most of these dates back several years and the museums may well have been redesigned in the meantime.

17. William M. Johnston, *Celebrations: The Cult of Anniversaries in Europe and the United States* (New Brunswick, NJ: Transaction Publishers, 1991).

18. D. J. Sherman, 'Objects of Memory: History and Narrative in French War Museums', *French Historical Studies* 19, no. 1 (1995): 49–74.

19. Beatrice Heuser, 'Culloden 1746: Six Myths and their Politics', in *Famous Battles, vol. 2*, eds. Heuser and Leoussi, 84–108.

20. Heinrich von Kleist, *Die Hermannsschlacht* (Leipzig: Insel Bücherei, originally published 1821).

21. Erich Loest, *Froschquartett* (Hamburg: Knaust, 1987).

22. Cf. Paul Fussell, *The Great War and Modern Memory* (New York & London: Oxford University Press, 1975); Michael Paris, *Warrior Nation: War in British Popular Culture, 1850–2000* (London: Redaktion, 2000).

23. The former called *Ekaterina* (2014), the latter *Velikaya* (2015).

24. Raymond Aron, *Les Guerres en Chaîne* (Paris: Gallimard, 1951).

25. Ibid., 398–402.

26. Beatrice Heuser, *NATO, Britain, France and the FRG: Nuclear Strategies and Forces for Europe* (London: Macmillan, 1997): 148–72.

27. Norman Angell, *The Great Illusion* (London: Heinemann, 1912).

28. Beatrice Heuser, 'Stalin as Hitler's Successor', in *Securing Peace in Europe, 1945–62: Thoughts for the Post-Cold War Era*, eds. Beatrice Heuser and Robert O'Neill (London: Macmillan, 1992): 17–40.

29. Cf. David Holloway, *Stalin and the Bomb: The Soviet Union and Atomic Energy, 1939–1956* (London & New Haven, CT: Yale University Press, 1994).

30. Beatrice Heuser, *Reading Clausewitz* (London: Pimlico, 2002): 143f.

31. Cf. the article by Andrey A. Kokoshin and Major-General Prof Valentin V. Larionov in 1991, 'Origins of the Intellectual Rehabilitation of A. A. Svechin', in *Aleksandr A. Svechin: Strategy*, ed. Kent Lee (Minneapolis: East View Information Service, 1992): 11–26.

32. Carl von Clausewitz, *On War*, 19th edition, ed. Werner Hahlweg (Bonn: Dümmler, 1991): VIII.3B, 972f.

33. Beatrice Heuser, 'NSC 68 and the Soviet Threat: A New Perspective on

pp. [281–284]   NOTES

Western Threat Perception and Policy Making', *Review of International Studies* 17, no. 1 (1991): 17–40.

34. The great classic being Robert Osgood's, *Limited War: The Challenge to American Strategy* (Chicago: University of Chicago Press, 1957).

35. F.L. Taylor, *The Art of War in Italy, 1494–1529* (Westport, CT: Greenwood, 1973); André Corvisier, *Armées et Sociétés en Europe de 1494 à 1789* (Paris: PUF, 1976); Frank Tallett, *War and Society in Early-Modern Europe, 1495–1715* (London: Routledge, 1992).

36. Quoted in Geoffrey Parker, *The Military Revolution: Military Innovation and the Rise of the West 1500–1800* (Cambridge: Cambridge University Press, 1988): 10. Max Booth also sees this campaign as the crucial turning point: Max Booth, *War Made New: Technology, Warfare, and the Course of History, 1500 to Today* (New York: Gotham Books, 2006): 1–6.

37. Simon Pepper, 'Castles and Cannon in the Naples Campaign of 1494–95', in *The French Descent into Renaissance Italy*, ed. David Abulafia (Aldershot: Variorum, 1995): 263–94.

38. Martin van Creveld, *The Transformation of War* (London: Brassey's, 1991): 192–223; Beatrice Heuser, 'La Guerre Asymétrique', in *Guerre et Politique*, ed. Jean Baechler (Paris: Editions Hermann, 2014): 107–21.

39. John F. Kennedy, 'United States Military Academy Commencement Address', *American Rhetoric Online Speech Bank*, https://www.americanrhetoric. com/speeches/jfkwestpointcommencementspeech.htm (last accessed 17 January 2020).

40. Herfried Münkler, 'Was ist Neu an Neuen Kriegen?', in *Den Krieg Überdenken*, ed. Anna Geis (Baden Baden: Nomos, 2006): 133–50.

41. Creveld, *The Transformation of War*. See also Martin van Creveld, 'The waning of major war', in Raimo Väyryen (ed.), *The Waning of Major War: Theories and Debates* (Abingdon: Routledge, 2006): 97–102.

42. Kersti Larsdotter, 'New Wars, Old Warfare?', in *Rethinking the Nature of War*, eds. Isabelle Duyvesteyn and Jan Angstrom (London: Frank Cass, 2005): 135–58.

43. Colin McInnes, 'A Different Kind of War?', in ibid., 109–34.

44. 'President Bush's radio address Sept.(sic) 15 on terrorist attacks', 15 September 2001, http://www.usembassy.org.uk/bush76.html

45. Mary Kaldor, *New and Old Wars: Organised Violence in a Global Era* (Cambridge: Policy Press, 1999); Mary Kaldor, 'Elaborating the "New War" Thesis', in *Rethinking the Nature of War*, eds. Duyvesteyn and Angstrom, 210–24.

46. Clausewitz, *Vom Kriege*, VIII. 3B.

47. 'The Secret Brilliance of "You go to war with the Army you have..."', *Carrying the Gun*, 9 December 2014, https://carryingthegun.com/2014/12/09/you-go-to-war-with-the-army-you-have/ (last accessed 14 May 2023).

NOTES pp. [285–288]

48. Vladimir V. Putin, 'On the Historical Unity of Russians and Ukrainians', *President of Russia*, 12 July 2021, http://en.kremlin.ru/events/president/news/66181 (last accessed 20 November 2022).

14. FORECASTING THE FUTURE OF WAR

1. Alexander Kott and Philip Perconti, 'Long-Term Forecasts of Military Technologies for a 20–30 Year Horizon: An Empirical Assessment of Accuracy,' *Technological Forecasting and Social Change* 137 (2018): 272–9.
2. Patricia Sullivan, *Who Wins? Predicting Strategic Success and Failure in Armed Conflict* (New York: Oxford University Press, 2012); David T. Mason, Joseph P. Weingarten Jr. and Patrick J. Fett, 'Win, Lose, or Draw: Predicting the Outcome of Civil Wars,' *Political Research Quarterly* 52, no. 2 (1999): 239–68.
3. Frank L Klingberg, 'Predicting the Termination of War: Battle Casualties and Population Losses,' *Journal of Conflict Resolution* 10, no. 2 (1966): 129–71; Scott D. Bennett and Allan C. Stam, 'Predicting the Length of the 2003 U.S.-Iraq War,' *Foreign Policy Analysis* 2, no. 2 (2006): 101–15.
4. Alberto Nardelli, Jennifer Jacobs and Nick Wadhams, 'U.S. Warns Europe that Russia May Be Planning Ukraine Invasion,' *Bloomberg*, 11 November 2021, https://www.bloomberg.com/news/articles/2021-11-11/u-s-warns-europe-that-russian-troops-may-plan-ukraine-invasion
5. For example, the initial review of US Department of Defense's policies and programs that ultimately sparked the development of the F-35 began in October 1993; the first F-35 reached operational capability nearly twenty-three years later, in August of 2016. Jeremiah Gertler, 'F-35 Joint Strike Fighter (JSF) Program,' *Report from the Congressional Research Service* (Washington, DC: 16 February 2012): 8–9, https://apps.dtic.mil/sti/pdfs/ADA590244.pdf; Colin Clark, 'Air Force Declares F-35A IOC; Major Milestone for Biggest U.S. Program,' *Breaking Defense*, 2 August 2016, https://breakingdefense.com/2016/08/air-force-declares-f-35a-ioc-major-milestone-for-biggest-us-program/. Even fielding the far less technologically advanced Mine Resistant Ambush Protected (MRAP) vehicle took more than three years after the US Marine Corps first identified 'an urgent operation need for armored tactical vehicles to increase crew protection and mobility'; see *Defense Acquisitions: Rapid Acquisition of MRAP Vehicles, Testimony before the House Armed Services Committee, Defense Acquisition Reform Panel*, 111th Congress (2009) (Michael J. Sullivan, Director Acquisition and Sourcing Management).
6. Paul McCleary, 'NATO Sets Sights on Rebuilding Ukraine's Defense Industry,' *POLITICO*, 12 October 2022, https://www.politico.com/news/2022/10/12/pentagon-chief-ukraine-support-00061387

pp. [289–290] NOTES

7. Jochem Wiers, 'Building a Bridge or Nurturing the Gap?' *Journal of Strategic Studies* 40, no. 1/2 (2017): 283–6; Jo Guldi and David Armitage, *The History Manifesto* (Cambridge: Cambridge University Press, 2014).

8. Erik Gartzke, 'War is in the Error Term,' *International Organization* 53, no. 3 (1999): 567–87.

9. 'French Military Spy Chief Quits After Failure to Predict Russian Invasion,' *France 24*, 31 March 2022, https://www.france24.com/en/france/20220331-french-military-spy-chief-quits-after-failure-to-predict-russian-invasion

10. Office of the President of Ukraine, 'Ukraine to Consider Severance of Diplomatic Relations with Russia and Other Effective Steps to Respond to Recent Events – Volodymyr Zelenskyy,' *President of Ukraine*, 22 February 2022, https://www.president.gov.ua/en/news/ukrayina-rozglyane-pitannya-rozrivu-diplomatichnih-vidnosin-73057

11. Philip E. Tetlock, *Expert Political Judgement* (Princeton, NJ: Princeton University Press, 2005).

12. Gartzke, 'War is in the Error Term,' 584.

13. Geoffrey Blainey, *The Causes of War* (New York: Free Press, 1973): 49.

14. Håvard Hegre, Håvard Mokleiv Nygård and Peder Landsverk, 'Can We Predict Armed Conflict? How the First 9 Years of Published Forecasts Stand Up to Reality,' *International Studies Quarterly* 65, no. 3 (2021): 660–8.

15. Jack A. Goldstone et al., 'A Global Model for Forecasting Political Instability,' *American Journal of Political Science* 54, no. 1 (2010): 190–208.

16. Hegre et al., 'Can We Predict Armed Conflict?' 7.

17. Drew Bowlsby, Erica Chenoweth, Cullen Hendrix and Jonathan D. Moyer, 'The Future Is a Moving Target: Predicting Political Instability,' *British Journal of Political Science* 50, no. 4 (2019): 1405–17.

18. Lars-Erik Cederman and Nils B. Weidmann, 'Predicting Armed Conflict: Time to Adjust Our Expectations?' *Science* 355, no. 6324 (2017): 474–6.

19. Nazli Choucri and Thomas W. Robinson, *Forecasting in International Relations: Theory, Methods, Problems, Prospects* (San Francisco: W.H. Freeman, 1978).

20. John R. Freeman and Brian L. Job, 'Scientific Forecasts in International Relations: Problem of Definition and Epistemology,' *International Studies Quarterly* 23, no. 1 (1979): 130.

21. Ted Robert Gurr and Mark Irving Lichbach, 'Forecasting International Conflict: A Competitive Evaluation of Empirical Theories,' *Comparative Political Studies* 19, no. 1 (1986): 33.

22. Scott D. Bennett and Allan C. Stam, *The Behavioral Origins of War* (Ann Arbor, MI: University of Michigan Press, 2004): 155.

23. Patrick T. Brandt, Vito D'Orazio, Latifur Khan, Yi-Fan Li, Javier Osorio and Marcus Sianan, 'Conflict Forecasting with Event Data and Spatio-temporal

NOTES pp. [290–293]

Graph Convolutional Networks,' *International Interactions* 48, no. 4 (2022): 814, Figure 8.

24. Gurr and Lichbach, 'Forecasting International Conflict,' 6–7.

25. Nils Petter Gleditsch, Steven Pinker, Bradley A. Thayer, Jack S. Levy and William R. Thompson, 'The Forum: The Decline of War,' *International Studies Review* 15, no. 3 (2013): 396–419.

26. Bear F. Braumoeller, *Only the Dead: The Persistence of War in the Modern Age* (New York: Oxford University Press, 2019).

27. Azar Gat, 'Is War Declining? Why and Where?' *Anali Hrvatskog Politoloskog Drustva* 16, no. 1 (2019): 201–8.

28. United Nations Development Programme, *Human Development Report 2021/2022* (New York: United Nations, 2022), https://hdr.undp.org/system/files/documents/global-report-document/hdr2021-22pdf_1.pdf

29. Kenneth E. Boulding, *Conflict and Defense* (New York: Harper and Row, 1962): 291.

30. Ibid., 65–6.

31. Paul Goodwin, and George Wright, 'The Limits of Forecasting Methods in Anticipating Rare Events,' *Technological Forecasting and Social Change* 77, no. 3 (2010): 355–68.

32. Martin van Creveld, 'Less Than Meets the Eye,' *Journal of Strategic Studies* 28, no. 3 (2005): 452.

33. Andrea Ichino, and Rudolf Winter-Ebmer, 'The Long-run Educational Cost of World War II,' *Journal of Labor Economics* 22, no. 1 (2004): 57–86.

34. Gerard Rocolato, 'The Character of War is Constantly Changing,' *Proceedings* 148/5/1,431 (2022), https://www.usni.org/magazines/proceedings/2022/may/character-war-constantly-changing

35. Bowlsby et al., 'The Future is a Moving Target.'

36. Cederman and Weidmann, 'Predicting Armed Conflict,' 475.

37. Deaths, for example. Lawrence Freedman, *The Future of War: A History* (New York: Hachette, 2017), 124–33.

38. Gurr and Lichbach, 'Forecasting International Conflict,' 5.

39. One commonly used threshold for the definition of war is 1,000 battle deaths. Melvin Small and J. David Singer, *Resort to Arms* (Beverly Hills: Sage, 1982): 56.

40. Hubert M. Blalock, *Causal Inferences in Nonexperimental Research* (Chapel Hill, NC: University of North Carolina Press, 1961).

41. R. Robert Huckfeldt, C.W. Kohfeld and Thomas W. Likens, *Dynamic Modeling: An Introduction* (Beverly Hills: Sage, 1982): 17–18.

42. Thomas Hobbes, *Leviathan* (Open Road Integrated Media, 2018 [1651]): 80–1.

43. Huckfeldt et al., *Dynamic Modeling*, 18.

NOTES

44. Alvin M. Saperstein, 'Predictability, Chaos, and the Transition to War,' *Bulletin of Peace Proposals* 17, no. 1 (1986): 87–93; Lewis F. Richardson, *Arms and Insecurity* (Chicago: Quadrangle Books, 1960).

45. Laurence H. Meyer, 'Remarks Before the Economic Strategy Initiative' (Washington, DC: 8 January 1998), https://www.federalreserve.gov/boarddocs/speeches/1998/19980108.htm

46. Bennett and Stam, *The Behavioural Origins of War*, 159.

47. Sean P. O'Brien, 'Anticipating the Good, the Bad, and the Ugly: An Early Warning Approach to Conflict and Instability Analysis,' *Journal of Conflict Resolution* 46, no. 6 (2002): 793.

48. Thomas Chadefaux, 'Conflict Forecasting and Its Limits,' *Data Science* 1 (2017): 13.

49. Boulding, *Conflict Defense*, 291.

50. Freedman, *The Future of War*, 12.

51. Kott et al., 'Long-term Forecasts.'

52. Ibid.

53. Michael O'Hanlon, *A Retrospective on the So-called Revolution in Military Affairs, 2000–2020* (Washington, DC: Brookings Institution, 2018), https://www.brookings.edu/wp-content/uploads/2018/09/FP_20181217_defense_advances_pt1.pdf

54. Martin van Creveld, *Technology and War: From 2000 B.C. to the Present* (New York: The Free Press, 1989/1991).

55. David Johnson, 'The Tank is Dead: Long Live the Javelin, the Switchblade, the …?' *War on the Rocks*, 18 April 2022, https://warontherocks.com/2022/04/the-tank-is-dead-long-live-the-javelin-the-switchblade-the/

56. Mark L. Haas, *The Ideological Origins of Great Power Politics, 1789–1989* (Ithaca, NY: Cornell University Press, 2005).

57. Such as with the Center for Systemic Peace's POLITY scores.

58. One factor is the pressure to conform. Ralph K. White, *Nobody Wanted War* (Garden City, NY: Anchor Books, 1970): 244–9.

59. Frank P. Harvey, *Explaining the Iraq War: Counterfactual Theory, Logic, and Evidence* (New York: Cambridge University Press, 2012).

60. Barry B Hughes, *International Futures: Building and Using Global Models* (Cambridge, MA: Elsevier, 2019).

61. We use probability here in the colloquial sense and as a useful shorthand, though not one above critique. Philip B. Stark, 'Pay No Attention to the Model Behind the Curtain,' *Pure and Applied Geophysics* 179 (2022): 4121–45.

62. Barry B. Hughes, *IFs Interstate Politics Model Documentation* (Denver, CO: Frederick S. Pardee Center for International Futures, Josef Korbel School of International Studies, 2014), https://pardeewiki.du.edu/index.php?title=Interstate_Politics_(IP)

NOTES pp. [297–299]

63. Glenn Palmer et al., 'The MID5 Dataset, 2011–2014: Procedures, Coding Rules, and Description,' *Conflict Management and Peace Science* 39, no. 4 (2022): 470–82.

64. Nazli Choucri and Marie Bousfield, 'Alternative Futures: An Exercise in Forecasting,' in *Forecasting in International Relations: Theory, Methods, Problems, Prospects*, eds. Nazli Choucri and Thomas W. Robinson (San Francisco: W.H. Freeman, 1978): 308–26.

65. For example, in Schelling's Nobel prize acceptance speech, he cited the following warning: 'In 1960, the British novelist C. P. Snow said on the front page of the *New York Times* that unless the nuclear powers drastically reduced their nuclear armaments, thermonuclear warfare within the decade was a 'mathematical certainty.' Nobody appeared to think Snow's statement extravagant.' Thomas C. Schelling, 'An Astonishing 60 years: The Legacy of Hiroshima' (Stockholm, Sweden: 8 December 2005), https://www.nobelprize.org/uploads/2018/06/schelling-lecture.pdf

66. An as example, O'Hanlon cites 'the remarkable chain of improvements in robotics technologies such as drones, and in other types of miniaturized systems such as satellites. This progress was probably greater than I had projected back in 2000—perhaps because the synergies created by greater computing power and miniaturization, tested and refined on the real battlefields of the Middle East, created innovative dynamics that were difficult to anticipate by looking at individual areas of technology.' The aforementioned 'death of the tank' argument is another example. O'Hanlon, 'A Retrospective,' 20.

67. Bennett and Stam, *The Behavioural Origins of War*, 159.

68. UN Department of Economic and Social Affairs, *World Population Prospects 2022: Summary of Results* (New York: United Nations, 2022), https://www.un.org/development/desa/pd/sites/www.un.org.development.desa.pd/files/wpp2022_summary_of_results.pdf

69. US Energy Information Administration, *International Energy Outlook 2021, With Projections to 2050* (Washington, DC: Department of Energy, 2021), https://www.eia.gov/outlooks/ieo/pdf/IEO2021_Narrative.pdf

70. Lowy Institute, 'Asia Power Index: 2023 Edition,' *Lowy Institute*, https://power.lowyinstitute.org/data/future-resources/defence-resources-2030/military-capability-enhancement-2030/ (last accessed 16 February 2023); 'IFs version 7.8,' *International Futures*, https://www.ifs.du.edu/IFs/frm_GraphicalDisplay/%22Government%20Consumption%20by%20Dest%2c%20Currency%2c%20Hist%20and%20Forecast%22/True/0/0-185/229/1/2100 (last accessed 16 February 2023).

71. Juliana Liu, 'US-China Trade Defies Talk of Decoupling to Hit Record High in 2022,' *CNN*, 8 February 2023; United Nations, 'UN Chief Raises Alarm

422

NOTES

Over 'Backsliding' of Democracy Worldwide,' *UN News*, 15 September 2022, https://news.un.org/en/story/2022/09/1126671; US National Intelligence Council, *Global Trends 2040: A More Contested World* (Washington, DC: National Intelligence Council, 2021), https://www.dni.gov/files/ODNI/documents/assessments/GlobalTrends_2040.pdf

72. US National Intelligence Council, *Global Trends 2040*, 103.

73. Daniel Kahneman and Amos Tversky, 'Prospect Theory: An Analysis of Decision Under Risk,' *Econometrica* 47, no. 2 (1979): 263–91.

74. Daniel W. Drezner, 'The Perils of Pessimism,' *Foreign Affairs* 101, no. 4 (2022), https://www.foreignaffairs.com/articles/world/2022-06-21/perils-pessimism-anxious-nations

75. Daniel L. Byman and Kenneth M. Pollack, 'Let Us Now Praise Great Men: Bringing the Statesmen Back In,' *International Security* 25, no. 4 (2001): 107–46.

76. Wendy Pearlman and Kathleen Gallagher Cunningham, 'Nonstate Actors, Fragmentation, and Conflict Process,' *Journal of Conflict Resolution* 56, no. 1 (2012): 3–15.

77. Sara McLaughlin Mitchell and John A. Vasquez, 'Norms and the Democratic Peace,' in *What Do We Know about War?* 2nd edition, ed. Sara McLaughlin Mitchell and John A. Vasquez (Lanham, MD: Rowman & Littlefield, 2012): 167–88.

78. Alexander Wendt, 'Anarchy Is What States Make of It: The Social Construction of Power Politics,' *International Organization* 46, no. 2 (1992): 391–425.

79. Stephen Biddle, *Nonstate Warfare* (Princeton, NJ: Princeton University Press, 2021).

80. Mark W. Zacher, 'The Territorial Integrity Norm: International Boundaries and the Use of Force,' *International Organization* 55, no. 2 (2001): 215–50.

81. Richardson, *Arms and Insecurity*.

82. Bennett and Stam, *The Behavioural Origins of War*, 47, 184.

83. Thomas J. Christensen and Jack Snyder, 'Chain Gangs and Passed Bucks: Predicting Alliance Patterns in Multipolarity,' *International Organization* 44, no. 2 (1990): 137–68; John Braithwaite and Bina D'Costa, *Cascades of Violence: War, Crime and Peacebuilding Across South Asia* (Acton, Australia: Australian National University Press, 2018).

84. Boulding, *Conflict and Defense*, 269.

85. Kristian Skrede Gleditsch, 'One without the Other? Prediction and Policy in International Studies,' *International Studies Quarterly* 66, no. 3 (2022): sqac036.

86. Michael D. Ward, 'Can We Predict Politics? Toward What End?' *Journal of Global Security Studies* 1(1): 80–91; Robert D. Kaplan, 'The Perils of Forecasting,' *Orbis* 65, no. 1 (2020): 3–7.

NOTES

87. Cederman and Weidmann, 'Predicting Armed Conflict,' 476.
88. Nazli Choucri, 'Forecasting in International Relations: Problems and Prospects,' *International Interactions* 1, no. 2 (1974): 70.
89. Francis G. Hoffman, 'The Future is Plural: Multiple Futures for Tomorrow's Joint Force,' *Joint Forces Quarterly* 88, 1 (2018): 4–13.
90. Frank Hoffman, 'American Defense Priorities After Ukraine,' *War on the Rocks*, 3 January 2023, https://warontherocks.com/2023/01/american-defense-priorities-after-ukraine/
91. Boulding, *Conflict and Defense*, 261.
92. For example, see Souva's measure of material military power (MMP). As of 2016, Russia had an MMP score twelve times that of Ukraine. Even accounting for the considerable improvements that Ukrainian military forces made between 2016 and 2022, it is difficult to imagine that Ukraine would have closed the power gap with Russia according to the MMP. Mark Souva, 'Material Military Power: A Country-year Measure of Military Power, 1865–2019,' *Journal of Peace Research* (2022): 1–8.
93. Tara Copp, 'How Ukraine War has Shaped US Planning for a China Conflict,' *AP News*, 16 February 2023, https://apnews.com/article/russia-ukraine-taiwan-politics-china-8a038605d8dd5f4baf225bdaf2c6396b
94. Volodymyr Zelenskyy, 'We Stand, We Fight and We Will Win. Because We Are United. Ukraine, America and the Entire Free World, Address before a joint meeting of the U.S. Congress,' *President of Ukraine*, 21 December 2022, https://www.president.gov.ua/en/news/mi-stoyimo-boremos-i-vigrayemo-bo-mi-razom-ukrayina-amerika-80017
95. Rosa Brooks, 'Can There be War Without Soldiers?' *Foreign Policy*, 15 March 2016, https://foreignpolicy.com/2016/03/15/can-there-be-war-without-soldiers-weapons-cyberwarfare/

15. THE WAR IN UKRAINE AND THE APOCALYPTIC IMAGINARY

1. Mircea Eliade, *Myths, Dreams and Mysteries: The Encounter between Contemporary Faiths and Archaic Realities* (New York: Harper & Row, 1975): 243.
2. Today, apocalyptic visions arguably have more dramatic impact upon non-state actors, as seen in the Salafi-jihadist and far-right accelerationist movements. This is of limited relevance to the topic of this book, however.
3. Frank Kermode, *The Sense of an Ending*, 2nd edition (New York: Oxford University Press, 2000).
4. Adam Roberts, *It's the End of the World: But What Are We Really Afraid Of?* (London: Elliott and Thompson, 2020): 23.
5. Kermode, *Sense*, 8.
6. Key works for understanding the field and its diversity of approaches include: Nick Bostrum, 'Existential Risks: Analyzing Human Extinction

pp. [306–311] NOTES

Scenarios and Related Hazards', *Journal of Evolution and Technology* 9, no. 1 (2002): 1–36; John Leslie, *The End of the World: The Science and Ethics of Human Extinction* (London: Routledge, 1996); Johan Rockström et al., 'Planetary Boundaries: Exploring the Safe Operating Space for Humanity', *Ecology and Society* 2, no. 32 (2009): 472–5; Phil Torres, 'Facing Disaster: The Great Challenges Framework', *Foresight* 21, no. 1 (2019): 3–34.

7. For example, see *WMO Global Annual to Decadal Climate Update* (Geneva: World Meteorological Association, 2022).

8. Cf. Shahar Avin et al., 'Classifying Global Catastrophic Risks', *Futures* 102 (2018): 20–6; Hin-Yan Liu, 'Governing Boring Apocalypses: A New Typology of Existential Vulnerabilities and Exposures for Existential Risk Research', *Futures* 102 (2018): 6–19.

9. Examples include the French initiative Red Team Defense (www.redteamdefense.org) and the US Army's Science Fiction Prototypes (https://threatcasting.asu.edu/scifi-prototypes).

10. The White House, *National Security Strategy* (Washington, DC: October 2022), https://www.whitehouse.gov/wp-content/uploads/2022/10/Biden-Harris-Administrations-National-Security-Strategy-10.2022.pdf

11. Nick Bostrum's distinction between 'existential risk' and 'global endurable risk' is typical; see Bostrum, 'Existential Risks'.

12. William M. Deneven (ed.), *The Native Population of the Americas in 1492* (Madison, WI: University of Wisconsin Press, 1992): xxix.

13. The most obvious objection to this definition of apocalyptic threat (AT) is that it risks stretching the concept too far, or that it encompasses such a huge range of threats as to impede strategic action. To counter this, I propose four levels of AT, which represent an ascending order of severity: AT I—the physical and/or social destruction of a distinct social community; AT II— the physical and/or social destruction of societies across a significantly large region; AT III—the physical and/or social destruction of societies across a continent or continents; and AT IV—the destruction of all (or virtually all) human life on the planet.

14. In a 2022 poll in the United States by Pew Research Center, 39 per cent of respondents agreed that 'we are living in the end times' and 10 per cent believed that the Second Coming of Jesus would occur in their lifetimes. Jeff Diamant, 'About four-in-ten U.S. adults believe humanity is "living in the end times"', *Pew Research Center*, 8 December 2022, https://www.pewresearch.org/fact-tank/2022/12/08/about-four-in-ten-u-s-adults-believe-humanity-is-living-in-the-end-times/ (last accessed 31 January 2023).

15. For example, see the classic vignette of Seneca's deluge: 'Waters will converge from the west and from the east. A single day will bury the human race': Lucius Annaeus Seneca, *Natural Questions*, trans. Harry Hine (Chicago: University of Chicago Press, 2010): 51.

425

NOTES pp. [311–321]

16. Daniel Wojcik, *The End of the World as We Know It: Faith, Fatalism and Apocalypse in America* (New York: NYU Press, 1997): 4.

17. Elizabeth Kolbert, 'The Climate of Man II: The Curse of Akkad', *The New Yorker*, 2 May 2005. This was long thought to be merely a fantastical recounting of Akkadian collapse, until scientists in the 1990s discovered that environmental disasters in the region at that time matched the poem's descriptions of deadly droughts and famines.

18. Thomas Moynihan, 'Existential risk and human extinction: An intellectual history', *Futures* 116 (2020): 1–13.

19. David Seed, 'The Course of Empire: A Survey of the Imperial Theme in Early Anglophone Science Fiction', *Science Fiction Studies* 37, no. 2 (2010): 233.

20. Gerry Canavan, 'Ice-Sheet Collapse and the Consensus Apocalypse in the Science Fiction of Kim Stanley Robinson', in *The Cambridge Companion to Literature and Climate*, eds. Adeline Johns-Putra and Kelly Sultzbach (Cambridge: Cambridge University Press, 2022): 179–90.

21. Wojcik, *End of the World*, 1; I.F. Clarke, *Voices Prophesying War: Future Wars, 1763–3749* (Oxford: Oxford University Press, 1992): 164–217.

22. It is important to note that all typologies of this nature are culturally relative, and this particular example reflects my perspective as a scholar of future war in a contemporary Anglo-American context.

23. Gerardo Ceballos, 'Biological Annihilation via the Ongoing Sixth Mass Extinction Signaled by Vertebrate Population Losses and Declines', *Proceedings of the National Academy of Sciences* 114, no. 30 (2017): 6089–96.

24. Qi Zhao et al., 'Global, Regional and National Burden of Mortality Associated with Non-Optimal Ambient Temperatures From 2000 to 2019', *The Lancet Planetary Health* 5, no. 7 (2021): 415–25.

25. For example, the Inupiat communities: see Emily Witt, 'An Alaskan Town Is Losing Ground – And a Way of Life', *The New Yorker*, 28 November 2022.

26. Christopher Williams, *Terminus Brain: The Environmental Threats to Human Intelligence* (London: Cassell, 1997).

27. Torres, 'Facing Disaster', 12.

28. Bostrom, *Superintelligence*.

29. 'Climate Action Tracker Emissions Gap', *Climate Action Tracker*, https://climateactiontracker.org/global/cat-emissions-gaps/ (last accessed 21 March 2023).

30. Robert Jervis, *The Illogic of Nuclear Strategy* (Ithaca, NY: Cornell University Press, 1984).

31. Torres, 'Facing Disaster'.

32. Aamer Madhani, Ellen Knickmeyer and Josh Boak, 'Biden's "Armageddon" Talk Edges Beyond US Intel', *Associated Press*, 7 October 2022, https://apnews.com/article/biden-nuclear-risk-1d0f1e40cff3a92c662c57f274ce0e25

pp. [322–324]   NOTES

33. Thomas C. Schelling, *Arms and Influence* (New Haven, CT: Yale University Press, 1966): 68.
34. Cited in 'War in Ukraine: Could Russia Use Nuclear Weapons?', *House of Lords Library*, 24 November 2022.
35. Originally published by RIA Novosti, 3 April 2022, https://web.archive.org/web/20220403212023/ https://ria.ru/20220403/ukraina-1781469605.html (last accessed 20 February 2023).
36. Article by Vladimir Putin, 'On the Historical Unity of Russians and Ukrainians', *President of Russia*, 12 July 2021, http://en.kremlin.ru/events/president/news/page/101 (last accessed 20 February 2023).
37. Oksana Dudko, 'A Conceptual Limbo of Genocide: Russian Rhetoric, Mass Atrocities in Ukraine, and the Current Definition's Limits', *Canadian Slavonic Papers* 64, no. 2/3 (2022): 133–45.
38. 'Address by the President of the Russian Federation', *President of Russia*, 24 February 2022, http://en.kremlin.ru/events/president/news/page/58 (last accessed 20 February 2023).
39. 'UN Annual Results Report 2022: Early Recovery Efforts in Ukraine', *United Nations*, 11 April 2023, https://ukraine.un.org/en/resources/publications?f%5B0%5D=resources_agency%3A28 (last accessed 21 February 2023); 'OSCE Special Monitoring Mission to Ukraine (closed)', *Organization for Security and Co-operation in Europe*, https://www.osce.org/special-monitoring-mission-to-ukraine-closed (last accessed 21 February 2023).
40. Transcript of initial hearing in *Ukraine vs Russia* (2022), International Court of Justice, 'Allegations of Genocide under the Convention on the Prevention and Punishment of the Crime of Genocide (Ukraine v. Russian Federation)', *International Court of Justice*, March 2022, www.icj-cij.org/public/files/case-related/182/182-20220307-ORA-01-00-BI.pdf (last accessed 21 February 2023).
41. A review of Russia's claims and their falsity can be found in the report of the Independent International Fact-Finding Mission on the Conflict in Georgia (the 'Tagliavini Report'). See Independent International Fact-Finding Mission on the Conflict in Georgia (IIFFMCG), 'Report: Volume 1', *BBC*, 19 September 2009, http://news.bbc.co.uk/1/shared/bsp/hi/pdfs/30_09_09_iiffmgc_report.pdf (last accessed 21 February 2023).
42. Available at: Felix Light and Olzhas Auyezov, 'Putin's address to Russia's parliament', *Reuters*, 21 February 2023, https://www.reuters.com/world/europe/putins-address-russias-parliament-2023-02-21/
43. I have borrowed this concept from the British sociologist Colin Campbell, who in 1972 introduced the concept of the 'cultic milieu' to describe the diverse range of movements comprising the cultural underground of that era. Colin Campbell, 'The Cult, the Cultic Milieu and Secularization', *A Sociological Yearbook of Religion in Britain* 5 (1972): 119–36.

427

NOTES   pp. [325–331]

44. IPSOS, 'Most Americans Agree the US Should Continue to Support Ukraine Despite Threat of Nuclear Weapons Use by Russia', *IPSOS*, 10 October 2022, https://www.ipsos.com/en-us/news-polls/americans-agree-us-continue-support-ukraine-despite-russia-threatening-use-nuclear; YouGov, 'Nuclear Fears Rise Among Extinction Worries Following Ukraine Invasion', *YouGov*, 10 March 2022, https://yougov.co.uk/topics/politics/articles-reports/2022/03/10/nuclear-fears-rise-among-extinction-worries-follow; Chicago Council on Global Affairs and Levada Center, 'Russians and Americans Sense a New Cold War', *Global Affairs*, 20 April 2022, https://globalaffairs.org/sites/default/files/2022-04/Final%20New%20Cold%20War%20Brief.pdf (all last accessed 21 February 2023).

45. American Psychological Association, 'Stress in America', *American Psychological Association*, March 2022, https://www.apa.org/news/press/releases/stress/2022/march-2022-survival-mode (last accessed 21 February 2023).

46. For example, the mass shootings in Christchurch, El Paso and Buffalo were perpetrated by self-described eco-fascists.

47. Polls by the Levada Center, the last independent polling centre in Russia. See Levada Center, 'Indicators', *Levada Center*, February 2023, https://www.levada.ru/en/ratings/ (last accessed 22 February 2023).

48. Quoted in 'What Secret Russian State Polling Tells Us about Support for the War', *Moscow Times*, 9 December 2022, https://www.themoscowtimes.com/2022/12/06/what-secret-russian-state-polling-tells-us-about-support-for-the-war-a79596 (last accessed 22 February 2023).

16. WAR AS BECOMING

1. Lawrence Freedman, *The Future of War: A History* (London: Penguin, 2018).

2. Michael P. Ferguson, 'The Nature of War is not Changing in Ukraine', *The Hill*, 16 June 2022, https://thehill.com/opinion/national-security/3526558-the-nature-of-war-is-not-changing-in-ukraine/

3. David Lonsdale, *The Nature of War in the Information Age: Clausewitzian Future* (London: Routledge, 2004).

4. 'Why Ukraine's Army Still Uses a 100-Year-Old Machinegun', *The Economist*, 11 May 2022, https://www.economist.com/the-economist-explains/2022/05/11/why-ukraines-army-still-uses-a-100-year-old-machinegun; David L. Stern, 'Russia Attacks Kyiv Overnight with Swarm of Self-Detonating Drones', *The Washington Post*, 19 December 2022, https://www.washingtonpost.com/world/2022/12/19/kyiv-drones-attack-belarus-putin/

5. Antoine Bousquet, 'War', in *Concepts in World Politics*, ed. Felix Berenskoetter (London: Sage Publications, 2016): 91–106.

pp. [331–338]   NOTES

6. John P. Abizaid and Rosa Brooks, *Recommendations and Report of the Task Force on US Drone Policy* (Washington, DC: Stimson Center, 2014): 35.

7. Carl von Clausewitz, *Vom Kriege* (Berlin: Ferdinand Dümmler Berlagsbuchhandlung, 1905): 627.

8. Ernst Jünger, 'Total Mobilization', in *The Heidegger Controversy: A Critical Reader*, ed. Richard Wolin (Cambridge, MA: MIT Press, 1992): 126.

9. Bernard Brodie, *The Absolute Weapon: Atomic Power and World Order* (New York: Harcourt, Brace and Company, 1946).

10. Norman Moss, *Men Who Play God: The Story of the Hydrogen Bomb* (London: Penguin Books, 1970): 351.

11. Tarak Barkawi, 'Decolonising War', *European Journal of International Security* 1, no. 2 (2016): 119–214.

12. Isabel V. Hull, *Absolute Destruction: Military Culture and the Practices of War in Imperial Germany* (Ithaca, NY: Cornell University Press, 2005); Manfred F. Boemeke, Roger Chickering and Stig Förste (eds), *Anticipating Total War: The German and American Experiences, 1871–1914* (Cambridge: Cambridge University Press, 1999): 399–471.

13. Jairus Grove, *Savage Ecology: War and Geopolitics at the End of the World* (Durham, NC: Duke University Press, 2019).

14. The enduring fascination with the writings of two obscure Chinese colonels on 'unrestricted warfare' in 1999 is symptomatic of an increasingly held conviction in war as 'an unlimited zone of conflict' that includes 'attacks on all elements of society'. David Barno and Nora Bensahel, 'A New Generation of Unrestricted Warfare', *War on the Rocks*, 19 April 2016, https://warontherocks.com/2016/04/a-new-generation-of-unrestricted-warfare/

15. Frank G. Hoffman, *Conflict in the 21st Century: The Rise of Hybrid War* (Arlington, VA: Potomac Institute for Policy Studies, 2007); Geraint Hughes, 'War in the Grey Zone: Historical Reflections and Contemporary Implications', *Survival* 62, no. 3 (2020): 131–58.

16. Antoine Bousquet, Jairus Grove and Nisha Shah, 'Becoming Weapon: An Opening Call to Arms', *Critical Studies on Security* 5, no. 1 (2017): 1–8.

17. Jairus Grove, 'An Insurgency of Things: Foray into the World of Improvised Explosive Devices', *International Political Sociology* 10, no. 4 (2016): 332–51.

18. Antoine Bousquet, 'A Revolution in Military Affairs? Changing Technologies and Changing Practices of Warfare', in *Technology and World Politics: An Introduction*, ed. Daniel R. McCarthy (London: Routledge, 2017): 165–81.

19. Roman Horbyk, '"The War Phone": Mobile Communication on the Frontline in Eastern Ukraine', *Digital War* 3, no. 1 (2022): 1–16.

20. It is worth noting, as evidenced by countless personal testimonies, that armed combat is one of the privileged sites in which individuals experience the intense depersonalization and deregulation of habitual frames of

reference in all their terror and sublimity, putting them into direct contact with the flow of becoming.

21. Colin Gray, *Modern Strategy* (Oxford: Oxford University Press, 1999): 362.

22. On the issues surrounding the Michael Howard and Peter Paret translation, see Jan Willem Honig, 'Clausewitz's *On War*: Problems of Text and Translation', in *Clausewitz in the Twenty-first Century*, eds. Hew Strachan and Andreas Herberg-Rothe (Oxford: Oxford University Press, 2007): 57–73; Hew Strachan, 'Michael Howard and Clausewitz', *Journal of Strategic Studies* 45, no. 1 (2022): 143–60.

23. Von Clausewitz, *Vom Kriege*, 20.

24. Carl von Clausewitz, *On War*, trans. Michael Howard and Peter Paret (Princeton, NJ: Princeton University Press, 1976): 89.

25. Antulio J. Echevarria, 'War's Changing Character and Varying Nature: A Closer Look at Clausewitz's Trinity', *Military Strategy Magazine* 5, no. 4 (2017): 15–20, https://www.militarystrategymagazine.com/article/wars-changing-character-and-varying-nature-a-closer-look-at-clausewitzs-trinity/

26. Translation: '*Die Aufgabe ist also, daß sich die Theorie zwischen diesen drei Tendenzen wie zwischen drei Anziehungspunkten schwebend erhalte*', in Clausewitz, *Vom Kriege*, 21.

27. Clausewitz, *On War*, 89.

28. Clausewitz, *Vom Kriege*, 78.

29. Translation: '*Die Kriegführung verläuft sich fast nach allen Seiten hin in unbestimmte Grenzen; jedes System, jedes Lehrgebäude aber hat die beschränkende Natur einer Synthesis*', in ibid., 77.

30. Ibid., 60.

31. Ibid., 61.

32. François Jullien, *Traité de l'Efficacité* (Paris: Grasset, 1996): 18 [my translation].

33. Helmut Satz, *The Rules of the Flock: Self-Organization and Swarm Structure in Animal Societies* (Oxford: Oxford University Press, 2020).

34. Alan Beyerchen, 'Clausewitz, Nonlinearity and the Unpredictability of War', *International Security* 17, no. 3 (1992): 59–90; Brian Cole, 'Clausewitz's Wondrous Yet Paradoxical Trinity: The Nature of War as a Complex Adaptive System', *Joint Force Quarterly* 96 (2020): 42–9; B.A. Friedman, 'War Is the Storm – Clausewitz, Chaos, and Complex War Studies', *Naval War College Review* 75, no. 2 (2022): 37–65.

35. Antoine Bousquet, *The Scientific Way of Warfare: Order and Chaos on the Battlefields of Modernity*, 2nd edition (London: Hurst Publishers, 2022).

36. H. Diels and W. Kranz, *Die Fragmente der Vorsokratiker* (Berlin: Weidmann, 1957): 53.

37. Ibid., 12.

pp. [343–353] NOTES

38. Nicholas Rescher, *Process Metaphysics: An Introduction to Process Philosophy* (Albany, NY: State University of New York Press, 1996).
39. Diogenes Laertius, *Lives of Eminent Philosophers*, trans. R.D. Hicks (London: William Heinemann, 1925): 415.
40. William James, *A Pluralistic Universe: Hibbert Lectures at Manchester College on the Present Situation in Philosophy* (Lincoln, NE: University of Nebraska Press, 1996): 117.
41. William James, *Essays in Radical Empiricism* (Cambridge, MA: Harvard University Press, 1976): 42.
42. Henri Bergson, *Creative Evolution* (London: Routledge, 20220); Alfred North Whitehead, *Process and Reality: An Essay in Cosmology* (New York: Free Press, 1978).
43. Gilles Deleuze and Felix Guattari, *A Thousand Plateaus: Capitalism and Schizophrenia* (Minneapolis, MN: University of Minnesota Press, 1987).
44. Antoine Bousquet, Jairus Grove and Nisha Shah, 'Becoming War: Towards a Martial Empiricism', *Security Dialogue* 51, no. 2/3 (2020): 99–118.
45. Bernard Brodie, 'The Atomic Dilemma', *The Annals of the American Academy of Political and Social Science* 249 (1947): 37.
46. Marcel Detienne and Jean-Pierre Vernant, *Cunning Intelligence in Greek Culture and Society* (Chicago: University of Chicago Press, 1991): 47.
47. Antoine Bousquet, *The Eye of War: Military Perception from the Telescope to the Drone* (Minneapolis, MN: University of Minnesota Press, 2018).
48. Christopher Coker, *Warrior Geeks: How 21st Century Technology is Changing the Way we Fight and Think About War* (London: Hurst Publishers, 2013).
49. Harun Farocki, 'Phantom Images', *public* 29 (2004): 15, 32–41.

AFTERWORD

1. Prior to his unfortunate recent death, Professor Coker prepared these notes as a review of this book for Hurst. In honour of his memory, we are republishing it here. 'Debating the Future of War: Change and Continuity After the Invasion of Ukraine', 11 September 2023, https://www.hurstpublishers.com/debating-the-future-of-war-change-and-continuity-after-the-invasion-of-ukraine/
2. Total number of wars calculated from two Wikipedia entries: for 1990–2002, https://en.wikipedia.org/wiki/List_of_wars:_1990%E2%80%932002; for 2003–present, https://en.wikipedia.org/wiki/List_of_wars:_2003%E2%80%93present
3. Margaret MacMillan, *War: How Conflict Shaped Us* (NY: Random House, 2020).
4. Edward N. Luttwak, 'Gory or glorious? The pains and pleasure of war,' *Times Literary Supplement*, Issue 6137, November 13, 2020.

431

INDEX

Note: Page numbers followed by *'n'* refer to notes, *'t'* refer to tables, *'f'* refer to figures.

3D printers, 17, 206–7, 253
4Chan, 149, 152
9/11 attacks, 48–9, 331, 338

A2/AD (Anti-Access/Area Denial)
  capabilities, 56, 62, 67, 164,
  219–20
'adaptation before fire', 197–9
'Afghan Model', 49, 53–4
Afghanistan, 2, 38, 50, 52, 53,
  131, 142, 143, 184, 283
AI. *See* artificial intelligence (AI)
Air Alert app, 198
air defence systems, 58, 63, 66, 67
AirLand Battle doctrine, 166
'air-power revolution', 43
Ajax Systems, 198
Alexander Nevsky (film), 274
Algerian National Liberation
  Front, 283
al-Qaeda, 49, 52, 338
  See also 9/11 attacks
Amazon Web Services, 196
Amazon, 98, 198

American Civil War, 201, 203
American Psychological
  Association, 325
Angell, Norman, 277, 285
Animal Farm (Orwell), 354
animal war, 225, 226
Anonymous (hacker group), 92, 97
Another Bloody Century (Gray), 47
'anticipatory intelligence', 271
anti-politics practices, 254–5
antithetical potentialities, 275–7
anti-war norms, 27, 29, 38
anti-war youth campaign, 81
apes, 226, 269–70
apocalyptic imaginary, 19, 305–27,
  315*t*, 427*n*13
  apocalyptic threats, defined,
    308–10
  biological, 316–17
  Box X: the unknown/
    unimaginable, 319–21
  brief history, 310–14
  categorizations, 310–11
  defined, 308–10

INDEX

'enduring nature' of, 306
environmental, 316
European tradition, 311–13
human-centred strain of, 313–14
perceptions on policy and
strategy, 307, 314–21, 315*t*,
321–6, 427*n*13
and Russo-Ukrainian War,
321–7
sacred, 310–11, 312, 313
secular, 311–13
societal, 317–18
space and planetary, 319
technological apocalyptic
threats, 318–19
war and genocide, 317, 321,
323
APT SandWorm, 97, 102–3
'Arc of Instability', 49, 59
armed conflict law, 26–7, 28–9
Aron, Raymond, 275, 276
Arquilla, John, 85
artificial intelligence (AI), 17, 58,
59, 62, 64, 119, 122, 164, 187,
193, 194, 195, 223–39, 241–2,
330
apocalyptic imaginary, 307
character of war, changing,
230–2
cognitive domain operations,
115
drones manufacturing, 206–7
'friction', reducing, 232–3
human minds, changing, 233–7,
235*f*
human nature and decision-
making, 224–5, 226
human nature of war and,
223–4
language models and decision-
making, 230, 237–9

semiconductors, 167, 173, 175,
399–400*n*48
Socratic model, 236–7
assemblage, 247–8, 249, 251, 252,
254, 257–8
Aurora Orion, 206
Australia, 73, 270
authoritarian-capitalist regime,
81–2
authoritarianism, 54, 62
autocratic regimes, 25
Autonomous Weapon Systems
(AWS), 58, 59, 62, 64, 65, 167
AWACS (Airborne Warning and
Control Systems), 187
Awad, Edmund, 229

Baghdad, 153, 168
Balkan crisis, 44
ballistic missiles, 170, 172, 209
'Banana Wars', 36
barbarism, sophisticated, 60
Bard, 224
bargaining-versus-brute force
model, 33
Barkawi, Tarak, 241
battalion tactical groups (BTG),
63, 67
Beevor, Antony, 101
Belgium, 72, 270–1
Berlin Airlift (1948–9), 270
Biden, Joe, 321
biological weapons, 188–90
biotechnology, 189–90, 195
Black Death, 316
Blue Helmets, 45
Book of Revelation, 305, 311
Bookchin, Murray, 152
Bosnia, 45–6
Boston Dynamics, 250
Bostrom, Nick, 319

434

INDEX

'botnets', 96
Boulding's general theory of
conflict, 291, 301
Brexit, 277
Brighton, Shane, 241
Britain, 79, 81, 267, 271, 277
Cold War-era civil defence, 161
'dehousing' campaign, 219
food security, 156
information infrastructure,
158–9
intercommunal violence, 153
Brodie, Bernard, 344
BTG. *See* battalion tactical groups
(BTG)
Bush, George W., 283

C4ISR (command, control,
communications, computers
intelligence, surveillance and
reconnaissance), 122
Cameron, David, 150
Canada, 72, 155
Capella Space, 204
capitalism, 42, 82, 144
Carter, Ashton, 187, 194
Catherine the Great, 274
Cebrowski, Arthur, 43
Central Intelligence Agency, 186
Chalmers, David, 234, 236, 237
Charles VIII, 281–2
Chechnya, 64, 277
Chiang, Ted, 237
China, 3, 61, 71, 72, 73, 82, 83,
144, 165, 170, 172, 175, 176,
220, 246, 270, 298, 302
A2/AD capabilities, 164
cognitive domain operations,
114–15
commercial espionage, 112
cross-domain synergy, 120

cyber activity, 57
investment in AI, 167, 188
'offset strategy' against, 187–8
PMCs usage, 118
proxy wars, 117, 118
'system destruction warfare',
121
Chips Act, 175
Churchill, Winston, 184, 266
civil wars, 15, 147–61
causes of, 149
chances for, 148
character, 147, 148
civil defence, 161
economic factor, 150–1
guardedness of, 149
information infrastructure
attacks, 158–9
peace operations in, 44–6
as 'political war', 160
primary objective, 157
strategy, 160–1
trust levels in society, 149–50
weapons availability, 151–2
Western states and, 147
See also urbicide
civilian technological innovation,
195–7
Clash of Civilizations, The
(Huntington), 48
Clausewitz, Carl von, 19, 33, 125,
126, 142, 191, 223–5, 226,
279, 281, 283–4, 302, 330, 332
'apocalyptic imaginary', 305,
339–42
future of war, 280
on ideal commander, 225
'mystery' of war, 351
on people's war, 127, 128–9
See also war's nature and
character

435

INDEX

climate change, 18, 268–9, 313, 316
coercion, 14, 113
   See also strategic coercion
cognitive warfare, 14, 112–16
   definition of, 113–14
Cohen, Eliot, 42, 43
Cold War, 7, 29, 38, 42, 44, 48, 56, 127, 141, 149, 161, 163, 184, 188, 244, 278, 279, 282, 322, 333
Collapse of Complex Societies, The (Tainter), 317
'collateral damage', 47
command and control (C2), 65
commemoration and mourning, 271–9
commercial espionage, 112
commercial satellites, 205
communism, 48, 79, 140, 144
compellence, 30, 32–3, 34
'comprehensive approach', 50
conflict and mediation event observations (CAMEO), 290
Congress of Vienna, 133
conscription, 126, 134, 135–6, 137, 140, 143
Continental European countries, 267
conventional warfare, 14, 118–22, 141, 297, 331
   'cross-domain synergy', 119–20
   'system destruction warfare', 121–2
cool war, 61, 63
counterinsurgency (COIN), 51, 53, 60, 280, 282
counterproliferation tools, 188–90
counterterrorism, 48–9
Covid-19, 316, 324

Crimea, annexation of (2014), 3, 37, 54, 56, 89, 142, 249, 277, 279
cruise missiles, 59, 63, 67, 209, 210, 220
Cuban missile crisis, 278–9
cultural anthropology, 228
cultural evolution, 227–9
cultural psychology, 228
'culture wars', 147
'cunning intelligence', 345
Curse of Akkad, The (poem), 311
cyber espionage and sabotage, 170–1, 214
cyber weapons, 88, 100–1, 103
cyberattacks, 57–8, 60, 61, 62, 63, 65, 86, 307, 317. *See* cyberwar
CyberBerkut, 89
'cybernetics', 232
Cybersecurity and Infrastructure Security Agency (CISA), 110, 111
cyberspace, 52, 58, 85–6
   offense dominance, 214–15
   See also cyberattacks; cyberwar
cyberwar, 85–104
   anti-cyber options, 215
   characteristics, 87–8
   cyber operations, expectations and reality, 97–101
   digital intelligence operations, 88
   evolution of, 92–7
   hard-cyber operations, 88, 89, 90, 97, 98, 99, 101
   malicious software (malware), 88, 90
   pitfall, 86
   soft-cyber operations, 88, 89, 90–1, 97
   term, 87

INDEX

Da Jiang Innovations, 250
Damen Shipyards Group, 248
Darwinist nationalism, 277
DDoS. *See* Distributed-Denial-of-Service (DDoS) attacks
'decline of war' thesis, 290
deep fakes, 59
Defence Innovation Agency, France, 9, 353
defense dominance. *See* offense–defense balance
Defense Innovation Accelerator for the North Atlantic (DIANA), NATO, 242–3
Defense Innovation Competition (DIC), The Netherlands, 256
Defense Innovation Unit (DIU), United States, 194, 195
Deleuze, Gilles, 343, 344
Delft Dynamics, 257
democracies, 74, 140, 143
  COIN campaigns and, 53
  new technologies, exploitation of, 115–16
  spread of, 25, 79–80
democratic constitutional reform, 133
'de-Nazification', 322–3
denial. *See* economic and military denial
depopulation policy, 33
Der große König (film), 274
deterrence, 12, 30–3, 123, 140
  by denial, 30–1, 32
  by punishment, 30
  rediscovering, 56–9
Die Geschwister Oppermann (Feuchtwanger), 274
Die Hermannsschlacht (play) (von Kleist), 274
disinformation, 55, 61, 89, 114

Distributed-Denial-of-Service (DDoS) attacks, 89, 91, 92, 93
diversity of war, 9–11
DNA, 189–90
DoD. *See* US Department of Defense (DoD)
Donbas, 25, 37, 66, 89, 93, 142, 323
drones, 53–4, 60, 62, 64–5, 68, 185, 193, 196, 197, 198, 221, 224
  commercial drone sector, 206
  high-altitude drones, 213
  ISIS, 186
  MoD's drones purchasing, 250
  MQ-9 Reaper drones, 255
  naval drones, 209
  'Project PARIS', 251, 256, 257
  proliferation of, 59
  rise of, 205–6
  SkyHive system, 256, 257, 259–60
  SPEAR (Swarm-based Persistent Autonomous Reconnaissance), 245, 247, 257–8
  See also swarm technology
dual-use technologies, 59, 164, 169, 175
Dutch military–industrial–commercial complex, 244–62
  assemblage, 247–8, 249, 251, 252, 254
  commonalities and critiques, 254–5
  new alliances, forging, 250–4
  re-assemblage, 257–8
  shift in threat perceptions and technical solutions, 248–9
  swarm technology, 246–8

437

INDEX

tensions, contradictions and
failures management, 258–60
valley of death, bridging the,
255–7
Dutch Ministry of Defense (MoD),
244–5, 248–9
anti-politics practices, 254–5
Defense Innovation Strategy
(2018), 259
Defensity College program,
250–1, 256
efforts to forge alignments,
255–7
employees, 252
'Frontdoor' (website page), 252
MQ-9 Reaper drones
acquisition, 255
off-the-shelf technologies,
buying, 250
'Project PARIS', 251, 256, 257
Robots and Autonomous
Systems (RAS) units, 247,
251, 252–3, 254, 257
startups, collaboration with,
251–2, 255–7
See also Dutch military–
industrial–commercial
complex
Dutch Royal Military Police, 250
dynamite, 185, 186

economic and military denial,
163–79
combining possibilities, 170–3
current trends, 164–5
globalization, 165–7
impact of, 173–9, 177–8t
inequalities, deepening, 168–70
maritime costs, decline in,
165–6
working of, 168

economic crisis (2007–8), 82, 148,
151, 267, 268
economic integration, 26, 27–8
economic modernization, 74
economic openness, 78, 79
Edelman Trust Barometer, 149–50
Elbit Systems, 246
electronic warfare (EW), 216–17,
371n24
EMP systems, 62, 212, 216
EMS domain, 216–17
End of History and the Last Man, The
(Fukuyama), 48
Ender's Game (Card), 353
Enlightenment Now (Pinker), 349
enmity, 24, 39, 128, 129, 133,
140, 339
'environmentally-mediated
intellectual decline' theory,
316–17
ESET, 92, 94
Esper, Mark T., 214
Estonia cyberattack (2007), 57,
350
European army, 276
European Coal and Steel
Community, 276
European colonialism, 312
European Space Agency, 319
European Union (EU), 25, 45, 50,
61
Evera, Stephan van, 219
existential risk ('X-risk'). *See*
apocalyptic imaginary
*Expulsions: Brutality and Complexity
in the Global Economy* (Sassen),
261
'extra-judicial killing', 53

Facebook, 93, 98,
far-right 'extremism', 160

INDEX

fascists, 272
Fedorov, Mykhailo, 92, 93, 241
Feuchtwanger, Lion, 274
Flame attack (2012), 57
flu pandemic, 316
forecasting, future war, 287–303
  conventional war, probability
    of, 297
  invasion of Ukraine and
    calculation changes, 299–303
  prediction of war, possibilities,
    289–91
  See also war and patterns of (un)
    knowability
four faces of warfare. *See* warfare,
  four faces of
'fourth-generation warfare', 48, 60
France, 76, 79, 130, 134–6, 153,
  269, 276, 277, 353
Franco-Prussian War, 135–6
free trade, 13, 78, 79, 83
Freedman, Lawrence, 4, 30, 44
French Revolution, 125–6, 127,
  129–30, 131, 132, 143, 145,
  268, 272, 283
Fukuyama, Francis, 48, 62
'futurology', 149

Gaddafi's regime, 53
Gartzke, Erik, 289, 294
general purpose technologies,
  188–90
genocide, 76, 95, 153, 311, 317,
  321–4
geopolitical rivalry, 61
George, Kennan, 25–6
Georgia, war against (2008), 278,
  324
Gerasimov, Valery Vasilyevich, 265,
  266
German armed forces, 136–7

German Democratic Republic
  (GDR), 274
Germany, 75, 76, 79, 81, 135, 269
  economy, 78
  military leadership, 137–8
  Nazi political leadership,
    139–40
  Project Cassandra, 353
global positioning system (GPS),
  206, 213
Global War on Terror (GWOT),
  49, 147, 168, 331, 350
globalization, 165–7, 331
Goebbels, Joseph, 139–40
Google, 112, 196, 198
GPT, 17, 233, 236
  ChatGPT exchange, 223, 235f
  GPT-4, 224
'gray zone', 25, 29
Gray, Colin, 47, 339
Graziano, Michael, 234
Great Depression, 79
great powers, long peace between,
  74–5, 75f, 77–9, 334
Greenfield, Susan, 353
Guattari, Félix, 343–4
Guicciardini, Francesco, 282

Hädicke, Martijn, 254
Harari, Yuval, 349
Hedges, Chris, 48
Henry V (film), 274
Heraclitus, 342–3
HermeticWiper, 92, 94
Hezbollah, 51, 52, 117
high-mobility artillery rocket
  systems (HIMARS), 65, 66
Hitler, Adolf, 267, 277, 278
Holland, 72
How to Lose the Information War
  (Jankowicz), 350

439

INDEX

Howard, Michael, 68
Hughes, Wayne, 210
'human machine team', 237–9, 320
'human security', 309
'humane warfare', 47
humanitarian wars, 44–7, 59
Hungary, 268
Huntington, Samuel, 48
Hussein, Saddam, 49, 142
hybrid conflict, 51, 54–6, 60, 102, 336
hypersonic missiles, 58, 59, 62, 63, 64, 67, 188
Hyten, John, 216

ICANN, 98
ICJ. *See* International Court of Justice (ICJ)
ICT. *See* information and communications technology (ICT)
ideology, 62
Iliad (Homer), 346, 351
immaculate war, 60, 64
improvised explosive device (IED), 160, 186, 221, 338
India, 298
Industrial Control Systems (ICS), 90
industrial growth, 77–9
Industrial Revolution II, 185, 186–7, 204
Industrial Revolution III, 197
Industrial Revolution IV, 16
'Industroyer2' ICS malware, 94
Inflation Reduction Act, 175
'influence operations' doctrines, 52
information and communications technology (ICT), 43, 44, 51, 85, 99, 101, 103

'information revolution', 43
innovation. *See* military-technological innovation
Innovative Technologies Shaping the 2040 Battlefield, 350–1
Instagram, 87, 196
'intelligentized warfare', 121
International Atomic Energy Agency, 215
International Committee for the Red Cross, 54
International Court of Justice (ICJ), 323–3
International Criminal Court, 27
international law, 26–7, 28–9, 136–7
International Monetary Fund, 26, 61
International Organization for Migration, 38
Internet of Things, 198
Internet Research Agency (aka Glavset), 102–3
internet, 51–2, 97
Iran, 117, 164, 168–9, 171, 184, 214
Stuxnet attack (2010), 215
UAV production, 169, 172
Iraq, 2, 38, 43, 52, 142, 143, 184
civil war, 50
post-Saddam, 159
urban counterinsurgency operations, 153
Irish Republican Army, 283
irregular warfare, 51–2
IsaacWiper, 94
ISIS (Islamic State in Iraq and Syria), 6, 51, 52, 53, 60, 186
ISR capabilities, 174–5
Israel, 51, 52, 246
IT-Army of Ukraine, 93

440

INDEX

James, William, 343, 353
Japan, 72, 79, 140, 175
Joint Concept for Access and Maneuver in the Global Commons (US), 174
Jullien, François, 341

Kagan, Robert, 47
Kaldor, Mary, 44–5, 47, 59–60
Kaspersky Lab, 90
Kennedy, John F., 282
Kermode, Frank, 306
Kilcullen, David, 51, 60, 157
'killer robots', 58, 59, 62, 64, 65, 167
Kokoshin, Andrey, 279
Kolberg (film), 274
Korean War, 281, 282
Kosinski, Michal, 236
Kosovo, 46, 270
Kuhn, Thomas, 36

L'action française (periodical), 138
Landwehr, 131
Lebanon War (2006), 51
Leicester, intercommunal violence (Sep 2022), 153
Levada Centre, 325
'liberal peace security culture', 59–60
liberalism, 42, 54, 62, 79, 82
Limits of the City, The (Bookchin), 152
Lind, Bill, 48
Link Magazine, 252
Linkola, Pentti, 160–1
London, urban riots (2011), 156–7
Long Peace, 74–5, 75*f*, 77–9, 334
Louis XVI, 130
low-risk warfare, 53–4
Ludendorff, Erich, 139

Luhansk, 25

machine learning, 115, 187, 231, 242, 244
Maersk, 111
Mahabharata, 351
Major Accident Hazard Pipelines (MAHP), 154–5
major war, decline of, 24–9, 40
anti-war norms, spread of, 27, 29, 38
democracies, spread of, 25
economic integration, increasing, 26, 27–8
international law and the law of armed conflict, 26–7, 28–9
mass destruction weapons, proliferation of, 25
multilateral institutions, growth of, 25–6
Mali, 69
malware, 88, 90, 94
Mandiant, 92
Mao 2 (DeLillo), 351
Marconi, Guglielmo, 185
Marcus, Gary, 234, 235, 236
Marine Corps (US), 205–6
Maritime Research Institute Netherlands (MARIN), 248
Mariupol, 66, 68
martial empiricism, 337, 342–5
mass destruction weapons, 12, 25, 298
McFate, Sean, 60
media, 36
Medvedev, Dmitry, 54, 297, 324
Melvin, Mungo, 280
Merkel, Angela, 150
Meta, 196
metis, 345
Meyer, Laurence, 293

441

INDEX

Microsoft, 91, 92, 93–4, 98, 112, 196
middle powers, 164, 176, 177–8*t*
migration, waves of, 268–9
'Military Innovation by Doing' (MIND), 253
military–industrial–commercial complex. *See* Dutch military–industrial–commercial complex
military-technological innovation, 16–18, 183–200, 333
  'adaptation before fire', 197–9
  biological weapons, 188–9
  civilian technological innovation, 195–7
  counterproliferation tools, 188–90
  destructive use of inventions, 186
  late nineteenth century, 185
  lethal open technologies, 186
  pioneers, 183
  as primarily military, 193–5
  professional military education (PME), 190–3, 199
  views, 10–11
  See also artificial intelligence (AI)
Milley, Mark, 207
Miloševic, Slobodan, 46
mobilization, 2, 8, 31, 56, 131–2, 133–4, 139–40, 203, 330, 332, 344
'Modernization Peace', 78
Mongol conquests, 76
Moskva (Russia's Black Sea Fleet flagship), 96
MQ-9 Reaper drones, 255
Mueller, John, 44, 61
multiculturalism, 150
multidomain operations (MDO), 119–20

multilateral institutions, growth of, 25–6
museums, 272–4, 275
Musk, Elon, 92

N PRGRSS (magazine), 259
Nagorno-Karabakh War, 209, 246
Napoleon III, 134, 135
Napoleonic Wars, 269, 282, 305, 331–2
NASA, 319
National Transmission System (UK), 154
nationalism, 79, 133–4, 281
NATO. *See* North Atlantic Treaty Organization (NATO)
Nazis, 139–40
Near-Earth Objects Coordination Centre, 319
Nelson, Horatio, 209
Netherlands Organization for Applied Scientific Research (TNO), 248
the Netherlands, 17, 175, 244, 245, 247
  'start-up ecosystems', 253
  See also Dutch military–industrial–commercial complex; Dutch Ministry of Defense (MoD)
'network-centric warfare', 43–4
'New Cold War', 57
New START treaty, 321
'New World Order', 42
New Zealand, 73, 270
'new' wars, 281–4
Nobel, Alfred, 185
non-state actors, 36, 38, 41, 48, 51, 54, 164, 169
  cyber-enabled influence operations, 94–5, 97–8

442

INDEX

cyberspace, 87–8
drones usage, 59
sophisticated barbarism, 60
United States and, 168
North Atlantic Treaty Organization
(NATO), 7, 25, 29–30, 34, 38,
45, 46, 50, 63, 99, 176, 268,
281, 288, 307, 320, 324, 326
annexation of Crimea
responses, 56
conscription, 140
conventional defence, 140–1
defense dominance, 220–1
Innovation Fund, 261
Military Committee, 270
military spending, 242–3
multidomain operations
(MDO), 119, 120
nuclear war, fear of, 321–2
Russian cyberattacks, 112
Warsaw Summit (2016), 57
North Korea, 72, 73, 117, 214
Northern Alliance, 49
NotPetya ransomware attack
(2017), 86, 90, 100, 111,
214–15
nuclear peace, 74
nuclear war, 29, 307, 313, 317,
320, 321–2, 326–7
nuclear weapons, 13, 57, 59,
83, 140, 141, 184, 186, 187,
188, 193, 276, 281, 284, 313,
321–2, 326–7
decline of major war, 79–80
Nye, Joseph, 58

O'Brien, Philips, 171
offense–defense balance, 201–21,
221t, 303
air domain, 211–12
cyber domain, 214–15

dominance vs. temporary
advantage, 201
EMS domain, 216–17
interaction between domains,
217–18
land domain, 208–9
sea domain, 209–10
shifting balance in history,
203–8
space domain, 213
strike warfare, 218–19
the West, 219–21, 221t
On War (Clausewitz), 39, 129, 191,
302, 339, 340–1
OpenAI, 224, 233
GPT-4, 224
Operation Allied Force, 46
Operation Deliberate Force, 46
Operation Desert Storm, 42
Operation Enduring Freedom, 49,
53
Operation Iraqi Freedom, 49, 53
Operation Orchard, 86
Operation Unified Protector, 53
Oracle, 198
Orwell, George, 354
Ottoman Empire, 272

Palantir Technologies, 243
Panetta, Leon E., 57, 214
Paradise Built in Hell (Solnit), 352
Parry, Chris, 280
Payer, Friedrich von, 143
Peloponnesian War (431–04 BCE),
75, 271–2
'the people', 125–45
as a legitimate actor in war,
127–32
as a limited resource for
technocratic war, 133–7
total war, 137–42

443

INDEX

People's Liberation Army (PLA), 114
  'system destruction warfare', 121
  'three battles', 115
people's war. *See* 'the people'
  Prussia, 125–6
  Ukraine, 126, 142, 143
pervasive surveillance, 204–5, 209
Pinker, Steven, 290, 349, 353–4
Planet Lab, 204
Planetary Defense Coordination Office (US), 319
PME. *See* professional military education (PME)
Poland, 268, 277, 278
'political extremism', 157
'political polarization', 150
Poroshenko, Petro, 89
ports, mining of, 210
post-human condition, 352–3
post-Qaddafi Libya, 159
presentism, 18–19, 42
private military companies (PMCs), 117–18
professional military education (PME), 190–3, 199
Project Cassandra, 9, 353
protectionism, 79, 82–3
proxy wars, 14, 53–4, 116–18, 123, 320
Prussia, 75, 125–6, 129–130, 135, 280
Putin, Vladimir, 24, 27, 28–9, 64, 66, 97, 143, 277, 285, 291, 294, 321, 324, 325–6
  NATO as a threat, 25
  Russian economy, 26
  'special military operation' speech (24 Feb 2022), 323

See also Crimea, annexation of (2014); Russia; Russo-Ukrainian War; Ukraine, invasion of (2022)
Putnam, Robert, 150

quantum computing, 62, 188
Quds Force, 117

RAND, 115
RAS. *See* Robots and Autonomous Systems (RAS) units
revolution in military affairs (RMA), 44, 45, 46, 49, 62, 67
Richards, David, 57–8
RIPE, 98
Ritter, Gerhard, 135
robotics, 58, 119, 185, 188, 193, 194, 206
Robots and Autonomous Systems (RAS) units, 247, 251, 252–3, 254, 257
Ronfeldt, David, 85
Royal Netherlands Aerospace Centre (NLR), 248
Royal United Services Institute (RUSI), 207, 211
Rumsfeld, Donald, 284
Rusk, Dean, 333
Russia
  authoritarian and nationalist character, 82
  critical infrastructure, cyberattacks on, 110–12
  cyber disruption attacks, 214–15
  cyber operations against Ukraine, 89–92
  defense industry, 172–3
  democracy, failure of, 268
  Estonia cyberattack (2007), 57, 350

INDEX

GDP per capita, 290
geopolitics, 61
Georgia, war against (2008),
   278, 324
influence campaigns, 114
international trade, 83
New START treaty withdrawal,
   321
nuclear missiles, 57
offense as dominant, 204
proxy wars, 117
reliance on PMCs, 117
semiconductors, inflow of, 173
strike warfare, 219
See also Crimea, annexation of
   (2014); Russo-Ukrainian War;
   Ukraine, invasion of (2022)
Russian economy, 26
   sanctions and embargoes, 28,
      30, 34–5, 279
Russian Military Intelligence
   (GRU), 90, 91, 93
Russian military
   air power, 66–7
   'Anti-Access/Area Denial' (A2/
      AD) capabilities, 56
   cross-domain integration, 120
   loss of tanks, armoured vehicles
      and soldiers, 66
   professionalism and discipline,
      56
Russophobia, 95
Russo-Ukrainian War, 20, 62–9,
   71, 123, 142, 265–6, 277, 278,
   334
   AI usage, 232–3, 241–2
   air superiority, 211–12
   apocalyptic imaginary, 321–7
   assumptions, 1–2
   cyber operations, expectations
      and reality, 97–101

defensive positions gaps, 202
denial and countering denial,
   170, 173
denial capabilities, 164, 165,
   172
drone tech in, 186, 207,
   211–12
failures, 65, 66–7, 68, 96
'the Ghost of Kyiv', 96
impact of, 12–14
pervasive surveillance, 209
social media and, 95–6
tech platforms and, 196, 243
troops withdrawal, 66
UAVs, 169
weaponization, 338
See also cyberwar

Sapolsky, Robert, 269–70
Sassen, Saskia, 261
satellites, 17, 62, 92, 184, 193,
   204, 208, 233
   active satellites, 205
   Synthetic Aperture Radar
      (SAR), 194, 204, 205, 206
   war in space, 213
   See also Starlink
Savunma Teknolojileri Mühendislik
   (STM), 246
Schelling, Thomas, 26, 30, 33, 37,
   322
science fiction, 8–9, 306, 353
Scientific American (journal), 185
Second Punic War, 75
Security Service of Ukraine (SBU),
   90, 96
semiconductors, 167, 173, 175,
   399–400n48
Sense of an Ending, The (Kermode),
   306
Sergeytsev, Timofey, 322–3

445

INDEX

Shaw, Martin, 47
SkyHive system, 256, 257, 259–60
small powers, 164, 169, 174, 177–8t
Smith, Adam, 83
Smith, Rupert, 24, 44
'war amongst the people' paradigm, 24, 35–9
social media, 52, 55, 60, 61, 95, 97
exploitation of, 114, 115
Zelensky's use of, 64, 95
societal warfare, 14, 109–12
'soft power', 44
SolarWinds hack (2020), 111, 215
Solnit, Rebecca, 352
South China Sea, 3, 71, 83, 220
South Korea, 72–3, 175
Soviet Union (USSR)
collapse of, 126, 166, 283
covert biological weapons program, 189
NATO vs., 7, 140–1
recreating, 268
World War II casualty rate, 76
space surveillance, 205
SpaceX, 196
Starlink, 64, 92, 98, 196
Spain, 76, 153
SPEAR (Swarm-based Persistent Autonomous Reconnaissance), 245, 247, 257–8
Spetsnaz units, 56
Spot (dog-like robot), 250
Srebrenica massacre (Jul 1995), 46, 274
Stalin, Joseph, 278
Stand Out of Our Light (Williams), 353
Starlink, 64, 92, 98, 196

Starship Troopers (Heinlein), 353
startups
fear of, 254
Dutch MoD's collaboration with, 251–2, 255–7
valley of death, 255–7
State Service of Special Communications and Information Protection, Ukraine, 94
state-building, 50
strategic bombing, 110, 202
strategic coercion, 24, 30–5, 40
definition, 30
irrational actor versus the ignorant actor, 31–2
Stuxnet attack (2010), 57, 86, 215
submarine technology, 170
Sudan civil war, 69
Sun Tzu, 114–15
surface-to-air missile (SAM), 63, 65, 66
Surkov, Vladislav, 114
'surrogate warfare', 53, 60–1, 123
surveillance
pervasive surveillance, 204–5, 209
sea domain, 209
space surveillance, 205
V-BAT 128 surveillance drone, 206
Svechin, Aleksandr Andreyevich, 266, 279
swarm technology, 246–8
SPEAR (Swarm-based Persistent Autonomous Reconnaissance), 245, 247, 257–8
Tective Robotics, 256–7
See also drones
Sweden, 269

INDEX

Symantec, 94
Synthetic Aperture Radar (SAR), 194, 204, 205, 206
'system destruction warfare', 121–2

tactical defense. *See* offense–defense balance
Tainter, Joseph, 317
Taiwan, 71, 73, 278, 302
Taliban, 49, 50, 51, 52, 142–3
technological apocalyptic threats, 318–19
technological innovation. *See* military-technological innovation
Tective Robotics, 256–7, 259–60
Telegram, 93, 95
The Future of War: A History (Freedman), 4
'theory of mind', 226, 231, 236
Thirty Years War (1618–48), 76
This England (film), 274
Threat Analysis Group (TAG), Google, 112
'Thucydides trap', 80–1
TikTok, 115
total war, 137–42
Tournear, Derek, 218
Transformation of War, The (van Creveld), 60
Trojan.Killdisk, 94
troll armies, 55, 61
Turkey, 246, 269

Ukraine
  cyber infrastructure, 98
  denial, 179
  drones, 197, 198, 205
  Kyiv, 28, 31, 63–4, 112, 198, 322

power grid attack, 86, 90, 94, 97
presidential election (May 2014), 89
Russian cyber operations against, 89–97, 110–12
Russian cyberattacks, countering, 91, 92–3, 96
Russian troops withdrawal, 66
strategies of control, 33–4
tech capacity, 197–8
'total defense' policy, 30–1
use of EMS, 216
West's support to, 221
Ukraine, invasion of (2022), 1, 3, 11, 23–40, 63, 163, 280, 285, 303
  cyberspace activities, 85, 86, 87
  material capabilities, 302
  plight of refugees, 38
  Putin's justifications, 40
  sanctions and embargoes, 28, 30, 34–5, 279
  strategic coercion, 24, 30–5, 40
  US intelligence community's predictions, 288, 289
  'war amongst the people' paradigm, 24, 35–9
  Western materiel support for, 34, 35
  See also major war, decline of; Russo-Ukrainian War; war's nature and character
Ukrainian Armed Forces, 2, 35, 97, 169, 262
Ukrainian hackers, 198
UN Safe Areas, 45
(un)knowability characteristics. *See* war and patterns of (un) knowability
unconventional terrorism, 81

447

INDEX

Union of Concerned Scientists, 205
United Kingdom (UK)
  'dehousing' campaign, 219
  gas networks, 154–5
  London, urban riots, 156–7
United Nations (UN), 25, 45, 50, 61, 298
United States (US), 23, 61, 71, 72, 79, 168, 246, 266, 271, 282
  9/11 attacks, 48–9, 331, 338
  A2/AD system, 219, 220
  American Civil War, 201, 203
  armed forces, 2–3
  civil war and, 147
  'command of the commons', 164
  debt–GDP ratio, 150–1
  Global War on Terror (GWOT), 49, 147, 168, 331, 350
  globalization, benefits, 166–7
  hegemony, 71, 74, 81
  military power, 42–4
  monitoring PLA, 115
  National Defense Strategy (2018), United States, 118, 192, 307
  offense/defense competition, 201–2
  'Offset X', 187–8
  PME system, 192–3, 199
  power projection, 168–9, 170, 174, 175, 176
  presidential election (2016) and cyberattack, 86
  proxy wars, 117
  Roosevelt's 'New Deal', 268
  structural inequalities, use of, 165
  war participation, 36–7
(un)knowability. *See* war and patterns of (un)knowability

unmanned aerial vehicles (UAVs), 166, 167, 169, 173, 246
unmanned aircraft system (UAS), 216
Unrestricted Warfare (1999 book), 56
urbicide, 152–61
  aims, 152
  definition, 152
  gas networks as target, 154–5
  outflows of people, 156–7
  security services, 157
  tactics amongst anti-status quo, 154
  transportation and logistics infrastructure as target, 156
  See also civil wars
US Cyber Command Vision, 214
US Department of Defense (DoD), 11–12, 157, 187, 194–5, 199, 338
US Energy Information Administration, 298
US National Intelligence Council, 299
US Navy, 43, 168, 184, 205–6

van Creveld, Martin, 47, 48, 60, 283, 291
Vann, John Paul, 160
V-BAT 128 surveillance drone, 206
Vertical take-off and landing (VTOL), 206
Viasat, 92, 196, 215
Vietnam, 3, 117, 282, 283
VKontakte, 90, 95, 197
Vodafone, 93, 98
von Kleist, Heinrich, 274

Wagner Group, 61, 64, 69, 117
Wall Street Crash (1929), 267, 268
Walters, Barbara F., 150

448

INDEX

Walzer, Michael, 349

'war amongst the people' paradigm, 24, 35–9

war and patterns of (un)knowability, 287, 288, 292–6, 295–96t, 303
  complexity, 292
  equilibration, 293
  measurability, 292–3
  stochasticity, 294
  tractability, 294, 295

war as becoming, 329–47
  against an essence of war, 334–8
  becoming Clausewitz, 339–42
  martial empiricism, 342–5
  (no-)futures of war, 345–7
  perpetual transformation of war, 331–4
  See also war's nature and character

War in the Air, The (Wells), 312

War is a Force That Gives Us Meaning (Hedges), 48

war memorials, 272–4, 275

War of the Worlds, The (Wells), 313

war's nature and character, 19–20, 24, 33, 39–40
  cultural evolution, 227–9
  human nature and decision-making, 224–5, 226
  human nature of war, 223–4
  implications of, 229
  language capability, 227
  metaphysical dimension of war, 351–2
  See also artificial intelligence (AI); war and patterns of (un)knowability; war as becoming

warfare, four faces of, 107–24, 109f
  framework, 108–9, 109f

risk assessment, 122–3

Warsaw Pact, 29, 166, 282

'Washington Consensus', 50

Web 2.0, 51–2

Weimar Germany, 267

Wells, H.G., 149, 312

WhatsApp, 87

Williams, Christopher, 316–17

'wiperware'-attack, 91, 92, 94, 112

Work, Robert O., 187, 188

World Bank, 61

World Set Free, The (Wells), 312

World War I, 2, 36, 37, 75, 79, 137, 143–4, 186, 201, 208, 269, 272–3, 277, 281, 282, 284, 332

World War II, 27, 36, 37, 56, 75, 76, 139–40, 172, 184, 186, 187, 201, 219, 269, 270, 274, 281, 282, 332–3

World War III, 282, 313, 325

Wright, Orville, 185

Wright, Wilbur, 185

X (formerly known as Twitter), 93, 98, 196, 234

Yandex, 197

Yarosh, Dmytro, 89

Yudin, Grigory, 326

Yugoslav civil war, 45, 153

Zelensky, Volodymyr, 27, 64, 97, 143
  appeal to the US Congress, 302
  'total defense' policy, 30–1
  See also Russo-Ukrainian War; Ukraine; Ukraine, invasion of (2022)

Zmiinyi Island ('Snake Island'), 96

'zone of peace' / 'zone of war', 72, 83

449